AF334398

FRONTIERS OF FUNDAMENTAL PHYSICS

To learn more about AIP Conference Proceedings,
including the Conference Proceedings Series, please visit the webpage
http://proceedings.aip.org/proceedings

FRONTIERS OF FUNDAMENTAL PHYSICS

Eighth International Symposium

FFP8

Madrid, Spain   17 – 19 October 2006

EDITORS

Burra Gautam Sidharth
Birla Science Centre, Hyderabad, India

Antonio Alfonso-Faus
Polytechnic University of Madrid, Madrid, Spain

Màrius Josep Fullana i Alfonso
Polytechnic University of València, València, Spain

All papers have been peer reviewed.

SPONSORING ORGANIZATIONS
Polytechnic University of Madrid (UPM)
Polytechnic University of Valencia (UPV)
Community of Madrid Government Council
Generalitat Valenciana
Ministry of Education and Science
INTA
SENER
CIEMAT
Santander Group
Iberdrola
CS Ingenieros de Minas
Madrid Emprende
Iberia
Conde-Duque Hotel
NH Hotel

Melville, New York, 2007

AIP CONFERENCE PROCEEDINGS ■ VOLUME 905

Editors

Burra Gautam Sidharth
Birla Science Centre
Hyderabad, India
E-mail: iiamisbgs1@yahoo.com

Antonio Alfonso-Faus
E.U.I.T. Aeronáutica
Plaza del Cardenal Cisneros, s/n
28040 Madrid, Spain
E-mail: aalfonsofaus@yahoo.es

Màrius Josep Fullana i Alfonso
Universitat Politècnica de València
València, Spain
E-mail: mfullana@mat.upv.es

Cover image courtesy of the *Mass Boom* by Antonio Izquierdo Ortega, www.estudiodearte-aizquierdo.com.

L.C. Catalog Card No. 2007924742
ISBN 978-0-7354-0412-0
ISSN 0094-243X

Printed in the United States of America

CONTENT

I. ASTROPHYSICS, RELATIVITY, AND COSMOLOGY

II. QUANTUM PHYSICS

III. FOUNDATIONS OF PHYSICS AND SCIENCE IN GENERAL

IV SCIENCE AND SOCIETY

V. POSTERS

PREFACE

For the past decade the International Symposium series, Frontiers of Fundamental Physics has attracted brilliant physicists from all over the world. The broad objective of the series has been to enable scholars working in slightly different areas to meet on a single platform and exchange ideas and status reports. It has also been the intention to provide the participants an opportunity to meet eminent physicists. The areas covered in each symposium have included Astronomy and Astrophysics, Particle Physics, Theoretical Physics, Gravitation and Cosmology and even Computational Physics.

The symposia have been held in India, Europe and Canada. The eminent physicists who have delivered special lectures over the years, sometimes with multiple participation, have included Nobel Laureates Professors Gerard 't Hooft, Stephen Chu, Charles Townes, Klaus von Klitzing, Pierre De Gennes, Douglas Osheroff, Yuval Ne'eman, Jogesh Pati, Ashoke Sen and several other prominent scholars. The Proceedings of all except the third symposium in the series have been or are being published by the Universities Press (Orient Longman), Kluwer Academic, Springer and now the American Institute of Physics.

The eighth symposium in this series in Madrid has been no exception. There have been three special symposium lectures delivered by professors G 't Hooft, De Gennes and Douglas Osheroff (professor Carlo Rubbia, the fourth Nobel Laureate to participate, could not assist due to health problems). This apart there have been invited talks, contributed papers and some posters—nearly one hundred participants in all. This edition has been organized by the Technical University of Madrid (UPM) and the Technical University of València (UPV).

It is worthwhile to comment that in the present edition a space for the discussion of the role of science and its connection with society has been supplied. This has been finally organized as parallel talks where scientists and engineers have given their point of view on this subject. The continuation of this idea could supply a promising site of discussion and could be the impellor of a debate to extend among workers of science and all the society. One of the ideas that appeared is the possibility of changing the way of doing science in order to contribute to improve the society in a fairer social system, a science for all the people and all the peoples of the world. It also appeared the opinion that alternative scientific ideas should be taken more into account among the scientific community. This fact always helps the progress of the status of science in every moment. Furthermore, a debate about ethics implications on the new advances of science and technology has been done.

The symposium would not have been possible but for the kind cooperation and help of Gonzalo Milans del Bosch, María Cecilia Estefan and María Jesús Otero; and the next institutions: the two organizer universities UPM and UPV, Gobierno de la Comunidad de Madrid, Generalitat Valenciana, Ministerio de Educación y Ciencia, INTA, SENER, CIEMAT, Grupo Santander, Iberdrola, CS Ingenieros de Minas, Madrid Emprende (Ayuntamiento de Madrid), Iberia, Hotel Conde-Duque and NH Hoteles.

These proceedings have been mainly financed by the "Programa de Recolzament a la Investigació i el Desenvolupament de la Universitat Politècnica de València (PAID-03-06)". Special mention has to be done to the hard and well done work of María Jesús Otero in the preparation of this book. One of us, Màrius J. Fullana i Alfonso, wish to thank his wife, Maria Teresa Jordan i Pla and their daughter, Ariadna Mariola, his sisters, Pepa and Sònia, and his niece, Laia, for her help and patience, I love you. Last but not the least thanks are also due to Dr. Maya Flikop, Dr. Kristen Girardi and the American Institute of Physics team but for whose initiative and hard work these proceedings would not have been a reality.

The Editors
Hyderabad, Madrid and Burjassot
February 2007

LECTURERS

ALFONSO Y GARCIA, JAVIER DE
ALFONSO-FAUS, ANTONIO
ALFONSO-GARCIA, MIGUEL
AMAYA Y GARCIA DE LA ESCOSURA, JOSE MANUEL
ANABALON, ANDRÉS
ARRANZ, F.J
ASHTEKAR, A
ASKEROV, B. M.
BENITO, R.M.
BORONDO, F
BROEKAERT, JAN
CABRERA SOLÉ, JOSE RICARDO
CARBONELL, EUDALD
CARBONELL, M.V.
CHUBYKALO, ANDREW
DE GENNES, PIERRE,
DEGROOTE, EUGENIO
DELGADO-BARRIO, GERARDO
DOMÉNECH, J.
DOMÉNECH-MANTECA, G.F.
DUQUE, PEDRO
EL NASCHIE, M.S .
ELIZALDE, EMILIO
ELVIRA, R.
ENDERS, PETER
ESPINOZA, AUGUSTO
FERNÁNDEZ-CABRERA, J.
FIGAROV, V.R
FIGAROVA, S.R.
FINKELSTEIN, DAVID
FLÓREZ, M.,
FULLANA i ALFONSO, MÀRIUS
FULLANA i ALFONSO, SÒNIA
GAITE, JOSE
GALAN, PABLO
GARCÍA PALOMO, MIGUEL
GIRALDA, C.G.
GÓMEZ GONZÁLEZ, A.
GONZALO, JULIO
GREINER, WALTER J.W.
GUIJARRO, JOSE FRANCISCO

GUSEINOV, G.I.
GULSHANI, PARVIZ
HADLEY, MARK
HARTNETT, JOHN
HOOFT, G. 'T
IVANOV, R.
JOURNEAU, PHILIPPE
KHOLMETSKII, ALEXANDER L.
KLEINERT, H.,
KOSMIEDER, LOTHAR-
KROEGER, H,
LARA CASTELLS, M.P. DE
LIM, S.C.
LÓPEZ DURAN, D.
LOSADA, J.C
LUEPKE ESTEFAN, XARLES ERIK
MANTECA, CONSOLACIÓN
MISSEVITCH, O.V.
MIZUSHIMA, MASATAKA-
MONTERO, JOSÉ DE LA LUZ
MONZÓN, A.
MORALES, J.A
NIKULOV, ALEXEY
NOTTALE, L.
NOVOA BLANCO, J.F.
OHIRA, TORU
OKON GURVICH, ELIAS
OSHEROFF. D.D.
PALAZZI, PAOLO
PROSMITI, R.
PUJALS, E.R.
RIESGO, I
RUBBIA, CARLO,
SAEZ MILAN, DIEGO P.
SALAS PERALTA, PEDRO J.
SIDHARTH, B.G. –
SMIRNOV-RUEDA, ROMAN
VALDÉS,A
VILACOBA RAMOS, ANDRES –
VILLARREAL, P
WISNIACKI,.D.A
ZAGRAVAEV, V.

His Majesty the King of Spain Juan Carlos I.

HONORARY COMMITTEE

President:

S.M. el Rey D. Juan Carlos I

Members:

Excma. Sra. Dña Esperanza Aguirre
Presidenta de la Comunidad de Madrid

Excma. Sra. Dña. Mercedes Cabrera Calvo-Sotelo
Ministra de Educación y Ciencia

Excmo. Sr. D. Alberto Ruiz Gallardón
Alcalde-Presidente del Ayuntamiento de Madrid

Excmo. Sr. D. Javier Uceda Antolín
Rector Magnífico de la Universidad Politécnica de Madrid

Excmo. Sr. D. Francisco Juliá Igual
Rector Magnífico de la Universidad Politécnica de Valencia

Excmo. Sr. D. Ángel Gabilondo Pujol
Rector Magnífico de la Universidad Autónoma de Madrid

Excmo.Sr. D. Carlos Berzosa Alonso Martínez
Rector Magnífico de la Universidad Complutense de Madrid

Excmo. Sr. D. Ernesto Martínez Ataz
Rector Magnifico de la Universidad Castilla-La Mancha

Excmo. Sr. D. Emilio Botín Sanz de Sautuola
Presidente del Grupo Santander

Excmo. Sr. D. Enrique Alarcón
Presidente de la Real Academia de Ingeniería

Excmo. Sr. D. Alberto Galindo Tixaire
Presidente de la Real Academia de las Ciencias Fxactas, Físicas y Naturales

Excmo. Sr. D. Gerardo Delgado-Barrio
Presidente de la Real Academia de Físicos de España

Excmo. Sr. D. Carlos Andradas
Presidente de la Real Sociedad de Matemáticas Española

Excmo. Sr. D. Luis Giménez-Cassina Basagoiti
Presidente del Instituto de Ingeniería de España

XV

Excmo. Sr. D. Francisco Pardo Piqueras
Presidente del Instituto Nacional de Técnica Aeroespacial

Excmo. Sr.D. José Luis Quintanilla
Presidente del CIEMAT

Excmo. Sr. D. Fernando de Cuadra
Rector de la ETSI ICAI de la Universidad Pontificia de Comillas

INTERNATIONAL ORGANIZING COMMITTEE:

CHAIRMAN: Dr. Sidharth, B.G

Prof. Alfonso-Faus, A – Technical University Madrid(UPM)
Prof. El Naschie, M. S.- Physics Dpt.-University of Alexandria
Prof. Finkelstein, David – Georgia Inst. Of Technology, USA
Prof. Fullana i Alfonso, M.J.- Universitat Politécnica Valencia
Prof. Greiner, Walter – Goethe Universität, Germany
Prof. Kroeger, H.-University of Laval, Canada
Prof. Lai, C.H.- National University of Singapore

Prof. B. G. Sindharth   Prof. A. Alfonso-Faus

LOCAL ORGANIZING COMMITTEE:

Chairman:	Prof. Antonio Alfonso-Faus
Special Advisor:	D. Gonzalo Milans del Bosch
Public Relations/Coordinator:	Dña María Cecilia Estéfan
Secretary:	Prof. Màrius J. Fullana
Administrative Secretary:	Dña. María Jesús Otero

Dr. Azcárraga Arana, A
Dr. Chicot Urech, J.M.
Prof. Fernandez de la Bastida, J.M.
Prof. Fullana i Alfonso, M.J
Dr. Gopegui Palacios, L
Prof. Herrero Debón, A
Prof. Lara Sáez, A
Prof. López Ruiz, J.L.
Prof. Masegosa Fanego, R.
Prof. Ruiz Delgado, M
Prof. Sáez Milán, D.P.
Dr. Sidharth, B.G.

Prof. M. J. Fullana

I. ASTROPHYSICS, RELATIVITY, AND COSMOLOGY

Lorentz-Poincaré type aspects of the matter Lagrangian in General Relativity Theory

Jan B. Broekaert

Philosophy of Physics Group, University of Oxford (temp. affil.) & CLEA-FUND, Vrije Universiteit Brussel

Abstract. It is well known that the solution to the Einstein Field Equation, $g_{\mu\nu}$, can be either interpreted as the metric tensor itself or the mere gravitational field, their geochronometric correspondence being assured by the Equivalence Principle (e.g. Brown 2005). Within the field interpretation, which allows emphasis on physical effects of gravitation on microphysical constituents of matter, we expose gravitational Lorentz-Poincaré type properties of the relativistic gravitational matter Lagrangian. The Weierstrass parametrization (Johns 2005) of the matter Lagrangian L_M in the explicit Lorentz-Poincaré type gravitation model is shown to render it equal to the standard matter Lagrangian —which can be reduced to the proper time invariant $(g_{\mu\nu}dx^\mu/d\lambda\,dx^\nu/d\lambda)^{1/2}$ for the geodesic motion (Stephani 2004)— in GRT. As such the GRT matter Lagrangian can be interpreted to result —following a Legendre transformation— from the energy of the matter fields obtained from Gravitationally Modified Lorentz Transformations (Broekaert 2005). The resultant correspondence between matter Lagrangians exposes explicitly the Lorentz-Poincaré type features such as (coordinate) spatially-variable speed of light, $c(r) = c'\Phi^2$, partial Machian mass induction $m(r,v) = m'_0\gamma\Phi^{-3}$ and gravitational affecting of space and time observations in local coordinates in GRT. These features are only apparent relative to the coordinative manifold, while locally and in physical coordinates the effects all vanish in concordance with the local Minkowski metric.

Keywords: Lorentz-Poincaré, matter lagrangian
PACS: 04.20.-q, 04.50.+h

THE LP-TYPE INTERPRETATION OF GRT

The LP gravitational model emphasizes physical effects of gravitation on matter, other than mere attraction. Principally we intend the effects of gravitational length shortening ("rod contraction") and time dilation ("clock slowing") as we understand them in the Lorentz-Poincaré interpretation of Special Relativity. A recent discussion of this interpretation of SRT underpins "dynamically" the foundations of SRT; the Lorentz-covariance of the electromagnetic interaction of the micro-constituents in rods and clocks are considered the primary effect, only subsequently resulting in the Minkowski "spacetime" [7, 1]. The LP interpretation distinguishes gravitationally affected observers —denoted S'— and gravitationally unaffected "observations", similar to "unrenormalized" and "renormalized" coordinates which are distinguished in field approaches [12, 8, 9, 13]. The unaffected perspective —here denoted S_0— corresponds to the *coordinate* space description in GRT. The gravitational effects on space and time observations were developed as gravitationally modified Lorentz Transformations (GMLT) for space and time intervals and, energy and momentum. It was also shown that the

CP905, *Frontiers of Fundamental Physics (FFP8), Eighth International Symposium*
edited by B. G. Sidharth, A. Alfonso-Faus, and M. J. Fullana
© 2007 American Institute of Physics 978-0-7354-0412-0/07/$23.00

elimination of the unaffected perspective from the GMLT between two local observers restores the local Lorentz covariance of the relations even if in S_0 the spatial-variable speed of light is present [4]. The fundamental time-space and energy-momentum transformations conveying the LP properties are;

$$\begin{pmatrix} \delta t' \\ \delta \mathbf{x}' \end{pmatrix} = \Lambda_{tr}(\mathbf{u},\mathbf{r}) \begin{pmatrix} \delta t \\ \delta \mathbf{x} \end{pmatrix}, \quad \begin{pmatrix} E' \\ \mathbf{p}' \end{pmatrix} = \Lambda_{Ep}(\mathbf{u},\mathbf{r}) \begin{pmatrix} E \\ \mathbf{p} \end{pmatrix} \tag{1}$$

with

$$\Lambda_{tr} \equiv \begin{pmatrix} \gamma\Phi & -\mathbf{u}c^{-2}\gamma\Phi \\ -\mathbf{u}\gamma\Phi^{-1} & \mathbf{1}\Phi^{-1} + \frac{u_i u_j}{u^2}(\gamma-1)\Phi^{-1} \end{pmatrix}, \Lambda_{Ep} \equiv \begin{pmatrix} \gamma\Phi^{-1} & -\mathbf{u}\gamma\Phi^{-1} \\ -\mathbf{u}c^{-2}\gamma\Phi & \mathbf{1}\Phi + \frac{u_i u_j}{u^2}(\gamma-1)\Phi \end{pmatrix}$$

and where $\Phi = \Phi(\mathbf{r})$ is related to the *scalar* part of the gravitational field, $c = c'\Phi^2$ is the *coordinate* spatially variable speed of light and $\gamma = (1 - u^2/c^2)^{-1/2}$ the kinematical affecting factor. In the case of a non-stationary source, the GMLT must take into account the *vector* potential field $\mathbf{w}$ — the *induced velocity* — caused by source movement [6]. An additional "Galilean" relation in coordinate space, according a local translation by the field $\mathbf{w}$ must then be applied to the $\delta t, \delta x$ according;

$$\delta\mathbf{x}_0 = \delta\mathbf{x}_w - \mathbf{w}\delta t_w \quad , \quad \delta t_0 = \delta t_w \tag{2}$$

The ensuing Hamiltonian formalism verifies till 1-PN correctly the gravitational phenomenology of GRT [3, 4, 5, 6].

THE MATTER LAGRANGIAN IN GRT AND LP-TYPE MODEL

In order to allow comparison we specify in GRT an *isotropic* metric with, at least till 1PN, the reciprocity between space and time components: $g_{00} = -g_{ii}^{-1}$ (or in general for the Yilmaz-Rosen metric). In GRT, "geodesic motion" follows form the variational principle extremizing the spacetime interval s between two points: $\delta s = 0$. With $ds = (g_{\mu\nu}dx^\mu/d\lambda\, dx^\nu/d\lambda)^{1/2}d\lambda$ the method corresponds to Hamilton's Principle with the "Lagrangian":

$$L = \left(g_{00}U_0^2 + g_{ii}\mathbf{U}^2\right)^{1/2} \quad (GRT) \tag{3}$$

where $U_0 = dt/d\tau$ and $\mathbf{U} = d\mathbf{x}/d\tau$.

In the LP model the Hamiltonian is obtained by transforming the free energy $m'c'^2$ in S' of the free-falling particle to E in S_0 by Λ_{Ep};

$$E = \left(m_0(\mathbf{x})^2 c(\mathbf{x})^4 + p^2 c(\mathbf{x})^2\right)^{1/2} = mc^2 \tag{4}$$

with $m = m_0\gamma(p,x) = m_0'\gamma(p,x)\Phi^{-3}$. Subsequently the Lagrangian becomes via the Legendre transformation:

$$L = \mathbf{p}.\mathbf{v} - H = -m_0 c^2\gamma^{-1} \tag{5}$$

let us now express the corresponding differential action in a covariant manner, using the hybrid —with variables of S_0 and S'— four velocities $(U_0, \mathbf{U})$ (Weierstrass parametrization [10]);

$$Ldt \;=\; -m_0 c^2 \gamma^{-1} dt \;=\; -m_0' c'^2 \left(\Phi^2 u_0^2 - \Phi^{-2} \mathbf{u}^2 / c'^2 \right)^{1/2} d\tau \quad (LP) \qquad (6)$$

Under the condition of the isotropic metric and reciprocity of space and time part of the metric, the Lagrangian forms of GRT and the LP-model are equivalent. The heuristic to obtain the Lagrangian in each is different; in the LP-model the Lagrangian associated to a particle comes about in a classical manner proper to Hamilton mechanical description of a test particle, while in GRT the "Lagrangian" of a test particle is obtained by formal correspondence in the variational principle for geodesic motion.

Directly posing the "Lagrangian" of a test particle in GRT appears therefor artificial, e.g. by Stephani ([11], section 14.2); "The Lagrangian L, which for force-free motion it is identical to the kinetic energy":

$$L \;=\; \frac{1}{2} m v^2 \;=\; \frac{1}{2} m g_{\alpha\beta} \dot{x}^\alpha \dot{x}^\beta \qquad (7)$$

with v a hybrid velocity ds/dt and m the rest mass. This "most quick" derivation of the equations of motion hinges on the ansatz of a *non-relativistic* energy expression followed by a substitution with a relativistic hybrid velocity. In this case the Lorentz-Poincaré interpretation of GRT and its 1PN implementation in the LP-model shows a clear heuristic advantage: the Lagrangian is an immediate consequence of the Hamiltonian which itself is directly obtained through the gravitationally modified Lorentz transformations. The principal aim of developing a Lorentz-Poincaré type interpretation is precisely to enhance a "classical" physical heuristic while maintaining the phenomenology of GRT.

Acknowledgments

This work was supported by FWO–Vlaanderen project F6/15-VC. A87.

REFERENCES

1. Bell J. S., *Speakable and Unspeakable in Quantum Mechanics*, Cambridge University Press, 1987
2. Brans C. and Dicke R.H., *Physical Review* **124**, 925, 1961.
3. Broekaert J., A Lorentz-Poincaré type interpretation of the WEP, gr-qc/0604107 (to appear in *IJTP*)
4. Broekaert J., in Proc. ESA SP-605 November 2005, (ed.) M. Cruise, 2005 a, gr-qc/0510017 (to appear)
5. Broekaert J., *Foundations of Physics*, **35**, 5, 839-864, 2005 b, gr-qc/0309023.
6. Broekaert J., A spatially-VSL gravity model with 1-PN limit of GRT, 2004, gr-qc/0405015 (submitted)
7. Brown H.R., Physical Relativity, Space-time structure from a dynamical perspective, Oxford UP, 2005.
8. Dehnen H., Hönl H., Westpfahl K., *Annalen der Physik*, 7, 6, 370-406, 1960.
9. Dicke R.H., *Reviews of Modern Physics*, **29**, 363-376, 1957.
10. Johns O., *Analytical Mechanics for Relativity and Quantum Mechanics*, Oxford UP, 2005.
11. Stephani H., *Relativity. An Introduction to Special an General Relativity*, Cambridge UP, 2004.
12. Thirring W.E., *Annals of Physics*, **16**, 96-117, 1961.
13. Wilson H.A., *Physical Review*, 17, 54-59, 1921.

Is it realistic to assume the same cosmic equation of state prior to and after atom formation?

N. Cereceda, M. I. Marqués, G. Lifante and J. A. Gonzalo

Facultad de Ciencias, C-IV
Universidad Autónoma de Madrid
28049-Madrid, Spain

Abstract Previous work (Acta Cosmologica XXIV-3, Krakov 1998) did show the close coincidence of two times: (1) the matter/radiation density equality time and (2) the atom formation time. This coincidence was obtained substituting observed data into the standard Friedmann-Lemaître solutions (with $\Lambda=0$). This work anticipated precisely the present "age" of the universe [$t_0 \sim (13.7\pm0.2)\times10^9$ years] and the present value of the local Hubble parameter [$H_0 \sim (65\pm2)$ km s/Mcp] given in February 2003 by NASA's WMAP satellite. For the cosmic epoque considered in this previous work, the cosmic equation of state was $RT \sim$ constant, consistent with a transparent universe made of atoms. Present work shows that a better coincidence between calculated and observed cosmic time (t_{ns}) and density (ρ_{ns}) is obtained for primordial nucleosynthesis with a "plasma" equation of state $RT^{4/3} \sim$ constant. This is perfectly consistent with a change in equation of state from atom formation time to present resulting in a t_0 and a ρ_0 in agreement with WMAP observations.

Keywords: Cosmic equation of state, nucleosynthesis, atom formation
PACS: 98.80.-k,98.80.Bp,98.80.Ft

1.- Introduction

At the time that explicit open solutions of Einstein's cosmological equation were worked out by Friedmann and Lemaître [1], the existence of a cosmic blackbody background radiation (CBR) was not known. Only some years later [2] the possible existence of such background radiation was predicted. Subsequently, Penzias and Wilson [3] detected a highly isotropic microwave 3 K radiation, latter identified with the CBR. More recently, [4] this radiation was characterized with the outmost precision by the team of researchers responsible for the successful COBE mission. They established its nearly perfect blackbody character as well as the very small anisotropies signalled by galaxy precursors.

The Friedmann-Lemaître open solutions relate parametrically the scale factor (R) and the cosmic time (t), but they do not specify the relationship between scale factor and background temperature (T). In fact the temperature is a much more convenient parameter to characterize cosmic events through the history of the universe.

Since the early days of the hot big-bang proposal, it has been assumed that the relative evolution with time of matter mass density (ρ_m) and radiation mass density (ρ_r) was the same at very early times as it is now.

At present (t_0) the radiation mass density ($\rho_{r0}=\sigma T_0^4/c^2$) decreases faster than the matter mass density ($\rho_{m0}=3M/4\pi R_0^3$). Consequently, the ratio ρ_r/ρ_m has been steadily

CP905, *Frontiers of Fundamental Physics (FFP8), Eighth International Symposium*
edited by B. G. Sidharth, A. Alfonso-Faus, and M. J. Fullana
© 2007 American Institute of Physics 978-0-7354-0412-0/07/$23.00

decreasing from earlier times to present times. In fact, now, $\rho_{r0}<<\rho_{m0}$. At an early time (t_x) at which $\rho_r(t_x)\cong\rho_m(t_x)$, which happens to coincide (Cereceda el al. 1998) with the atom formation time $(t_x=t_{equality}=t_{atom\ formation})$, a transition must have occurred from a previous opaque plasma universe, to the present transparent universe made up overwhelmingly of bound atoms.

Under present conditions (transparent universe), as it is well known, the relationship RT=constant between scale factor and background temperature describes well cosmic evolution. But, is it justified to use RT=constant prior to decoupling (t_x), when the universe was opaque to the expanding radiation? Using RT=constant prior to this time, as currently done, implies that the cooling rate of expanding sphere of radiation commoving with matter was the same as it is now, taking no notice of the transition from opacity to transparency at atom formation time. On the other hand, $RT^{4/3}$=constant, prior to decoupling, which directly implies $\rho_r(t)=\rho_m(t)$ all the way from $t<<t_x$ to t_x (i.e. at all times at which the universe was opaque to the expanding radiation), is compatible with a change in cooling rate at $t=t_x$ (decoupling) such that at $t\geq t_x$, when the universe became transparent, the relationship RT=constant begun to hold.

Prior to decoupling, $RT^{4/3}$=constant, which implies $\rho_r(t)/\rho_m(t)=1$ (constant), requires a steady entropy increase per particle from early times to atom formation, in contrast to RT=constant, which implies that the entropy per particle remained constant from very early times to present. We may note, however, that while it is reasonable to expect little scattering in a basically transparent universe (after decoupling) it is not so to assume zero scattering in an opaque plasma universe if this is the case (prior to decoupling). Large scale scattering is an irreversible process, and therefore, the entropy per particle should be expected to increase with time (prior to decoupling). To assume that violent interactions (scattering) are continuously taking place between the constituents (photons and ions/electrons) in an isolated system (the universe) and that, at the same time, the entropy per particle does not increase, appears most unlikely (see below).

A careful comparison of the predictions prior to decoupling $(t\leq t_x)$ obtained under both assumptions, $RT^{4/3}$=constant (radiation/matter balanced expansion, i.e. $\rho_r(t)=\rho_m(t)$), and RT=constant (radiation dominated expansion, i.e. $\rho_r(t)=(T/T_x)\rho_m(t)$), is therefore worthy of examination.

2.- Friedmann-Lemaître solutions

In order to make quantitative comparisons prior to decoupling we need to put the Friedmann-Lemaître solutions in proper form [5] and to use reliable values for H_0 (Hubble's constant) [6], t_0 (age of the universe) [7], and T_0 (present background radiation temperature) [8] , to get well defined values for temperature, time and density at decoupling/atom formation time. As explained in detail in reference [5] these values can be obtained in a straightforward manner as:

$$T_x\cong3.86\text{x}10^3\ K$$
$$t_x\cong1.47\text{x}10^{13}\ s = 4.66\text{x}10^5\ yrs$$

$$\rho_r(t_x) = \rho_r(t_x) \cong 1.88 \times 10^{-21} \ \text{g/cm}^3$$

In what follows we will make a quantitative comparison of results obtained under:

(1) $RT^{4/3}$=constant at $t \leq t_x$ (matter/radiation balanced expansion)   (1)

and

(2) RT=constant at $t \leq t_x$ (radiation dominated expansion)   (2)

at primordial nucleosynsthesis. A close examination of what is to be expected at still earlier times is left for further consideration elsewhere.

The general open ($k<0$) solution of Einstein's equation can be written in a convenient parametric form [5] as:

$$R = R_+ \sinh^2 y = R_+ y^2 \qquad\qquad (y<<1) \qquad\qquad (3)$$

with "y" being the usual parameter describing parametrically the Friedmann-Lemaître solutions (Peebles,1993).

$$t = \frac{R_+}{|k|^{1/2} c} \left(\sinh y \cosh y - y \right) \cong \frac{R_+}{|k|^{1/2} c} \frac{2}{3} y^3 \qquad\qquad (4)$$

for $y<<1$, which is applicable at times equal to or prior to atom formation, t_x. In general, for a relationship in the form of $(R/R_x)=(T_x/T)^n$, we get:

$$t_x \left(\frac{R}{R_x} \right)^{3/2} = t_x \left(\frac{T_x}{T} \right)^{3n/2} \qquad\qquad (5)$$

$$\rho_m = \rho_{mx} \left(\frac{R_x}{R} \right)^3 = \rho_{mx} \left(\frac{T}{T_x} \right)^{3n} \qquad\qquad (6)$$

Then for (1) $RT^{4/3}$=constant (matter/radiation balanced expansion) i.e. $n=4/3$ we obtain:

$$t_1 = t_x \left(\frac{T_x}{T} \right)^2 \qquad\qquad (7)$$

$$\rho_{m1} = \rho_{mx} \left(\frac{T}{T_x} \right)^4 \qquad\qquad (8)$$

and for (2) RT=constant (radiation dominated expansion), i.e. $n=1$, we get:

$$t_2 = t_x \left(\frac{T_x}{T} \right)^{3/2} \qquad\qquad (9)$$

$$\rho_{m2} = \rho_{mx}\left(\frac{T}{T_x}\right)^3 \qquad\qquad (10)$$

Since we know [5] T_x, t_x and ρ_{mx} at atom formation, to compare the results obtained under assumptions (1) and (2), all we need is using equations (7), (8) and (9)-(10), respectively. A direct check on how realistic or unrealistic are the predictions obtained under the above assumptions is therefore straightforward prior to decoupling.

3.- Nucleosynthesis cosmic time (t_{ns}) and density (ρ_{ns})

We take T_{ns} as the temperature at which primordial nucleosynthesis of ^{4}He stops. Fusion data [9,10] indicate that, for ignition to stop, the lowest value of the triple product $n_{io}T_{io}\tau_E$, where n_{io}=ion density, T_{io}=ion temperature and τ_E=confinement time, is given by:

$$n_{io}T_{io}\tau_E = 7x10^{21}\ \text{KeV s/m}^3 = 4.37x10^{30}\ \text{erg s/cm}^3$$

which corresponds to a lowest value for the temperature:

$$T_{io} = T_{ns} = 30\ \text{KeV} = 3.47x10^8\ \text{K} \qquad\qquad (11)$$

The expected time at nucleosynsthesis, on the other hand, can be estimated from the neutron to proton ratio (n/p), which became frozen thereafter [11]:

$$n/p \cong \frac{\frac{1}{2}Y_p}{1-\frac{1}{2}Y_p} = 0.131 \qquad\qquad (12)$$

being Y_p=0.232±0.008 the Helium abundance.

and from the neutron lifetime [12]:

$$\tau_n = 14.78\ \text{min} = 887\ \text{s} \qquad\qquad (13)$$

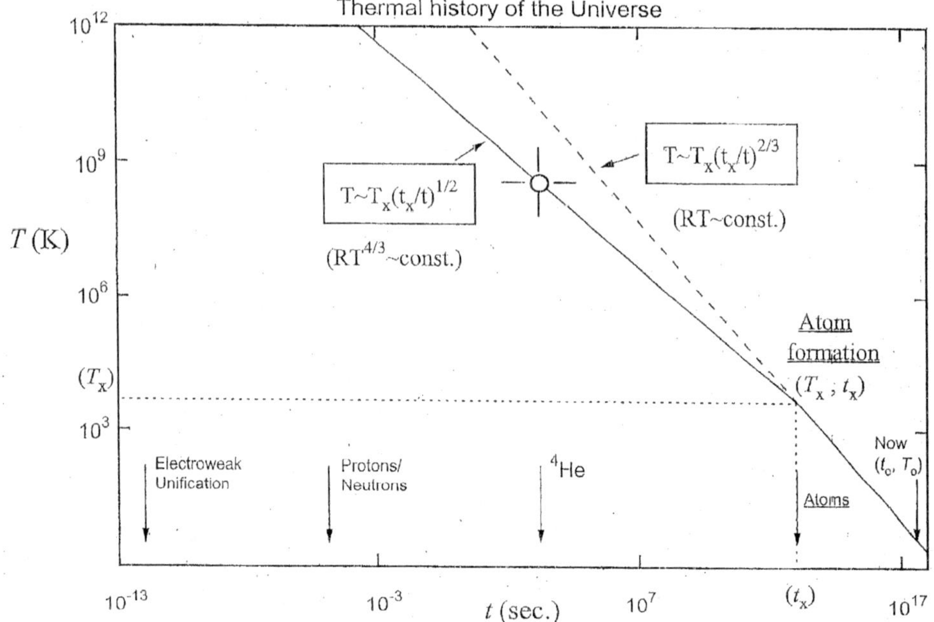

FIGURE.1 Cosmic temperature vs. Cosmic time: at $t \geq t_x$ (decoupling), $RT\sim$constant, resulting in $T\approx T_x(t_x/t)^{2/3}$, describes well the evolution of a transparent universe; at $t \leq t_x$ (prior to decoupling) $RT^{4/3}\sim$constant, resulting in $T\approx T_x(t_x/t)^{1/2}$, describes much better cosmic evolution at nucleosynthesis. The circle marks the expected values $t_{ns}=1.8\times10^3$ sec, $T_{ns}=3.47\times10^8$ K, from ^{4}He abundance observations, and fusion temperature estimation (see text) in good agreement with an equation of state $RT^{4/3}\sim$constant prior to decoupling.

These two values imply that:

$$n/p \cong e^{-t_{ns}/\tau_n} \tag{14}$$

i.e. $t(T_{ns})=t_{ns} = 1803$ s

It may be noted that current estimates of time, density, temperature at nucleosysnthesis, based on entirely different arguments, may give results at variance with equation (14).

Under assumption (1), according to equation (7), at any $t<t_x$, $T>T_x$ (i.e. before atom formation), using $t_x=1.47\times10^{13}$ s and $T_x=3.86\times10^3$ K we have:

$$tT^2 = t_xT_x^2 = 2.19\times10^{20} \text{ sK}^2 \tag{15}$$

and, in particular, at nucleosynsthesis ($t_{ns}=1803$ s, $T_{ns}=3.47\times10^8$ K) we obtain:

$$t_{ns}T_{ns}^2 = 2.18\times10^{20} \text{ sK}^2 \tag{16}$$

in excellent agreement with the value (1) predicted in equation (15).

Under assumption (2), after equation (9), before atom formation, using again the values for t_x, and for T_x previously given, we get:

$$tT^{3/2} = t_x T_x^{3/2} = 3.52 \times 10^{18} \text{ sK}^{3/2} \tag{17}$$

which, at nucleosynthesis (t_{ns} and T_{ns} as above), leads to:

$$t_{ns} T_{ns}^{3/2} = 1.17 \times 10^{16} \text{ sK}^{3/2} \tag{18}$$

in clear disagreement with the value (2) predicted in equation (17).

The fusion research data [9,10] appear to support physical conditions at nucleosynsthesis in consonance with a radiation/energy balanced cosmis expansion (assumption (1)). This is not the case with the currently assumed radiation dominated cosmic expansion (assumption (2)).

Therefore a radiation dominated cosmic expansion (assumption (2)) is unrealistic, or at least questionable, because either t_{ns} is required to be much larger than 1.8×10^3 s, resulting in an unacceptable low ^{4}He abundance, or T_{ns} is forced to be much larger than 3.47×10^8 K, which is in conflict with the information provided by fusion data for a lower limit on the ignition/fusion temperature.

4. Photon to particle ratio before and after atom formation

Let us examine briefly the behaviour of the proton (entropy)/particle ratio through decoupling (atom formation) under the assumption of a coupled expansion of matter and radiation prior to atom formation and an uncoupled expansion thereafter.

The entropy per particle (ion/electron pair prior to decoupling and bound atom after) is given by:

$$\sigma_{pp} = \frac{\rho_r / 2.8 k_B T}{\rho_m / mc^2} = \frac{\rho_r}{\rho_m} \frac{T_b}{T} \tag{19}$$

with $T_b = m_b c^2 / 2.8 k_B = 3.88 \times 10^{12}$ K

Then we have:

    - Prior to decoupling ($t \leq t_x$, $(\rho_r/\rho_m)=1$): $\sigma_{pp}(T)=(T_b/T)$ (20)

    - After decoupling ($t \geq t_x$, $(\rho_r/\rho_m)=T/T_x$): $\sigma_{pp}(T)=(T_b/T) \sim 10^9$ (21)

In other words, the entropy per particle increases from about $\sigma_{pp}(T_{ns}) \cong (3.88 \times 10^{12}/3.48 \times 10^8) \cong 1.11 \times 10^4$ at nucleosynthesis to about $\sigma_{pp}(T \geq T_x) \sim 10^9$ (constant) at atom formation and thereafter. This can be viewed as describing a period of energy transfer from radiation field to plasma particles (supplying to them kinetic energy, and increasing the entropy per particle of the system) followed by a period of zero energy transfer to the gas of formed atoms, later clustered into galaxies (not contributing to the entropy per particle)

We conclude that assuming the same cosmic cooling rate (RT~constant) prior to and after atom formation, which is currently the standard assumption, is inadequate to describe properly physical conditions at nucleosysnthesis as given by fusion research data. A cosmic cooling rate ($RT^{4/3}$~constant) prior atom formation (t_x) leading $\rho_r(t<t_x)= \rho_m(t<t_x)$, $\rho_r(t)= \rho_m(t)$ for $t\leq t_x$ describes properly conditions at nucleosysnthesis.

With this cooling rate for the cosmic plasma the coincidence previously noted between atom formation time, t_{af}, and decoupling time, t_x, maximum time at which $\rho_r=\rho_m$, is automatically guaranteed.

The most precise observational information on cosmic quantities after atom formation is presently given by reference to WMAP's data [13].

ACKNOWLEDGMENTS

One of us (JAG) would like to thank J.C.Matter for encouragement and helpful correspondence

REFERENCES

1. See f.i. P.J.E. Peebles, "Principles of Physical Cosmology" (Princeton University Press, 1993).
2. R.A. Alpher and R.C. Herman, Nature 162, 774 (1948).
3. A.A. Penzias and R.W. Wilson, Astrophys. J. 142, 419 (1965).
4. J.C. Mather et al. Astrophys J. 354, L37 (1990).
5. N. Cereceda, G. Lifante and J.A. Gonzalo, Acta Cosmologica XXIV-3, Krakov (1998).
6. W.L. Freedman et al. Nature 371, 757 (1994).
7. M.Botte and C.J. Hogan, Nature 376, 399 (1995).
8. J.C. Mather et al. Astrophys. J. 429, 439 (1994).
9. J.G. Cordey, R.J. Goldston and R.R. Parker, Phys. Today (Jan 1992), pp.22-30 and references therein.
10. J.D. Callen, B.A. Carreras and R.D. Stanbaugh, Ibid, pp. 33-42, and references therein.
11. K.A. Olive and G. Steigman, Astrophys. J. Suppl. Ser. 97, 49 (1995)
12. See f.i. E.J. Pagel, Physica Scripta, T36, 7 (1991).
13. See f.i. C.L. Bennett et al., ApJS **148**, 1 (2003).

Status of CMB Radiation

Màrius J. Fullana i Alfonso[1]* and Diego Sáez Milán[†]

*Institut de Matemàtica Multidisciplinar, Universitat Politècnica de València,
Camí de Vera s/n, 46022 València

†Departament d'Astronomia i Astrofísica, Universitat de València,
Dr. Moliner 50, 46100 Burjassot (València), Spain

Abstract. A brief review on the current status of cosmic microwave background researches is presented. First, a description of its discovery and nature is given and, then, its importance in Cosmology is pointed out. COBE and WMAP satellite observations are considered. The main information obtained from these space experiments is analyzed. Perspectives on the field are outlined.

Keywords: Cosmic microwave background, Cosmology, Large-scale structure of universe
PACS: 98.70.Vc, 98.80.-k, 98.65.Dx

INTRODUCTION

The Cosmic Microwave Background (CMB) radiation is a photon distribution, which has been always present in the Universe. Its discovery is a curious example of the intricate paths that science can take. Lemaître [1] was the first scientist to speculate about the possible remnants of the very early stages of the universe. This scientist imagined a hot beginning of the expansion and a relic background of cosmic rays. Tolman [2] introduced the idea of the thermal history of an expanding universe. He showed that expansion cools the black-body radiation while keeping a thermal spectrum. Alpher and Herman, [3] and Gamow [4] predicted a universal radiation background –remnant of the hot Big Bang– with a temperature of a few Kelvin degrees.

Penzias and Wilson [5], two researchers of the Bell Laboratories at Holmdel, detected a certain noise excess in their radiotelescope measurements. Dicke, Peebles, Roll & Wilkinson (Princeton University team) were required to explain the observed noise [6]; thus, the CMB was first detected as an almost isotropic antenna temperature excess $T_0 = 3.5 \pm 1.0K$ at a wavelength of 7.35 cm. Penzias and Wilson were awarded with the Nobel Prize (1978) for their discovery. Since then, many observations have been performed to describe the CMB with more and more accuracy. Among the most important of them, we can cite those of COBE [7]–[9] and WMAP [10]–[12] space missions.

Researches on the CMB have played a primordial role in Cosmology. Many advances in our understanding of the universe have been motivated by CMB observations and related theoretical predictions. Here, the information we have got during a long period of CMB researches is included in a short review on this subject (see also [13]). First of all, the CMB origin, nature, and properties are described. The so-called C_ℓ coefficients play

---

[1] E-mail. mfullana@mat.upv.es

CP905, *Frontiers of Fundamental Physics (FFP8), Eighth International Symposium*
edited by B. G. Sidharth, A. Alfonso-Faus, and M. J. Fullana
© 2007 American Institute of Physics 978-0-7354-0412-0/07/$23.00

a crucial role because they describe the angular correlations of the CMB temperature distribution induced by physical processes in the past. We focus our attention on satellite experiments. The first satellite designed to study the CMB was COBE [7]. We present its main results. Afterwards, a theoretical description of some important concepts related with the CMB is given and, then, the current status of the CMB researches is outlined by writing a brief summary of WMAP results and implications [10]. Finally, some perspectives are highlighted.

CMB ORIGIN AND RELEVANCE

There existed a primeval plasma with free protons and electrons (not confined in atoms). The mean free path of the CMB photons –through this hot plasma– was very small as a result of frequent interactions with free electrons. No information could be carried, the universe was opaque. The CMB thermalized in that epoch (black-body spectrum) due to its coupling with matter. Since the universe cooled as a result of expansion, its temperature became low enough to allow the formation of stable neutral hydrogen atoms (recombination), this process led to a transparent universe because free electrons disappeared and, consequently, matter and radiation decoupled (at $z \sim 1100$). From decoupling to present time, CMB propagation has been almost free, the form of the spectrum has not changed, and its temperature has decreased. The CMB photons reaching our detectors come from a narrow region between two close surfaces (perturbed spheres). If its thickness is neglected, one can define the so-called Last Scattering Surface (LSS), which would be a perfect sphere in a Friedmann-Robertson-Walker (FRW) universe, where the CMB would be absolutely isotropic.

Small CMB temperature fluctuations, anisotropies, have been found. The detection of these anisotropies has strongly enhanced the importance of the CMB; since they are due to inhomogeneities, the analysis of the observed anisotropy supplies substantial information about the composition and the large-scale structure of the universe at high redshift. A complete theory about the formation and evolution of FRW perturbations is then necessary. Three different kinds of linear FRW fluctuations exist: (1) Scalar perturbations: energy density fluctuations leading to the observed cosmological structures. (2) Vector perturbations: divergenceless velocity fields which do not appear in most scenarios, and (3) Tensor perturbations: pure metric fluctuations describing the propagation of gravitational waves in the universe. Scalar (tensor) modes appear in any (many) admissible inflationary model (models).

Observational constraints on the deviations of the CMB spectrum with respect to a perfect black-body one set constraints on physical processes generating this kind of deviations. Those producing too great deviations must be rejected. Observational constraints on the CMB anisotropies severely test theories of large structure formation. Any of these theories producing too large anisotropies must be ruled out.

CMB ANISOTROPIES

FRW fluctuations break the homogeneity of the universe producing CMB anisotropies through different physical processes. Anisotropies created near the LSS are called *primary*, whereas those produced long after decoupling (far from the LSS) are called *secondary*. The latter ones are subdominant.

Primary anisotropies

(1) The Doppler effect produced by peculiar velocities on the LSS.
(2) Thermal anisotropies: initial fluctuations of the photon gas near the LSS.
(3) The Sachs-Wolfe effect due to the peculiar gravitational potential at the LSS.

Secondary anisotropies

(1) The Rees-Sciama effect: caused by the time variation of the peculiar gravitational potential created by nonlinear structures located close to the photon trajectories [14].
(2) Weak lensing: integrated deviations of the photon propagation directions produced by the gradient of the peculiar gravitational potential created by cosmological inhomogeneities. These deviations deform the angular distribution of CMB temperatures and, consequently, they modify the Angular Power Spectrum (APS, see below).
(3) The Integrated Sachs-Wolfe (ISW) effect: associated to the time variation of the peculiar gravitational potential created by big linear inhomogeneities along the null geodesics. This effect does not appear in flat universes without cosmological constant [15].
(4) The Sunyaev-Zel'dovich effect: inverse Compton scattering between CMB photons and hot electrons located inside galaxy clusters.
(5) Effects produced during reionization: Path mixing and Vishniac anisotropy
(5a) Path mixing damps pre-reionization anisotropies. As a result of Thompson scattering in a reionized universe, CMB photons change directions and, consequently, the photons reaching our detectors in a certain direction come from different points of the LSS (anisotropy suppression due to averaging).
(5b) The Vishniac anisotropy is due to a second order term, which couples density contrast and peculiar velocity. This term is relevant in a reionized universe and it produces small angular scale anisotropy.
(6) Anisotropies produced by primordial gravitational waves.

CMB ANGULAR POWER SPECTRUM

The APS describes the second order (two directions) angular correlations of the CMB temperatures. It plays a fundamental role in Cosmology. If the temperature distribution is Gaussian, this spectrum completely describes the CMB anisotropy from a statistical point of view; if not, it contains very important information, although a complete description requires angular correlations for three and more directions.

In standard inflationary models, the FRW fluctuations are Gaussian and remain so during linear evolution. Only nonlinear evolution produces deviations from Gaussianity, but these deviations are expected to be small. In some models for the generation of cosmological perturbations, primordial deviations from Gaussianity exist.

It is commonly assumed that CMB maps are Gaussian, homogeneous, and isotropic statistical distributions of temperatures. Under this assumption, the following definitions and equations hold.

The two directions correlation function is:

$$C(\theta) = \left\langle \frac{\delta T}{T}(\vec{n}_1) \frac{\delta T}{T}(\vec{n}_2) \right\rangle \tag{1}$$

where $\vec{n}_1 \cdot \vec{n}_2 = \cos\theta$ and the average is to be done on many CMB sky realizations.

The C_ℓ quantities, *angular power spectrum*, are:

$$C_\ell = \frac{1}{2\ell+1} \sum_{m=-\ell}^{m=\ell} \left\langle |a_{\ell m}|^2 \right\rangle \tag{2}$$

where $a_{\ell m}$ are the coefficients of the spherical harmonic expansion of $\delta T/T$. The mean is performed using the $a_{\ell m}$ quantities of many CMB sky realizations.

Furthermore, one easily proves the relations:

$$C_\ell = 2\pi \int_0^\pi C(\theta) P_\ell(\cos\theta) \sin\theta \, d\theta \tag{3}$$

$$C(\theta) = \frac{1}{4\pi} \sum_0^\infty (2\ell+1) C_\ell P_\ell(\cos\theta) , \tag{4}$$

from which one easily see that the C_ℓ multipole is mainly due to the correlations at angular scales close to $\theta = \pi/\ell$.

Only one realization of the CMB sky is available; therefore, quantities $C(\theta)$ and C_ℓ obtained from this limited statistical sample have a certain unavoidable error. The estimated angular spectrum C_ℓ^{est} deviates with respect to the true spectrum C_ℓ [16] and

$$\left\langle \left(C_\ell^{est} - C_\ell \right) \left(C_{\ell'}^{est} - C_{\ell'} \right) \right\rangle = \frac{2}{2\ell+1} C_\ell^2 \delta_{\ell\ell'} \tag{5}$$

hence, the tipical deviation of C_ℓ^{est} is $\Delta C_\ell = \left(\frac{2}{2\ell+1}\right)^{1/2} C_\ell$.

The distribution of C_ℓ^{est} around $C_\ell = \langle C_\ell^{est} \rangle$ is not Gaussian, but a $\chi^2_{2\ell+1}$ distribution which is almost Gaussian for $\ell > 15$.

COBE MAIN RESULTS

Accurate observations of the CMB spectrum and anisotropies were performed with FIRAS and DMR devices on board of COBE satellite. Mater and Smoot have been awarded with the Nobel Prize (2006) for leadering that project, whose main results (see [7] – [9]) can be summarized as follows:

The best-fit black-body temperature is (with 95% CL):

$$T_\gamma = 2.728 \pm 0.002K \,, \tag{6}$$

so, the number of CMB photons per unit of volume, n_γ, and the energy density, ρ_γ, are

$$n_\gamma \simeq 413cm^{-3}, \qquad \rho\gamma \simeq 4.68 \times 10^{-34} gcm^{-3} \tag{7}$$

The distortion parameter, y, with respect to a blackbody spectrum and the chemical potential of the CMB photons, μ_0, are constrained as follows:

$$|y| < 1.5 \times 10^{-5}, \qquad |\mu_0| < 9 \times 10^{-5} \tag{8}$$

The best-fit dipole (related with C_1) and quadrupole (related with C_2) amplitudes were found to be:

$$D_{obs} = 3.357 \pm 0.001 \pm 0.023 mK, \qquad Q_{rms-PS} = 18 \pm 1.6 \mu K \tag{9}$$

The C_ℓ quantities were estimated for ℓ (θ) values smaller (greater) than ~ 26 ($\sim 7°$). Two directions forming an angle $\theta > 1°$ subtend a distance on the LSS, which is greater than the effective horizon at decoupling H_d; therefore, in the standard FRW model without inflation, COBE should have observed vanishing C_ℓ coefficients. By this reason, COBE observations imply causal correlations at distances greater than H_d, which is only possible in some models as, e.g., the inflationary ones. Since inflation was known from 1980, the observed COBE correlations (for $\theta > 7°$) were not a surprise. COBE verification of these nonvanishing correlations gave strong support to inflation. It was one of the main achievements of that space mission.

SOME GENERAL CONSIDERATIONS

Let us now discuss some concepts related to the interpretation of CMB observations.

Concordance Model: Several recent observations: SN Ia, CMB, Galaxy Surveys, and so on strongly support the so-called *concordance model*, which is a flat universe with: a cosmological constant plus cold dark matter (ΛCDM model), a Gaussian distribution of adiabatic energy density fluctuations with a power law spectrum close to a Zel'dovich one, and a certain reionization. WMAP data alone lead to a similar model with slightly different parameters. It is hereafter called the WMAP ΛCDM model.

Reionization: After decoupling (at $z \simeq 1100$), matter and radiation may partially couple again as a result of a reionization produced by high frequency radiation from big (third generation) stars (at high redshift). During reionization, path mixing and Vishniac effects (see above) change previous anisotropy. Previous CMB polarization is also modified as a result of new Thompson interactions.

Inflation. The inflationary paradigm solve some problems of the standard FRW universe (flatness, horizon, monopoles and so on). It is a period of accelerated expansion produced by some scalar field with an appropriate effective potential. During inflation, quantum fluctuations of the inflationary field and the metric produce scalar and tensor perturbations, respectively. In the standard inflationary models, the resulting density inhomogeneities are Gaussian and nearly scale invariant (Zel'dovich spectrum) explaining the main aspects of structure formation for an appropriate spectrum normalization. During inflation, the FRW fluctuations are causally generated inside the effective horizon, but they rapidly become greater than this horizon (fast expansion) and, then, they keep outside it at $z \sim 1100$, which explains the existence of correlations for $\theta > 7°$ (in regions greater than H_d) observed by COBE.

Dark Energy: Supernova observations supply the most direct evidence for this kind of energy, which is also compatible with CMB observations and galaxy surveys. Its nature is a mystery. The presence of a residual vacuum energy density or cosmological constant, Λ, is the most common explanation; nevertheless, fine-tuning and coincidence troubles arise in trying to explain the 120 orders of magnitude discrepancy between the theoretically predicted energy density of Λ and the observed dark energy density. Several alternative theories have been proposed (quintessence).

Polarization: The CMB polarization appears as a result of the interaction between electrons and photons through Thompson scattering [17]. Polarization appears because the cross section of Thompson scattering is proportional to $|\varepsilon \cdot \varepsilon'|^2$, where ε and ε' are the incident and scattered polarization directions, respectively. Polarization can only appear in an inhomogeneous universe. During the recombination decoupling process, there are quadrupolar components of the CMB intensities (produced by previous evolution of the FRW perturbations) and, as a result of the angular dependence of the Thompson cross section, these quadrupoles lead to polarization. Scalar, vector and tensor FRW perturbations produce different quadrupolar CMB intensities leading to distinct polarization patterns.

Since Thompson polarization is linear, the Stokes parameter V vanishes. Symbol T stands for the temperature associated to I (the total intensity), and parameters Q, U are not the best ones for cosmological considerations. The transformation laws –under rotations– of quantities $Q \pm iU$ allow their expansion in terms of the so-called spin spherical harmonics. In these expansions appear the coefficients E_ℓ and B_ℓ from which the correlations $C_\ell^{TT}, C_\ell^{TE}, C_\ell^{EE}$, and C_ℓ^{BB} can be easily defined [18] (any other crossed correlation vanishes). It can be proved that scalar (vector) inhomogeneities produce linear polarization with $B = 0$, $(E = 0)$, whereas tensor fluctuations have $E \neq 0$ and $B \neq 0$; therefore, in the absence of vector modes, the detection of any B component implies the presence of a background of gravitational waves (tensor modes), at least, if it proved that this component is not produced either by contaminants or by lensing.

Gaussianity: The detection of primordial non-Gaussian fluctuations in the CMB would have a deep impact on our understanding of the physics of the early universe. Many plausible mechanisms have been proposed. The usual study parameterizes deviations with respect to Gaussianity by writing: $\Phi(\vec{x}) = \psi_L(\vec{x}) + f_{NL}\psi_L^2(\vec{x})$, where ϕ_L is the Bardeen curvature potential, $\psi_L(\vec{x})$ is a Gaussian field, and $|f_{NL}|$ is a parameter whose value measures deviations from Gaussianity. Minkowsky functionals and other techniques can be used to estimate deviations from Gaussianity.

PRESENT STATUS. WMAP RESULTS

The Wilkinson Microwave Anisotropy Probe (WMAP) has mapped the entire sky in five frequency bands between 23 and 94 GHz. More information can be found at `http://lambda.gsfc.nasa.gov`. Some cosmological implications of the WMAP measurements (see [10]–[12]) are now discussed taking into account the comments of previous section.

Constraining parameters in the WMAP ΛCDM model: According to [10], a simple cosmological model with only six parameters fits the three year WMAP temperature and polarization data, small angular scale CMB data from other experiments, light element abundances, large scale structure (lss) in big galaxy surveys, and the observed SN Ia magnitude-redshift relation. The best fit values obtained from WMAP data alone are: $h = 0.734^{+0.028}_{-0.038}$ for the reduced Hubble constant, $\Omega_m h^2 = 0.1268^{+0.0072}_{-0.0095}$ for matter density, $\Omega_b h^2 = 0.02233^{+0.00072}_{-0.00091}$ for baryon density, $\sigma_8 = 0.744^{+0.050}_{-0.060}$ for power spectrum normalization, $\tau = 0.088^{+0.028}_{-0.034}$ for the reionization optical depth, and $n_s = 0.951^{+0.015}_{-0.019}$ for the spectral index of the scalar FRW perturbations. The allowed volume in the six dimensional parameter space is small (good fit) and it decreases when all the available data sets are taken into account (concordance model). If additional data are considered, parameters $\Omega_b h^2$, n_s, and τ are almost unaltered, whereas the remaining three parameters undergo more important changes depending on the chosen data sets.

The WMAP ΛCDM model best fit value for the age of the universe is $t_0 = 13.73^{+0.13}_{-0.17} Gyr$, which is compatible with the estimated ages of globular clusters and white dwarf stars.

Inflation and Gravitational Waves: A model with seven parameters is considered [10] to constraint inflationary models using WMAP data. The new parameter is r, the ratio of the tensor to scalar power spectrum (at $k = 0.002\ Mpc^{-1}$). The predictions of inflationary models depend on the form of the inflation potential. For monomial potentials of the form $V \propto \phi^\alpha$, one has (i) $r \simeq 4\alpha/N$, where N is the number of e-folds of inflation between horizon crossing of relevant scales and the end of inflation, and (ii) $1 - n_s \simeq (\alpha + 2)/2N$; hence, for $N > 60$ and $\alpha \leq 4$, one finds $r < 0.27$. Since N could be much greater than 60 and α should not be much greater than 4, this rather general type of inflation (monomial potential) strongly suggests small r values and, consequently, great problems with the detection of primordial gravitational waves.

Tensor and scalar modes contribute to the APS for $\ell < 100$; hence, for a fixed APS (e.g., that obtained from WMAP data), the greater r, the smaller (greater) the amplitude of the scalar (tensor) modes. Since the amplitude of the scalar density perturbations cannot be too small to explain: (i) the APS for $\ell > 100$ and (ii) lss formation, parameter r cannot exceed a certain value. WMAP data alone bound the seventh parameter of the model to ensure compatibility with the full APS ($r \leq 0.55$ at 95% CL). The upper limits obtained from WMAP and different lss data sets are more restrictive (see [10]), e.g., WMAP plus SDSS (Sloan Digital Sky Survey) give $r \leq 0.28$ at 95% CL, in agreement with inflationary bounds. A running n_s spectral index reduces these r bounds leading to $r \leq 1.5$ at 95% CL for WMAP alone and $r \leq 0.67$ at the same CL for WMAP plus SDSS.

Dark energy properties: Let us first assume a flat universe with an homogeneous dark energy component (without perturbations), whose equation of state is $P_{DE} = W\rho_{DE}$ with constant negative W ($W = -1$ for vacuum energy). This component alters: (1) the evolution of the scale factor $a(z)$ and, consequently, the distance to the LSS, and (2) the linear evolution of density fluctuations, namely, the growing mode $D(z)$. In a flat universe without dark energy, the relation $D(z) \propto a(z)$ holds and no ISW effect is produced (see above); however, for the function $D(z)$ corresponding to the flat model with dark energy –under consideration– an ISW contribution to the APS arises which depends on the W value. This contribution is particularly relevant for large angular scales (small ℓ values) and, consequently, it enhances the CMB multipoles for small ℓ values. The smaller the value of W, the greater the enhancement. Too small W values cannot be accepted because they lead to a too great ISW effect, which would account for a too great part of the WMAP multipoles for small ℓ (e.g., for $\ell = 2$); in this situation, the Sachs-Wolfe dominant contribution to C_2 would be too small and, consequently, parameters σ_8, h and so on would be incompatible with the lss data sets and with the full APS obtained from WMAP. In the case of WMAP plus SDSS, Spergel et al. [10] concluded that the inequality $W > -0.69^{+0.19}_{-0.18}$ must be satisfied, whereas for WMAP plus 2dFGRS (two degree Field Galaxy Redshift Survey) the resulting W lower limit is $-0.877^{+0.094}_{-0.110}$. In any case, values $W < -1$ are forbidden. If fluctuations of the dark energy are allowed, the ISW is smaller and the W bounds are less restrictive. Finally, if small deviations from flatness are considered, dark matter with $W \sim -1$ is still preferred by the data (see [10]).

Constraints on neutrino properties: Atmospheric and solar neutrino experiments show they are massive. These experiments give $m^2_{\nu_i} - m^2_{\nu_j}$. Cosmological measurements lead to complementary constraints on $\sum_i m_{\nu_i}$ (hereafter $\sum m_\nu$). Since light massive neutrinos are hot dark matter (HDM) they do not cluster as efficiently as CDM on small scales and, consequently, their presence alters lss evolution lowering the σ_8 parameter by a factor proportional to $\sum m_\nu$. This factor cannot be too great to prevent too low σ_8 values. For an effective number of neutrino species $N_\nu = 3.02$, WMAP alone, WMAP plus SDSS, and WMAP plus 2dFGRS give the following upper limits at 95% CL: $\sum m_\nu < 2.0 \, eV$, $\sum m_\nu < 0.91 \, eV$, and $\sum m_\nu < 0.87 \, eV$, respectively. Furthermore, the lss and CMB properties in a universe with stable light particles (HDM components) are the same as in a model with certain effective number of neutrino species N_ν^{eff} (see [19]). Since too high proportions of HDM are not compatible either with lss formation (small σ_8) or with the CMB APS, number N_ν^{eff} can be bounded. From WMAP plus SDSS (WMAP plus 2dF-GRS) one obtains $2.47 < N_\nu^{eff} < 6.17$ ($1.01 < N_\nu^{eff} < 2.94$), see [10]. The WMAP plus 2dFGRS bounds are very restrictive.

Gaussianity: After an appropriate cleaning of the WMAP three years maps (to avoid residual noise and contaminants), various methods have been used to look for primordial deviations from Gaussianity. The probability distribution function, the Minkowski funtionals, the bispectrum and trispectrum of the cleaned maps. The limits on primordial non-Gaussianity have improved from $-58 < f_{NL} < 137$ for one year WMAP data to $-54 < f_{NL} < 114$ (95% CL) for three years maps. These inequalities are compatible with Gaussianity ($f_{NL} = 0$) but also with deviations. Spergel et al. have detected a certain deviation from Gaussianity in the cleaned maps by using a certain method (modulation by an arbitrary function, see section 8.5 in [10]). It looks like deviations claimed in previous papers.

Polarization: WMAP has polarization sensitive radiometers. Significant levels of polarized foreground emission due to both Galactic synchrotron radiation and thermal dust emission must be subtracted to estimate intrinsic CMB polarization. In a ΛCDM model with optical depth $\tau = 0.09$ and $r = 0.3$, the CMB polarization levels are $\approx 0.3\mu K$ for the E-mode and $\approx 0.03\mu K$ for the B-mode.

In the three years WMAP foreground corrected maps, a signal with $\ell(\ell + 1)c^{EE}_{\ell=<2-6>}/2\pi = 0.086 \pm 0.029(\mu K)^2$ has been detected. This polarization could be produced by the Thompson scattering of CMB photons and free electrons released during an instantaneous reionization at $z_r = 10.9^{+2.7}_{-2.3}$. The optical depth of a reionization accounting for the observed EE correlations appears to be $\tau = 0.10 \pm 0.03$. If TT, TE and EE correlations are considered, one finds $\tau = 0.09 \pm 0.03$.

B-modes have not been detected with WMAP. Three years data only lead to the constraints $\ell(\ell + 1)c^{BB}_{\ell=<2-6>}/2\pi = -0.04 \pm 0.03(\mu K)^2$. Polarization signals alone leads to the inequality $r < 2.2$ (95% CL). These r values are close to the upper bounds predicted from monomial inflation effective potentials (see above).

Unexpected Anomalies and open problems: The C_ℓ multipoles measured by WMAP are too small for $\ell < 10$. Even if the cosmic variance is taken into account (see above), the observed value of C_2 appears to be very unlikely. The existence of multipole alignments, a certain asymmetry between the North and South hemispheres, and a big non-Gaussian cold spot have been reported. All these findings were not expected in the framework of standard cosmological models of structure formation (anomalies) and, consequently, they need further study. Moreover, there are some open problems as, e.g. the role of a running spectral index, the detection of gravitational waves, the fact that the primordial Li abundance predicted from WMAP data is different from the observed one. All these points also require more discussion.

SUMMARY AND PERSPECTIVES

WMAP has improved on: APS and parameter estimations in a ΛCDM model, polarization measurements (reionization), the study of deviations from Gaussianity, discussions about inflation, gravitational waves, dark energy, nucleosinthesis abundances and so on.

PLANCK ESA Probe Project will be launched in the near future. Its observations will improve on the sensitivity, the angular resolution and the frequency coverage with respect to WMAP; hence, it is expected: (1) a better subtraction of the galactic and extra-galactic contaminants, (ii) a more accurate estimation of the cosmological parameters, (iii) a better study of $\ell < 10$ multipoles and their alignments, (iv) a more detailed study of the deviations from Gaussianity, (v) a better description of the CMB polarization properties, (vi) new information on the Sunyaev-Zel'dovich effect, and so on. The detection of the B-polarization due to gravitational waves is very problematic, it will be only possible if parameter r takes on large enough values (other missions are being designed to measure this polarization mode). In the framework of PLANCK mission, we are studying: beam deconvolution of temperature maps, weak lensing by nonlinear structures, and the Rees-Sciama effect; see [20] [21] and [14] for interesting discussion on these topics.

ACKNOWLEDGMENTS

This work has been supported by the Spanish Ministerio de Educación y Ciencia, MEC-FEDER projects AYA2003-08739-C02-02 and FIS2006-06062. One of us, M.J. Fullana i Alfonso, wants to thank his wife Teresa Jordán i Pla and their daughter Ariadna Mariola for their help and patience in all the two year organization hard work. I love you. I also want to remember my parents Eugènia Matilde and Mariano, who recently died, for all their teaching and all the love they gave me. As workers, they defended their land and fought for a better world in a wonderful way. I carry you in my heart.

REFERENCES

1. A.G. Lemaître, *La reveu des Quetions Scientifiques, 4é série* **20**, 391 (1931).
2. R.C. Tolman, *Proc. Nat. Acad. Sci.* **20**, 169-176 (1934).
3. R. A. Alpher and R.C. Herman *Nature* **162**, 774 (1948).
4. G. Gamow, *Nature* **162**, 680-682 (1948).
5. A.A. Penzias and R.W. Wilson *ApJ Lett.* **142**, 419-421 (1965).
6. R.H. Dicke, P.J.E. Peebles, P.G. Roll and D.T. Wilkinson, *ApJ Lett.* **142**, 414-419 (1965).
7. G.F. Smoot et al., *Astrophys. J.* **396**, L1-L6 (1992).
8. C.L. Bennet et al., *Astrophys. J.* **396**, L7-L12 (1992).
9. E.L. Wright et al., *Astrophys. J.* **396**, L13-L18 (1992).
10. D.N. Spergel et al., *Astrophys. J.* (2006) submitted, (astro-ph/0603449).
11. L. Page et al., *Astrophys. J.* (2006) submitted, (astro-ph/0603450).
12. G. Hinshaw et al., *Astrophys. J.* (2006) submitted, (astro-ph/0603451).
13. M.J. Fullana, J.V. Arnau and D. Sáez, *Vistas in Astronomy* **41**, 467-492 (1998).
14. N. Puchades, M.J. Fullana, J.V. Arnau and D. Sáez, *MNRAS* **370**, 1849-1858 (2006).
15. M.J. Fullana and D. Sáez, *New Astronomy* **5**, 109-120 (2000).
16. L. Knox, *Phys. Rev.* **52D**, 4307 (1995).
17. D. Sáez, *Nuc. Phys. B (Proc. Suppl.)* **95**, 15-22 (2001).
18. W. Hu and M. White, *Phys. Rev. D* **56**, 596 (1997).
19. E.W. Kolb and M.S. Turner, *The Early Universe*, Addison-Wesley, New York, 1994.
20. C. Burigana and D. Sáez, *Astron. & Astrophys.* **409**, 423 (2003).
21. L. Antón, P. Cerdá-Durán, V. Quilis and D. Sáez, *Astrophys. J.* **628**, 1 (2005).

Scale invariance of dark matter clustering

José Gaite

Instituto de Matemáticas y Física Fundamental, CSIC, Serrano 113bis, 28006 Madrid, Spain

Abstract. The dark matter distribution is arguably scale invariant in a range of scales, so that it has a fractal geometry, observable in the clustering of galaxies and in *cosmic voids*. We review evidence of fractal geometry in recent observations, which shows, in particular, that a simple fractal model is not sufficient. Therefore, we propose a multifractal model of the dark matter distribution realized on fractal *halo* populations, in which voids appear naturally.

Keywords: Large-scale structure of the universe, Fractals
PACS: 98.65.Dx, 05.45.Df, 02.50.-r

INTRODUCTION

Scaling laws in the cosmic structure and fractal models of it arise from both ancient hierarchical models of the Universe and modern studies of the distribution of galaxies [1, 2]. Scaling is usually demonstrated by the appearance of power laws, which are found in the statistical analysis of the galaxy distribution, namely, in their correlation functions. Other scaling laws refer to geometrical features of the cosmic structure. For example, the counterpart of galaxy clusters are galaxy *voids*, which are large empty regions in the galaxy distribution. Fractal voids have scaling sizes [4, 5]. This law can be verified in galaxy surveys [6, 7].

However, most matter is dark, so the distribution of galaxies is only one aspect of the cosmic structure. The distribution of galaxies can be *biased* with respect to the full matter distribution and, in fact, different galaxy populations may have different distributions. Besides, galaxy voids may be filled with disperse dark matter.

At any rate, the dark matter distribution can be argued to be scale invariant on more general grounds. Cold dark matter (CDM) models predict *bottom-up* structure formation, such that, on small non-linear scales, the structures quickly become independent of the initial conditions and are dominated by the scale-invariant gravitational dynamics, namely, by *virial equilibrium*. Indeed, we shall show that the dynamics is dominated by a *multifractal attractor*. Multifractals are the most general scaling models and they were introduced in cosmology to describe "non-uniform" fractal clustering [3]. Non-linear CDM structure formation can be studied with N-body simulations. Their results agree with a multifractal model composed of fractal distributions of different dark-matter *halos* (mass concentrations) which can support different fractal galaxy populations [8].

We briefly review here basic notions of fractal clustering and the observational evidence of the fractal distribution of galaxies. This evidence, together with the analysis of results of cosmological simulations, leads us to introduce a multifractal model. Then we consider its implications for dark-matter halos and galaxy voids.

CP905, *Frontiers of Fundamental Physics (FFP8), Eighth International Symposium*
edited by B. G. Sidharth, A. Alfonso-Faus, and M. J. Fullana
© 2007 American Institute of Physics 978-0-7354-0412-0/07/$23.00

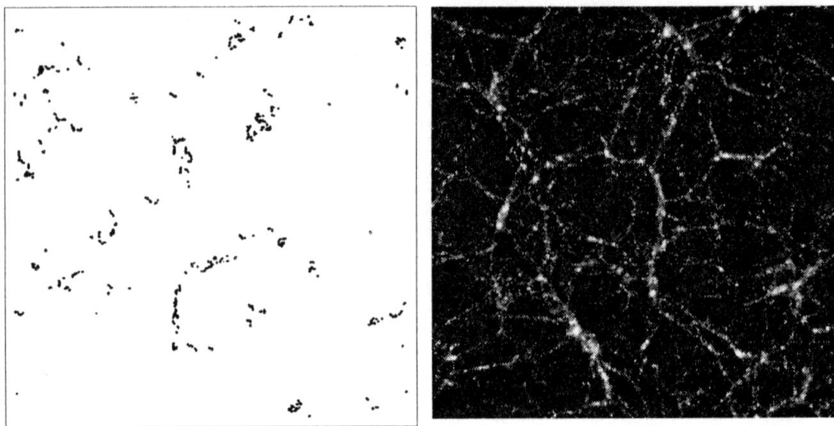

FIGURE 1. Random fractal with $D = 0.8$ and slice of "Cosmic Web" multifractal produced by a cosmological N-body simulation

FRACTAL GALAXY CLUSTERING

Hierarchical clustering can be described by means of fractal geometry [1], which deals with sets (or functions) that have *fine structure*, namely, detail at all scales. Usually, the fine structure of a fractal is due to its *self-similarity*, that is, to the set being similar to parts of itself, in a strict or approximate sense. Random fractals only have *statistical* self-similarity (Fig. 1), which implies that the correlation functions are power laws.

Random fractals can be defined by a power-law *number-radius* relation [1]. This relation expresses the number of points in a ball of radius r centered on one point and *averaged* over every point (the cumulative two-point correlation function). So $N(r) = Cr^D$, where D is the fractal dimension and C a constant. The distribution must have a transition to homogeneity at some scale r_0, such that $r \gg r_0 \Rightarrow D = 3$. Typical values of the fractal dimension and the scale of transition to homogeneity are, respectively, $D \simeq 2$ [2, 10] and $r_0 \simeq 15$ Mpc h^{-1} [10].

Scaling of voids

Voids scale if the number of voids with a given size is a power law of the size. Or if the cumulative count, namely, the number of voids $N(L > \ell)$ with linear size larger than a reference ℓ, fulfills $N(L > \ell) \propto \ell^{-D}$, where D is the fractal dimension [4, 5]. Since the cumulative count is the *rank*, the preceding law can also be expressed as a power-law dependence of size with rank (if size refers to volume, the exponent is $3/D$). A power-law rank order is *Zipf's law*.

Regarding voids in galaxy surveys, recent analyses find sets of convex-like voids that satisfy Zipf's law [6, 7]. The fractal dimension deduced from them, $D \simeq 2$, coincides with the dimension deduced from galaxy clustering. However, $D = 2$ is the dimension

24

of the boundary of voids and, therefore, the minimal value of D in the Zipf law for voids [11]. The actual fractal dimensions of the samples are probably smaller [6, 7]. The flattening of Zipf's law for large voids allows one to determine the scale of transition to homogeneity [5], which agrees with the one deduced from galaxy clustering [7].

Luminosity segregation

Different galaxy populations may have different statistical properties and, in particular, they may have different fractal dimensions. In fact, $D \simeq 2$ is typical, but other analyses find smaller values, and one can change D somewhat by selecting different galaxy populations. For example, analyses of galaxy populations in the Sloan Digital Sky Survey (SDSS) selected by luminosity show a decrease of fractal dimension with luminosity [9]. Therefore, it is preferable a *multifractal* model, which allows for various dimensions.

MULTIFRACTAL MODEL

Multifractals are the most general scaling mass distributions. They appear frequently as attractors of nonlinear dynamical systems. Multifractal measures represent highly irregular mass distributions, with mass concentrations of very different magnitude. This magnitude is defined by the *local* dimension $\alpha(x)$:

$$m[B(x,r)] \sim r^{\alpha(x)}, \tag{1}$$

where $m[B(x,r)]$ is the mass in the ball of radius r centered on x. In a regular mass distribution, $\alpha = 3$ (constant), so mass concentrations $\alpha(x) < 3$ are singularities. An ordinary fractal can be considered endowed with a *uniform* mass distribution over it, such that $\alpha < 3$ is the *constant* fractal dimension. Thus, in the context of multifractals, ordinary fractals are called *monofractals*. A full-fledged multifractal possesses a range of dimensions α, namely, $0 < \alpha_{\min} \leq \alpha \leq \alpha_{\max}$, and it can be considered as a set of interwoven monofractals. Every set of points in which α takes a definite value is a fractal set, with fractal dimension given by the *multifractal spectrum* $f(\alpha)$.

The multifractal spectrum is usually calculated by employing the statistical moment integrals $M_q(r)$, $-\infty < q < \infty$. Their scaling defines the function $\tau(q)$, such that

$$M_q(r) \sim r^{\tau(q)}. \tag{2}$$

$\tau(q)$ determines the multifractal spectrum through a Legendre transform [3, 8]: assuming $\alpha(q) = \tau'(q)$ to be monotone, $f(\alpha) = q(\alpha)\,\alpha - \tau[q(\alpha)]$.

The fractal distribution of halos and voids

We define halos as *singular* mass concentrations, with $\alpha(x) < 3$, such that the density given by Eq. (1) diverges as $r \to 0$. Note that scale invariance prevents us from assigning definite sizes or masses to these singularities. In order to do this, we may define a coarse-grained mass distribution using some small scale L. Then halos have equal size L but different masses. In CDM N-body simulations, the natural coarse-graining scale is the linear size of the volume per particle. Thus, initially and during the linear evolution, there is one particle per volume element, and halos only arise in the nonlinear stage, as some volume elements concentrate particles from other regions that become *voids*.

Since the the local dimension is related to the mass, namely, $\alpha \sim \log m / \log L$, every population formed by *equal-mass* halos is a monofractal of dimension $f(\alpha)$. This difference between the distribution of a particular halo population and the full dark matter distribution constitutes a multifractal type of *bias*.

In a multifractal analysis of N-body simulations [8], we have found fractal populations of halos of given mass. A particular simulation is shown in Fig. 1, on the right. The halo size is the box size/256. Halos have been assigned luminosity according to their mass, to make them play the role of parents of galaxies. Note that voids are filled with weak halos (dim galaxies), which are weakly clustered. The full dark matter distribution adopts a particular form that has been called the "Cosmic Web".

CONCLUSIONS AND OUTLOOK

Scaling in the dark-matter distribution is well supported by observations and CDM N-body simulations, but current galaxy surveys and simulations are insufficient to determine some details. The scale of homogeneity, which has been the subject of much controversy, seems to be in the range 10–20 Mpc/h. The distribution is multifractal, so its scaling properties are given by its multifractal spectrum, rather than by a unique fractal dimension. This spectrum can already be found with the help of simulations [8]. The determination of fine morphological features requires further analysis of the galaxy distribution and CDM simulations, with appropriate tools (statistical moments, Minkowski functionals, etc.). Among these morphological features are the voids. Scaling of voids is beginning to be observed, but deeper studies of voids will depend on improvement on their definition and, hence, detection [5].

REFERENCES

1. Mandelbrot B.B., *The fractal geometry of nature*, W.H. Freeman and Company, NY, 1977.
2. Sylos Labini F., Montuori M. and Pietronero L., *Phys. Rep.*, **293**, 61 (1998).
3. Pietronero L., *Physica A*, **144**, 257 (1987);
   Jones B.J., Martínez V.J., Saar E. and Einasto J., *Astrophys. J.*, **332**, L1 (1988).
4. Gaite, J. and Manrubia S. C., *Monthly Not. RAS*, **335**, 977 (2002).
5. Gaite, J., *Eur. Phys. Jour. B*, **47**, 93 (2005).
6. Tikhonov A V and Karachentsev I D, preprint `astro-ph/0609109`, to appear in *Astrophys. J.*.
7. Tikhonov A V, preprint `astro-ph/0610689`.
8. Gaite, J., *Europhys. Lett.*, **71**, 332 (2005);
   Gaite, J., preprint `astro-ph/0604202`.
9. Montuori M, Univ. of Rome report, unpublished;
   Tikhonov A.V., preprint `astro-ph/0610643`.
10. Tikhonov A.V. and Kopylov A.I., *Astrophys.*, **45**, 88 (2002).
11. Gaite, J., *Physica D*, **223**, 248–255 (2006).

Spiral galaxy rotation curves described using cosmological general relativity

John G. Hartnett

School of Physics, the University of Western Australia,
35 Stirling Hwy, Crawley 6009 WA Australia

Abstract. Spiral galaxy rotation curves are described using Carmeli's Cosmological General Relativity. A Tully-Fisher type relation results and rotation curves are reproduced without the need for non-baryonic halo dark matter. For accelerations larger than a critical value the Newtonian force law applies, but for accelerations less than the critical value the Carmelian force law applies.

Keywords: Cosmological General Relativity, Tully-Fisher, galaxy rotation curves
PACS: 98.52.Nr, 98.35.Df, 98.10.+z, 98.35.Df, 98.62.Ve

INTRODUCTION

Halo 'dark matter' [1] is used to explain lower than expected orbital speeds measured in the disk regions of spiral galaxies. Carmeli [3, 4] believes the usual assumptions in deriving Newton's gravitational force law from general relativity are insufficient, and successfully provided a theoretical description of the Tully-Fisher law. [2] This paper models the gravitational potential and the resulting forces in thin galaxy disks using cylindrical coordinates with exponential matter density distributions.

In the weak gravitational limit, where Newtonian gravitation applies, it is sufficient to assume the Carmeli metric with non-zero elements $g_{00} = 1 + 2\phi/c^2$, $g_{44} = 1 + 2\psi/\tau^2$, $g_{kk} = -1$, $(k = 1, 2, 3)$ in the lowest approximations in both $1/c$ and $1/\tau$, where τ is a universal constant $\approx 1/H_0$. The potential functions ϕ and ψ are determined by Einstein's field equations and from their respective Poisson equations, where $\psi = \phi/a_0^2$ and $a_0 = c/\tau$. In cylindrical coordinates (r, θ, z) the potential ϕ that satisfies the Poisson equation can be found from Ref. [5]. For a thin disk the z-dependence of the density can be integrated out and following from observation the density is best described by

$$\rho(r) = \frac{M}{2\pi a^2} e^{-r/a},\tag{1}$$

where a is a radial scale length and M is the mass of the galaxy.

EQUATIONS OF MOTION

The equations of motion are taken from equations B.62a and B.63a of Ref. [4]. Using B.62a with (1) and a cylindrical potential we get the standard circular motion equation

$$v^2 - \frac{GMr^2}{2a^3}\Pi, \quad \Pi - I_0\left(\frac{r}{2a}\right) K_0\left(\frac{r}{2a}\right) - I_1\left(\frac{r}{2a}\right) K_1\left(\frac{r}{2a}\right)\tag{2}$$

CP905, *Frontiers of Fundamental Physics (FFP8), Eighth International Symposium*
edited by B. G. Sidharth, A. Alfonso-Faus, and M. J. Fullana
© 2007 American Institute of Physics 978-0-7354-0412-0/07/$23.00

where G is the gravitational constant and I and K are standard Bessel functions. Equation (2) is the usual Newtonian result for the speed of circular motion in a cylindrical gravitational potential and is shown in curve 3 of figs 2(a).

Using $\psi = \phi/a_0^2$ in B.63a of Ref. [4] results in a new equation, which when integrated and solved for v as a function of r results in

$$v = \frac{2}{3}a_0 \frac{r^{3/2}}{\sqrt{GM}}. \tag{3}$$

To establish the combined result of equations (2) and (3), the simultaneous speed of test particles must be determined by the elimination of r between them. The Newtonian expression (2) describes motion under the central potential but assumes that spatial coordinates are fixed. Whereas the new equation (3) describes the expansion of space itself within a galaxy. Therefore we must find the combined (simultaneous) effect of these two. The result is a post-Newtonian equation,

$$v^{2/3} = \frac{(GM)^{5/3}}{(\frac{2}{3}a_0)^{4/3}2a^3}\Pi, \tag{4}$$

hereafter referred to as Carmelian, which cannot be solved analytically, but using the Mathematica software package can be solved numerically.

The result is plotted in curve 1 of fig. 1 where it has been assumed that $a = 1\ kpc$. This result indicates that the fourth order dependence on rotational speed (v) is directly proportional to mass (M) for large masses. Assuming that the mass of a galaxy is directly proportional to its luminosity, this dependence then becomes the Tully-Fisher relation.

By taking the $3/2$ power of (2) and multiplying it by (3) we can derive an equation describing the rotation curves in galaxies that is consistent with (4) in the limit of large r. [2], [3] The result is

$$v^4 = GM\frac{2}{3}a_0\left\{ \left(\frac{r}{2a}\right)^{9/2} 8\,\Pi^{3/2}\right\}, \tag{5}$$

remembering Π is a function of $r/2a$. It is is easily confirmed that as $r \to \infty$, the expression in the curley brackets in (5) tends to unity. Hence (5) then recovers the form of the Tully-Fisher relation.

By taking the 4th root of (5) we get an expression for the circular velocity of test particles as a function of their radial position r. That result has been plotted in curve 2 of fig 2(a) with a and M determined as fit parameters. At small values of r the rotation speeds determined from the Newtonian equation (curve 3) dominate.

The acceleration $\frac{2}{3}a_0$ in (3) is found to be a critical acceleration. When we compare the accelerations derived from the Newtonian equation (2) and the Carmelian equation (5) with this critical value we notice two regimes develop. See fig 2(b). Curve 2 represents the acceleration from (5) and curve 3 represents the acceleration from (2). For accelerations less than the critical value the Carmelian force applies and for accelerations greater than the critical value the Newtonian force applies. Note also that the Newtonian curve 3 has a r^{-2} dependence and the Carmelian curve 2 has a r^{-1} dependence above $10 kpc$. This theory also provides an understanding of the connection between the two regimes.

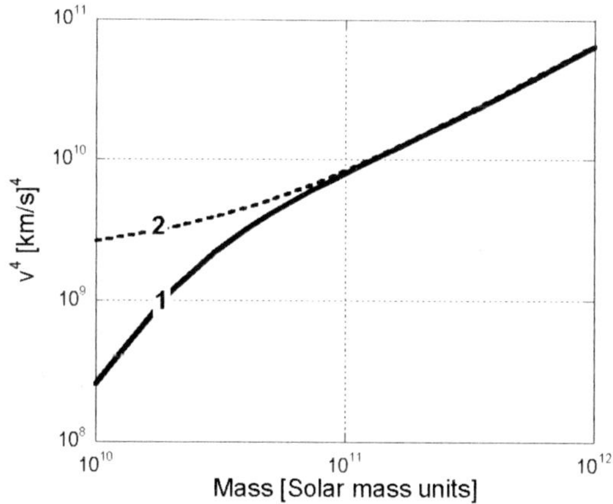

FIGURE 1.   Tully-Fisher law plotted on logarithmic axes. Curve 1 (solid line) represents the fourth order dependence of the rotational speeds of tracer gases in galaxies determined from the Carmelian equation (4). The masses are expressed in solar mass units of $M_\odot = 2 \times 10^{30} kg$. Curve 2 (broken line) represents the straight line $v^4 = 2 \times 10^9 + 0.064M$

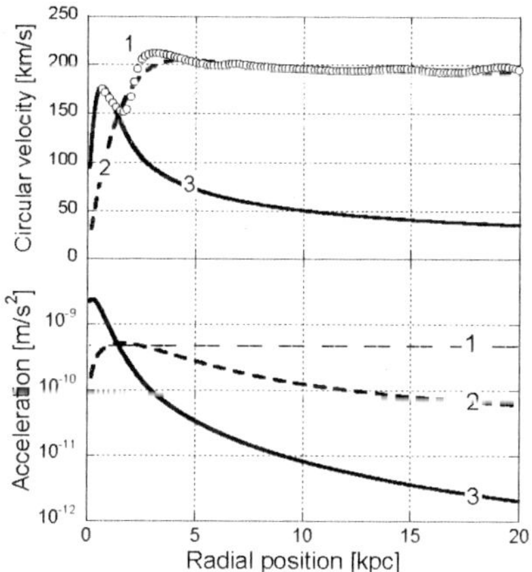

FIGURE 2.   (a) Above: The rotational speeds of tracer gases in NGC 2903 (Sc spiral) (circles - curve 1). Theoretical curve fits from the Carmelian equation (5) (curve 2) and from the Newtonian equation (2) (curve 3) (b) Below: The critical acceleration $\frac{2}{3}a_0 \approx 4.75 \times 10^{-10} m.s^{-2}$ (curve 1). The corresponding rotational accelerations determined from the Carmelian (curve 2) and the Newtonian (curve 3) equations

REFERENCES

1.   K.G. Begeman, *et al. M. N. R. A. S.* **249**, 523–537 (1991).
2.   M. Carmeli, *Int. J. Theor. Phys.* **37** (10), 2621–2625 (1998).
3.   M. Carmeli, *Int. J. Theor. Phys.* **39** (5), 1397–1404 (2000).
4.   M. Carmeli, *Cosmological Special Relativity.* World Scientific, Singapore, 2002.
5.   A. Toomre, *Ap. J.* **138**, 385–392 (1963).

Gravitational waves in Cosmological General Relativity

John G. Hartnett and Michael E. Tobar

*School of Physics, the University of Western Australia,
35 Stirling Hwy, Crawley 6009 WA Australia*

Abstract. The 5D Cosmological General Relativity theory indicates that gravitational radiation may not propagate as an unattenuated wave where effects of the Hubble expansion are felt. In such cases the energy does not travel over such large length scales but is evanescent and dissipated into the surrounding space as heat.

Keywords: Carmeli, cosmological relativity, gravitational waves, expanding universe
PACS: 04.30.-w, 04.25.Nx, 04.30.Nk, 95.85.Sz, 98.80.Jk

WAVE EQUATION IN CGR

In standard general relativity the expanding universe has no impact on the properties of gravitational waves. However, in Cosmological General Relativity (CGR)[1, 2] the velocity v of the expansion of the universe (or redshift of the gravitational wave) manifests as a fifth dimension [2] and in this paper we calculate possible consequences of that. In CGR the universe is represented by a 5-dimensional Riemannian manifold with a metric $g_{\mu\nu}$ and a line element $ds^2 = g_{\mu\nu}dx^\mu dx^\nu$, where $x^4 = \tau v$, and τ a universal constant. As usual $x^0 = ct$, where t is the time coordinate and $x^k, k = 1, 2, 3$, are spatial coordinates as in general relativity.

Now consider the wave equation [2]

$$\left(\frac{1}{c^2} \frac{\partial^2}{\partial t^2} - \nabla^2 \right) \bar{h}^{\mu\nu} = -\frac{1}{c^2 \tau^2} \frac{\partial^2}{\partial z^2} \bar{h}^{\mu\nu}, \tag{1}$$

where where $\bar{h}_{\mu\nu}$ is the trace-reversed $h_{\mu\nu}$ and the substitution $v/c \to z$ has been made. Here z is the redshift of the wave, and the substitution is valid where $z =< 0.1$, which is approximately $400\,Mpc$. We assume it is approximately valid beyond that. Now the solution to (1) is the sum of the solution to the homogeneous equation, which is the usual gravity wave solution in general relativity, and a particular solution of (1), which has a redshift dependent source term. In fact, in CGR, because the Hubble law is assumed *a priori*, the expansion velocity v (or gravitational wave redshift z) is not independent of r and depends on the matter density of the universe. Thus (1) can be written as

$$\left(\frac{1}{c^2} \frac{\partial^2}{\partial t^2} - \nabla^2 + \frac{1}{c^2 \tau^2} \frac{\partial^2}{\partial z^2} \right) \bar{h}^{\mu\nu} = 0, \tag{2}$$

CP905, *Frontiers of Fundamental Physics (FFP8), Eighth International Symposium*
edited by B. G. Sidharth, A. Alfonso-Faus, and M. J. Fullana
© 2007 American Institute of Physics 978-0-7354-0412-0/07/$23.00

with

$$\frac{\partial^2}{\partial z^2} = \left\{ \left(\frac{\partial r}{\partial z}\right)^2 \frac{\partial^2}{\partial r^2} + \frac{\partial^2 r}{\partial z^2}\frac{\partial}{\partial r} \right\}. \tag{3}$$

when the chain rule is applied. We look for a plane wave solution of the form

$$\bar{h}^{\mu\nu} = \varepsilon^{\mu\nu} \cos k_\alpha x^\alpha, \tag{4}$$

describing a wave propagating in the r direction which has $k^\alpha = (\omega/c, 0, 0, k_r)$. This means we only need to retain the r derivative in ∇^2 and effectively re-write (2) in spherical co-ordinates as

$$\left(\frac{1}{c^2}\frac{\partial^2}{\partial t^2} - \frac{\partial^2}{\partial r^2} - \frac{2}{r}\frac{\partial}{\partial r} + \frac{1}{c^2\tau^2}\frac{\partial^2}{\partial z^2} \right) \bar{h}^{\mu\nu} = 0. \tag{5}$$

Eqs (2) and (5) are only valid where the Hubble law applies. When it doesn't apply, that is, where $\partial r/\partial z = 0$ in (3), (2) becomes the normal wave equation for gravity waves in source free regions. However where the Hubble law is applicable,

$$\frac{1}{c^2\tau^2}\left(\frac{\partial r}{\partial z}\right)^2 - 1 + (1-\Omega)\frac{r^2}{c^2\tau^2}, \tag{6}$$

where $\Omega = \rho/\rho_c$ is the mass/energy density at some epoch expressed as a fraction of the 'critical' density, $\rho_c = 3/8\pi G\tau^2$. Substituting (6) into (5) with (3) we get

$$\left(\frac{1}{c^2}\frac{\partial^2}{\partial t^2} - \frac{2}{r}\frac{\partial}{\partial r} + \frac{1-\Omega}{c^2\tau^2}\left\{ r^2\frac{\partial^2}{\partial r^2} + r\frac{\partial}{\partial r} \right\} \right) \bar{h}^{\mu\nu} = 0. \tag{7}$$

This is new equation (7) depends on the surrounding matter density Ω.

When the gravity wave is very distant from the source and when $r \gg c\tau/\sqrt{1-\Omega}$ the second term of (7) is much smaller than the term in curly brackets, we assume the second term negligible. Therefore (7) can be approximated for large r and a solution obtained by separation of variables assuming $\bar{h}^{\mu\nu} \propto R(r)e^{i(\omega t + k_z)}$. Substituting back yields

$$\frac{\omega^2}{c^2} + \frac{\Omega-1}{c^2\tau^2} = 0, \tag{8}$$

with $k_z \approx 0$ and $R(r) = a_1 r^{-1}$ for $r \gg c\tau/\sqrt{1-\Omega}$. More generally $R(r) = \sum a_n r^{-n}$ a polynomial expression with an index $n > 0$. Equation (8) is a resonance condition. For this solution, which spans the whole extent of the Universe, the Universe acts like a resonant mode with a characteristic scale radius [3, 4] of

$$R_\Omega = \sqrt{|R_\Omega^2|} = \sqrt{\left|\frac{c^2}{\omega^2}\right|} = \frac{c\tau}{\sqrt{|1-\Omega|}}, \tag{9}$$

and resonance frequency

$$\omega = \frac{\sqrt{1-\Omega}}{\tau}. \tag{10}$$

For values of $\tau = 4.2 \times 10^{17}\, s$ and $\Omega = 0.02$ the scale radius is $R_\Omega \approx 4.13\,Gpc$ and the characteristic frequency is $\omega/2\pi \approx 3.66 \times 10^{-19}$ Hz.

When $r \approx< c\tau/\sqrt{|1-\Omega|}$ the second and fourth terms of (7) are much smaller than the third, and hence can be neglected. Therefore (7) can be approximated as

$$\left(\frac{1}{c^2}\frac{\partial^2}{\partial t^2} + \frac{1-\Omega}{c^2\tau^2}r^2\frac{\partial^2}{\partial r^2} \right)\bar{h}^{\mu\nu} = 0. \tag{11}$$

Now assuming $r/c\tau \approx z$, Eq.(11) becomes a wave equation with an approximate solution of the form $\bar{h}^{\mu\nu} \propto e^{i(k_r r + \omega t)}$, which results in the following dispersion relation

$$k_r^2 \approx \frac{\omega^2/c^2}{(\Omega-1)z^2}, \tag{12}$$

provided $z \ll 1$, otherwise (11) must be solved numerically. However this indicates the dependence. When $\Omega > 1$ the wave number is real and approximately $k_r = \omega/(cz\sqrt{\Omega-1})$ and hence gravity waves propagate yet are dependent on redshift. When $\Omega < 1$ the wave number is imaginary and the amplitude is attenuated with a decay constant $\kappa = ik_r = \omega/(cz\sqrt{1-\Omega})$.

CONCLUSION

On the local scale the Hubble law does not apply or is so insignificant as to be negligible. That is where GGR reduces to the usual special and general relativity theory and gravity waves propagate as is usually expected. On the cosmological scale the Hubble law is significant and there we expect a genuine modification to the usual 4D *spacetime* gravity wave equation found in general relativity textbooks. CGR predicts that gravitational waves from distant galaxies will be fully attenuated by the time they reach Earth.

Gravity waves that are generated within a galaxy quickly decay in the void between. Gravitational radiation therefore leaks into the surrounding space with attenuated amplitudes when Ω drops below unity. Gravity waves will not propagate far in an expanding universe. Therefore we would expect to see no stochastic gravity wave background spectrum. Instead we conjecture that the energy is deposited into space as heat. As a result they may contribute to the CMB blackbody temperature.

REFERENCES

1. M. Carmeli, *Cosmological Special Relativity*, World Scientific, Singapore, 2002.
2. M. Carmeli, "The Line Elements in the Hubble Expansion" in *Gravitation and Cosmology*, edited by A. Lobo et al, Universitat de Barcelona, Barcelona, 2003, [arXiv: astro-ph/0211043]
3. F.J. Oliveira, "Quantised intrinsic redshift in Cosmological General Relativity," aiXiv:gr-gc/0508094 (2005)
4. F.J. Oliveira, "Exact solution of a linear wave equation in Cosmological General Relativity," *Int. J. Mod. Phys. D* (in press) (2006) [arxiv:gr-qc/0509115]

Creation of Spiral Galaxies II

Masataka Mizushima

Department of Physics, University of Colorado, Boulder, Colorado 80309, U.S.A.

Abstract. Some discussions about quasars are presented. A gravito-radiative term is obtained as a correction term from the variational principle of Einstein General Relativity Theory. A collision between two massive black holes at the center of a quasar generates a gravito-radiative force. According to our theory, there cannot be any galaxies with an odd number of spiral arms, and that agrees with our observation. The size of the Milky Way galaxy was about 10 times the present size, and the galaxy must be shrinking now.

Keywords: spiral galaxies, galactic formation
PACS:

QUASARS

Hubble found that the estimated distance of a galaxy from us is proportional to the recession velocity of that galaxy from us. The proportionality constant, called Hubble's constant H, is about 77 km/s per Mpc, or 3 km/s per 10^{21} m. Thus the universe is expanding. The distance of galaxies from us, however, has an upper limit of about $d_{max} = 10^{25}$ m, and quasars are found beyond that distance with the same Hubble's constant [1]. Each quasar is regarded as an active galactic nucleus without spiral arms [2, 3]. We assume that a quasar has some black holes and that a highly condensed neutron (ylem) disk is orbiting around them. This disk developed into a spiral galaxy at $d_{max}/c = 10^{17}$ s $= 10^{10}$ years ago. As we are going to discuss in section 3, we assume that Earth was created at the same time as the Milky Way galaxy was created from a quasar, and the age of Earth is geologically known to be 4.6×10^9 years. Thus, we can assume that all galaxies were created from quasars here at the same time. We, however, do not know when and how the quasars were created.

GRAVITO-RADIATIVE FORCES

Einstein assumed that $(dx/dt)^2 = c^2 - (ds/dt)^2 \leq c^2$ for a particle moving in the x direction, and obtained the well-known expression of its energy as $E = mc^2/\sqrt{1 - (dx/cdt)^2}$. He [4] generalized this assumption into a general metric which describes gravity. Thus, in the general relativity theory, gravity also propagates with a finite speed equal to or less than the speed of light. In Einstein's general relativity theory, equations of motion under gravitational fields can be obtained from the variational principle $\delta \int ds = 0$ [4]. The first term, obtained in this way, is the Newtonian gravitational force, but as a correction term due to the Einstein principle, we obtain [5] a gravito-radiative term of the order of $(v/c)^4$,

CP905, *Frontiers of Fundamental Physics (FFP8), Eighth International Symposium*
edited by B. G. Sidharth, A. Alfonso-Faus, and M. J. Fullana
© 2007 American Institute of Physics 978-0-7354-0412-0/07/$23.00

$\mathbf{F}_{rad2}/m = -(4GM[\dot{\mathbf{v}}(\mathbf{V}\cdot\mathbf{v})+\mathbf{v}(\mathbf{V}\cdot\dot{\mathbf{v}})]/(c^4 r)$. Here, $\mathbf{v}$ and $\dot{\mathbf{v}}$ are the velocity and acceleration of the source particle of mass M, and $\mathbf{V}$ is the velocity of a test particle. A collision between two massive black holes at the center of a quasar generates a gravito-radiative force $\mathbf{F}_{rad2}$.

CREATION OF ENERGY

The total energy of a test particle of mass m and speed V under the gravitational field due to mass M at the origin is $\mathscr{E} = mc^2 + mV^2/2 - (GMm)/2$, to the first approximation in $(V/c)^2$ of static general relativity, and it agrees with the Newtonian theory. In the Newtonian theory we know that $d\mathscr{E}/dt = 0$, but in the general relativity theory, assuming that $dm/dt = 0$, we see that $d\mathscr{E}/dt = \mathbf{V}\cdot\mathbf{F}_E - GMmd(1/r)/dt = -8GMm(\mathbf{V}\cdot\mathbf{v})(\mathbf{V}\cdot\dot{\mathbf{v}})/(c^4 r)$, by taking $\mathbf{F}_E$ for $md\mathbf{V}/dt$. Thus energy is created by $\mathbf{F}_{rad2}$ by the collision of black holes at the ylem orbiting around the center. We assume that the ylem orbit around the black holes is a circular disk, and that the collision of two black holes takes place head-on in the ylem's plane. The equation above shows that only those ylems that are orbiting on opposite sides of a circle at which $\mathbf{V}$ was parallel to $\mathbf{v}$ or $\dot{\mathbf{v}}$ would gain energy from the gravito-radiative force. The ylem orbiting on the part of circle between these two parts are not touched [6]. The collision must be extremely relativistic. The velocity v must be close to c, and $\Delta t \dot{v}$ must also be close to c, where Δt is the duration time of the collision. Thus, the magnitude of the energy that part of the ylem disk (with mass m) gained must be about $\Delta\mathscr{E} \simeq 8GMm(\mathbf{V}\cdot\mathbf{n})^2/(c^2 r)$, where $\mathbf{n}$ is an unit vector in the direction of the (head-on) collision of black holes. Because the pulsar is very small, the orbiting speed of the ylem, V, must be close to c. The extra kinetic energy, each of these parts of the ylem disk that is orbiting in the direction of $\mathbf{n}$, or -$\mathbf{n}$ gain, is just enough to form a pair of arms extending to the edge of the galaxy [6]. If the original ylem disk was circular, the resulting two arms would be symmetric with a $180°$ rotation around the center. Now if the ylem is out of the strong gravitational compressing pressure, its constituents, the neutrons, will each start dissociating into an electron and a proton. The synthesis processes of chemical elements starts as Gamow assumed [7]. Because the expansion is so rapid, the processes end without completing, and the electromagnetic radiation created at the earliest stage of the galaxy creation remains as the 3K background. The two spiral arms still have the $180°$ rotational symmetry, in agreement with our observation of young galaxies such as M51, M74, M99, M100, and NGC1566. There may have been more than one black hole collision at the center of some galaxies, in which case the galaxy would have an even number of spiral arms. We see that galaxies NGC2997, NGC3310, M33, and M101 have two pair of arms, while galaxy M83 has three pair of arms. Our Milky Way galaxy has four pair of arms. But in all these galaxies, the $180°$ rotational symmetry is maintained [8]. According to our theory, there cannot be any galaxies with an odd number of spiral arms, and that agrees with our observation.

SPIRALITY OF MILKY WAY GALAXY

Take a star in the Milky Way galaxy at a radial distance r_1 from the center. If the average tangential speed was $<V_1>$ during the life of the galaxy, t_g, we see $<V_1>t_g = r_1\phi_1$, where ϕ_1 is the angular distance the star covered. Similarly, another star at a radial distance r_2 covered an angular distance ϕ_2 such that $<V_2>t_g = r_2\phi_2$. Let us take the largest and the smallest ends of the Orion arm of our Milky Way galaxy, so that $r_1 = 5.0 \times 10^{20}$ m and $r_2 = 3.3 \times 10^{20}$ m [1]. We know $\phi_2 - \phi_1 = \pi/3$. It is observed that the longitudinal speed is about 200 km/s and almost independent of the position of stars within the accuracy of 10 km/s. Taking $t_g = 5 \times 10^9$ yrs, assuming that the age of our galaxy is the same as that of Earth. The angular distance $\phi_1 + \phi_2$ is about 15, and that means $(<V_1> + <V_2>)/2 = 25$ km/s, which is about 1/8 of the present value of V. Because the angular momentum is conserved, the size of the galaxy was about 10 times the present size, and the galaxy must be shrinking now.

REFERENCES

1. W. K. Hartmann, *Astronomy: The Cosmic Journey*, Wadsworth Publishing, Belmont, CA, 1991.
2. M. Rowan-Robinson, *Cosmology*, Clarendon Press, Oxford, 1977, p. 61.
3. I. Robson, *Active Galactic Nuclei*, John Wiley & Sons, New York, 1996.
4. A. Einstein, *Ann. Phys.* **49**, 769 (1916).
5. M. Mizushima, *Hadronic J.* **18**, 577 (1995).
6. M. Mizushima, *Frontiers of Fundamental Physics 4*, edited by Sudharth and Altaisky, Kluwer Academic/Plenum Publishers, New York, 2001, pp. 179-188.
7. G. Gamow, *Phys. Rev.* **74**, 505 (1948).
8. M. Mizushima, *Frontiers of Fundamental Physics*, edited by Sudharth, Honsell, and de Angelis, Springer, The Netherlands, 2006, pp. 103-106.

Are the quasar polarization angles randomly distributed?

D. Sáez[1] and J.A. Morales[2]

*Departamento de Astronomía y Astrofísica, Universidad de Valencia,
46100–Burjassot, Valencia, Spain*

Abstract. Observations of the polarization angles of some quasar samples strongly suggest that these angles are not randomly distributed on Gpc scales. Large scale vector perturbations (vortical velocity fields) of the concordance cosmological universe produce rotations of the QSOs polarization directions; here, the amplitude and properties of these rotations are analyzed in the case of an unique vector mode. The resulting rotations depend on both the quasar redshifts and the line of sight and, consequently, they can alter a random initial distribution of QSO polarization angles. The question is: could this alteration explain correlations on Gpc scales? More work is necessary to answer this question starting from the conclusions of this preliminary paper.

Keywords: Cosmology: theory, Large-scale structure of universe—Quasars: general—Polarization
PACS: 98.65.Dx, 98.54.-h

INTRODUCTION

In the early eighties, P. Birch [1] claimed that the orientation of the quasar polarization vectors is not random. He observed a certain dipolar distribution of the polarization angles and proposed a rotating universe to explain it. Recently, some researchers have improved on both the observation techniques and the statistical methods to conclude that coherent orientations are very likely (see [2] and references cited therein).

Five decades ago, Skrotskii [3] used Maxwell equations and the metric describing a slowly rotating body (Minkowski perturbation) to prove that the polarization vector rotates as radiation propagates. This rotation (hereafter called the Skrotskii effect) also appears in other space-time structures. In any of them, the total variation, $\delta\psi$, of the polarization direction (Skrotskii rotation) can be calculated –from emission to observation– taking into account that the polarization vector is parallely propagated along the photon null geodesics and the coordinate basis is not. Here, Skrotskii rotations of the quasar polarization directions are calculated for vector cosmological perturbations (divergenceless peculiar velocities). Only linear vector modes (small enough perturbations of the concordance model) are considered.

A realization of the concordance model compatible with the analysis of three year WMAP data [4] is assumed to perform our calculations. In this model, the reduced Hubble constant is $h = 10^{-2} H_0 = 0.71$ (H_0 being the Hubble constant in units of

<hr>

[1] E-mail: diego.saez@uv.es

[2] E-mail: antonio.morales@uv.es

CP905, *Frontiers of Fundamental Physics (FFP8), Eighth International Symposium*
edited by B. G. Sidharth, A. Alfonso-Faus, and M. J. Fullana

$Km\ s^{-1}Mpc^{-1}$), and the density parameters of vacuum energy and matter (baryonic plus dark) are $\Omega_\Lambda = 0.73$ and $\Omega_m = 0.27$, respectively.

In this paper, Greek (Latin) indices run from 0 to 3 (1 to 3), and units are defined in such a way that $c = \kappa = 1$ where c is the speed of light and $\kappa = 8\pi G/c^4$ is the Einstein constant. Symbols a, η and x^i stand for the scale factor, the conformal time and the comoving spatial coordinates.

SKROTSKII ROTATIONS AND VECTOR PERTURBATIONS

Scalar, vector and tensor modes evolve separately during the linear regime; hence, vanishing scalar and tensor linear perturbations can be assumed to estimate the effects produced by linear vector modes. In this case, the relations $h_{00} = h_{ij} = 0$ define a certain gauge and, then, using the notation $h_{0i} = (h_1, h_2, h_3) = \vec{h}$, the line element can be written in the form $ds^2 = a^2(-d\eta^2 + 2h_i dx^i d\eta + \delta_{ij} dx^i dx^j)$. It is then easily proved that the polarization vector of any radiation emitted at time η_e (from a point with radial comoving coordinate r_e) rotates an angle

$$\delta\psi = -\frac{1}{2}\int_0^{r_e}(\vec{\nabla}\times\vec{h})\cdot\vec{n}\,dr \tag{1}$$

where $\vec{n} = \vec{r}/r = (\sin\theta\cos\phi, \sin\theta\sin\phi, \cos\theta)$ is the unit vector in the chosen radial direction (constant θ and ϕ spherical coordinates). Note that $\vec{\nabla}$ and the dot stand for the covariant derivative and the scalar product with respect the background flat 3-dimensional metric, respectively; hence, the Skrotskii rotation is obtained by integrating the curl of the vector perturbation $\vec{h}$ along the line of sight. Vector $\vec{h}$ and the peculiar velocity can be developed in terms of functions $Q^+(\vec{r},\vec{k}) = \vec{\varepsilon}^{\,+}(\vec{\kappa})\exp(i\vec{k}\cdot\vec{r})$ and $Q^-(\vec{r},\vec{k}) = \vec{\varepsilon}^{\,-}(\vec{\kappa})\exp(i\vec{k}\cdot\vec{r})$, $\vec{\kappa}$ being the unit vector $\vec{k}/k$. The definition of functions $\vec{\varepsilon}^{\,\pm}$ is given in [5]; the coefficients of the resulting expansions are $B^{\pm}(\eta,\vec{k})$ and $v^{\pm}(\eta,\vec{k})$, respectively. Quantities $v_c^{\pm} = v^{\pm} - B^{\pm}$ are gauge invariant [6]. It can be easily proved that, in the absence of anisotropic stresses, the equation $v_c^{\pm} = v_{c0}^{\pm}/a$ is satisfied, where $v_{c0}^{\pm}$ is the present value of $v_c^{\pm}$. Angle $\delta\psi$ is then given by the formula:

$$\delta\psi = 3H_0^2\Omega_m\int_0^{r_e}\frac{dr}{a^2(r)}[\vec{n}\cdot\vec{F}(\vec{r})], \tag{2}$$

where

$$\vec{F}(\vec{r}) = \int\frac{v_{c0}^+\vec{\varepsilon}^{\,+}(\vec{\kappa}) - v_{c0}^-\vec{\varepsilon}^{\,-}(\vec{\kappa})}{k}\exp(i\vec{k}\cdot\vec{r})\,d^3k. \tag{3}$$

For an unique vector mode $\vec{k}_0$, one can write $v_{c0}^{\pm}(\vec{k}) = v_{c0}^{\pm}\delta(\vec{k}-\vec{k}_0) - (v_{c0}^{\pm})^*\delta(\vec{k}+\vec{k}_0)$, where the complex numbers $v_{c0}^{\pm} = v_{c0R}^{\pm} + iv_{c0I}^{\pm}$ fix the mode amplitude and $\delta(\vec{k}-\vec{k}_0)$ and $\delta(\vec{k}+\vec{k}_0)$ are Dirac-distributions. From the above $v_{c0}^{\pm}$ distribution and Eqs. (2)–(3) one easily get:

$$\delta\psi = \frac{3H_0^2\Omega_m}{k_0}\vec{n}\cdot[v_{c0}^+\vec{\varepsilon}^{\,+}(\vec{\kappa}_0) - v_{c0}^-\vec{\varepsilon}^{\,-}(\vec{\kappa}_0)]\int_0^{r_e}\frac{\exp(i\vec{k}_0\cdot\vec{r})}{a^2(r)}\,dr. \tag{4}$$

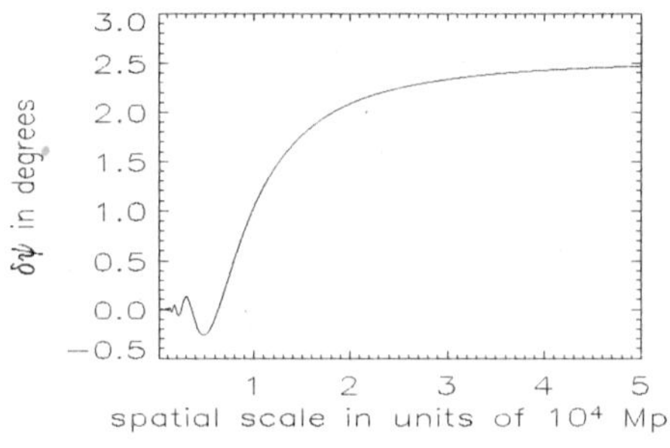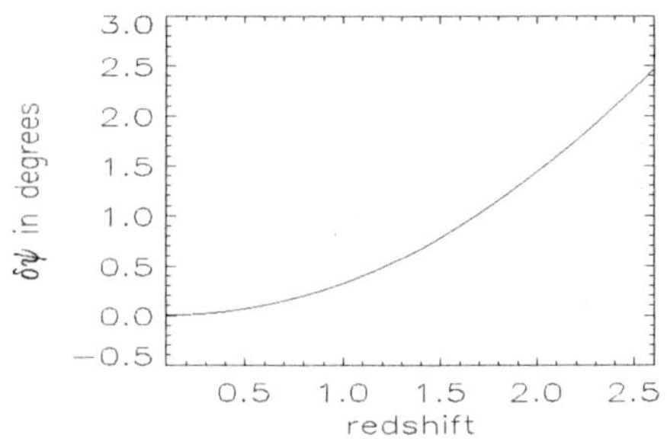

FIGURE 1. Left: Skrotskii rotation angle $\delta\psi$ as a function of the spatial scale in units of 10^4 Mpc for $z = 2.6$. Right: angle $\delta\psi$ in terms of the redshift for a scale of 5×10^4 Mpc.

After simple calculations based on the use of a basis (in momentum space) in which the components of vector $\vec{k}_0$ are $(k_0, 0, 0)$, the Skrotskii rotation can be written in the form $\delta\psi = \delta\psi^+ + \delta\psi^-$, where

$$\delta\psi^\pm = C\left[(I_s v_{c0I}^\pm - I_c v_{c0R}^\pm)\cos\theta - (I_s v_{c0R}^\pm + I_c v_{c0I}^\pm)\sin\theta\sin\phi\right], \tag{5}$$

with $C = 6H_0^2\Omega_m/k_0\sqrt{2}$, $I_s = \int_0^{r_e} a^{-2}(r)\sin\xi\,dr$, $I_c = \int_0^{r_e} a^{-2}(r)\cos\xi\,dr$, and $\xi = rk_0\sin\theta\cos\phi$. Equation (5) is used to estimate $\delta\psi$ for $v_{c0R}^\pm = v_{c0I}^\pm = 3 \times 10^{-3}$, $\theta = \pi/4$ and $\phi = \pi/2$. First, a fixed quasar redshift $z = 2.6$ (comoving distance of ~ 6000 Mpc) and variable k_0 values have been considered. Results are presented in the left panel of Fig. (1), where it is easily seen that the resulting values are not negligible for large spatial scales greater than $\sim 10^4$ Mpc, which are the scales we are interested in to avoid problems with CMB observations. Moreover, the angle $\delta\psi$ has been calculated for a fixed spatial scale of 5×10^4 Mpc and for variable redshifts (quasar positions), results are shown in the right panel of Fig. (1). Skrotskii rotations grow as z increases, reaching values of a few degrees for large enough redshifts.

SUMMARY AND PERSPECTIVES

Only very large spatial scales (a few times 10^4 Mpc) produce Skrotskii rotations of a few degrees for large enough redshifts $z > 2$. These angles are too small to explain observations [1], [2] about the statistical distribution of polarization angles. Moreover, we have verified that rotations of a few degrees are the largest ones which can be produced by linear vector modes. Greater angles are possible, but they would be produced by vector modes which require a nonlinear description.

More work is necessary to develop the following researches: (i) estimation of the Skrotskii rotations produced by appropriate superimpositions of linear vector modes, (ii) study of the effect generated by nonlinear vector perturbations, (iii) detailed calculation of the effects produced by large scale linear vector modes on the CMB, and so on.

ACKNOWLEDGMENTS

This work has been supported by the Spanish Ministerio de Educación y Ciencia, MEC-FEDER projects AYA2003-08739-C02-02 and FIS2006-06062.

REFERENCES

1. P. Birch, *Nature* **298**, 451 (1982).
2. D. Hutsemékers, R. Cabanac, H. Lamy and D. Sluse, *Astron. & Astrophys.* **441**, 915 (2005).
3. G. V. Skrotskii, *Sov. Phys.–Dokl.* **2**, 226 (1957).
4. D. N. Spergel et al., 2006, astro-ph/0603449
5. W. Hu and M. White, *Phys. Rev.*, **56D**, 596 (1997).
6. J.M. Bardeen, *Phys. Rev.*, **22D**, 1882 (1980).

II. QUANTUM PHYSICS

Relaxation-time and electrical-conductivity anisotropy of layered crystals at the scattering of charge carriers by impurity ions

. B.M.Askerov, S.R.Figarova, G.I.Guseinov, and V.R.Figarov

Baku State University, 23, Z.Khaliliov st., Baku, Az1148, Azerbaijan
e-mail:figarov@physics.ab.az

Abstract. The scattering of charge carriers by impurity ions in quasi-two-dimensional electronic systems with the cosine dispersion law is under investigation. It has been obtained analytical expressions for transverse and longitudinal components of the relaxation time against components of the wave vector. It has been shown that electrical conductivity is heavily anisotropic, the anisotropy magnitude depends upon the relation between the miniband width and Fermi level and is determined by the ratio of the screening radius to the lattice constant in the direction normal to the layer.

Keywords: Impurity scattering, relaxation time, quasi-two-dimensional electron system, electrical conductivity.

PASC : 72.15Lh;73.21.Cd.

The major key parameter in electronic transport phenomena studies is the relaxation time at different mechanisms of scattering. The relaxation-time magnitude and its dependence on wave vector components define kinetic properties of quasi-two dimensional electronic systems.

The object of the paper is the construction of a theory of the anisotropic scattering of charge carriers by impurity screening ions in quasi-two-dimensional electron systems, including also the purely two-dimensional case. The scattering anisotropy is taken into account through the introduction of the inverse relaxation time tensor in the Born approximation. For an electron gas it is assumed a cosine dispersion law that describes layered crystals and superlattices :

$$\varepsilon(k) = \hbar^2 k_\perp^{\,2} / 2m_\perp + \varepsilon_0 \left[1 - \cos(ak_z)\right], \quad (1)$$

here $k_\perp^{\,2} = k_x^{\,2} + k_y^{\,2}$, $k_\perp$ and k_z are longitudinal and transverse components of the wave vector, respectively, ε_0 is the one-dimensional conduction mini-band width, a is a lattice constant in

CP905, *Frontiers of Fundamental Physics (FFP8), Eighth International Symposium*
edited by B. G. Sidharth, A. Alfonso-Faus, and M. J. Fullana
© 2007 American Institute of Physics 978-0-7354-0412-0/07/$23.00

the direction perpendicular to the layer plane, $m_x^{-1} = m_y^{-1} = m_\perp^{-1}$ is the electron effective mass in the layer plane.

Considering explicit form of energy spectrum (1) in cylindrical coordinates for transverse, $\dfrac{1}{\tau_\perp}$, and longitudinal, $\dfrac{1}{\tau_{II}}$, components of the inverse relaxation time at the scattering of charge carriers by impurity ions after integrating in respect with φ and $\varepsilon_\perp$ with the aid of the δ-function yields

$$\frac{1}{\tau_\perp} = \frac{m_\perp CVa^2}{4\pi^2 h^2 k_\perp^2} \int_{-k_z}^{k_z} \frac{B^2 - 2k_\perp^2 a^2 A}{(A^2 + B^2)^{3/2}} \theta(\varepsilon - \varepsilon_z') dk_z', \quad (2)$$

$$\frac{1}{\tau_{II}} = \frac{m_\perp CVa^4}{4\pi^2 h^2 k_z} \int_{-k_z}^{k_z} |k_z - k_z'| \frac{A + 2k_\perp^2 a^2}{(A^2 + B^2)^{3/2}} \theta(\varepsilon - \varepsilon_z') dk_z', \quad (3)$$

where designations are used:

$$A = 2\gamma(\cos ak_z' - \cos ak_z) + (k_z - k_z')^2 a^2 + (a/r_0)^2,$$

$$B = 2k_\perp a\left[(k_z - k_z')^2 a^2 + (a/r_0)^2\right]^{1/2}, \gamma = m_\perp / m_{0z}, m_{0z} = h^2/\varepsilon_0 a^2,$$

$$\theta(\varepsilon_z - \varepsilon_z') = 1 \text{ as } \varepsilon_z - \varepsilon_z' > 0 \text{ and } \theta(\varepsilon_z - \varepsilon_z') = 0 \text{ as } \varepsilon_z - \varepsilon_z' < 0 \text{ is}$$

the Heaviside function, $C = \dfrac{2\pi}{h} \dfrac{N_i}{V} \left(\dfrac{4\pi e^2}{\infty}\right)^2$, N_i is the impurity concentration, χ is the electric permittivity.

Proceeding from the formulae, it has been performed a numerical computation of the dependence of longitudinal and transverse components of the relaxation time on the longitudinal and transverse wave vectors for $GaAs / Al_{0.36}Ga_{0.64}As$ superlattice parameters($a = 10nm$, $\chi = 13.18$, $\varepsilon_0 = 12.5meV$, $m_\perp = 0.06m_0$, $N = 10^{24}m^{-3}$) [1].

It is observed a growth of longitudinal and transverse components of the relaxation time with a rise in the transverse motion energy. The transverse relaxation time magnitude exceeds the longitudinal one, therewith the transverse relaxation time magnitude is close to the relaxation time of a bulk sample [2]. To the growth of $\tau_\perp$ with a rise in the transverse motion energy in the purely two-dimensional case ($\mu > 2\varepsilon_0$) it was noted in [3]. The mean value of the transverse relaxation time for the layered InP, computed by the formulae , gives the value of 0.71ps, which qualitatively agrees the experimental value of 0.57ps with the next parameters: the lattice constant $a = 50\mathring{A}$, the screening radius $r_0 = 18\mathring{A}$[3]. In the dependence of longitudinal and transverse relaxation time on the longitudinal wave vector it is evident a substantial distinction.

As the longitudinal wave vector increases the longitudinal relaxation time is reduced while the transverse one grows and at large magnitudes they approach an equal value. It might be point out that when the impurity concentration increases, distinctions in the dependence of the relaxation time on the wave vector longitudinal component are significant.

In the approximation when the effective mass in the layer plane is many fewer than the effective mass along the layer, having regard to the weak screening of Coulomb impurity atoms, in the Debye approximation for components of the inverse relaxation time one gets analytical expressions for longitudinal and transverse components of the relaxation time in the shape handy for further application in kinetic coefficient calculations :

$$\frac{1}{\tau_\perp} = \frac{1}{\tau_0} \frac{\ln 4k_z r_0}{(2k_\perp r_0)^3} , \quad (4)$$

$$\frac{1}{\tau_{II}} = \frac{1}{\tau_0} \frac{1}{4k_\perp k_z r_0^2} . \quad (5)$$

From Formulae (4) and (5) it follows that the transverse component of the inverse relaxation time at the weak screening strongly depends on the transverse component of the wave vector and weakly does on the longitudinal component of the wave vector. When passing to the purely two-dimensional case the dependence of $\tau_\perp$ on $k_\perp$ coincides with that in [2]. The longitudinal component of the inverse relaxation time in this limit in the same way, namely inversely, is dependent on the both longitudinal and transverse components of the wave vector.

From the expressions for components of the inverse relaxation time tensor, by the Boltzmann kinetic equation in the approximation of the anisotropic relaxation time via the methods of the approximate integration, it is calculated the longitudinal σ_{II} and transverse $\sigma_{\perp}$ components of the tensor of the electrical conductivity of a degenerate gas :

$$\sigma_{\perp} = 4\sigma_0 \frac{\mu}{\varepsilon_0} \frac{\mu}{k_0 T} (\frac{2m_{\perp}k_0 T}{\hbar^2 r_0^{-2}})^{1/2} \frac{(\mu^* - 2\varepsilon_0^* \sin^2 \frac{z_0}{4})^{1/2}}{\ln(4\frac{z_0}{a} r_0)} [z_0 + \frac{2\varepsilon_0}{\mu}(z_0 - \sin z_0) +$$

$$+ \frac{\varepsilon_0}{\mu}(\frac{3}{2} z_0 - 2\sin z_0 + \frac{1}{4}\sin 2z_0)], \quad (6)$$

$$\sigma_{II} = 4\sigma_0 \frac{r_0}{a} \frac{m_{\perp}}{m_{z_0}} (\frac{2m_{\perp}k_0 T}{\hbar^2 r_0^{-2}})^{1/2} (\mu^* - 2\varepsilon_0^* \sin^2 \frac{z_0}{4})^{1/2} (z_0 -$$

$$- \sin 2z_0 + 2\frac{\sin^2 z_0}{z_0}), \quad (7)$$

where $\quad \tau_0 = \frac{(m_{\perp}\chi)}{8\pi N e a^{3/2}}, \sigma_0 = \frac{e^2 \tau_0 n_0}{m}, \quad r_0^{-2} = \frac{4\pi e^2}{\chi} \frac{m_{\perp}}{\pi^2 \hbar^2 a} z_0,$

$$n_0 = \frac{m_{\perp}(\mu - \varepsilon_0)}{\pi^2 \hbar^2 a}, \quad \mu^* = \frac{\mu}{k_0 T}, \quad \varepsilon_0^* = \frac{\varepsilon_0}{k_0 T}, \quad z_0 = \pi \text{ if } \mu > 2\varepsilon_0,$$

$$z_0(\varepsilon) = \arccos\left(1 - \frac{\mu}{\varepsilon_0}\right) \quad \text{as} \quad \mu < 2\varepsilon_0; \text{ the rest are universally}$$

accepted.

Then, the ratio of electrical conductivity reads :

$$\frac{\sigma_{\perp}}{\sigma_{II}} = 8\frac{r_0}{a} \frac{1}{\ln(4\frac{r_0}{a}\pi)} \frac{\frac{3}{2} - \frac{2\sin z_0}{z_0} + \frac{\sin^2 z_0}{4z_0}}{z_0 - \sin 2z + 2\frac{\sin^2 z_0}{z_0}}. \quad (8)$$

From this it is inferred that, in contrast to the phonon scattering where the ratio of electrical conductivity is proportional to the ratio of the effective mass [4], on the ion impurity scattering the anisotropy coefficient is dictated by a ratio of the screening radius to the lattice constant across the layer.

It should be emphasized that electrical conductivity is heavily anisotropic, the anisotropy magnitude depends upon the relation between the miniband width and Fermi level.

REFERENCES

1. S.I. Borisenko, Semiconductors **37**, Issue 5, 569 and Issue9, 1093 (2003).

2. A.Gold, Physical Review B **35**, 723 (1987).

3. A.B. Henriques, Physical Review B **64**, 045319 (2001).

4. B.M. Askerov *et al*, Journal of Physics: Condensed Matter, 7, 843 (1995).

Dynamics and Potential Energy Surfaces for small to medium size He$_n$-dihalogen clusters

G. Delgado-Barrio, Á. Valdés, D. López-Durán, M.P. de Lara Castells, R. Prosmiti, and P. Villarreal

Instituto de Matemáticas y Física Fundamental C.S.I.C.
C/Serrano 123, 28006, Madrid, Spain

Abstract. The intermolecular forces between atoms and molecules are of great importance in studies of solids, liquids and clusters. Our current studies serve to bridge the gap between small cluster and large cluster limit.

Keywords: Foundations of quantum mechanics
PACS: 03.65.Ta

The intermolecular forces between atoms and molecules are of great importance in studies of solids, liquids and clusters. Over the last decade, enormous advances have been made in understanding of interaction potentials of small molecules. In particular, up to now accurate potentials have been obtained for triatomic van der Waals (vdW) systems either by fitting the potential parameters to experimental data obtained from high resolution spectroscopic methods or by performing state-of-the-art electronic structure calculations. Studies of larger species are more complex and the construction of ab initio potential energy surfaces (PESs) for such polyatomic molecular systems (including heavy atoms in our case) is a computationally prohibitive task. Thus, given the high quality *ab initio* PESs calculated for triatomic species, attempts are made using pairwise 3-body interaction potentials to represent PESs of larger Rg$_n$XY complexes [1],[2].

In the present study the pairwise additivity of the 3-body interactions, derived from the He$_2$ and He-dihalogen CCSD(T) potentials, is investigated by means of ab initio electronic structure and quantum-mechanical calculations [3],[4]. The intermolecular interactions and structural properties of these clusters, which consist of homopolar and heteropolar halogens, are analyzed and the importance of additional effects, such like introducing electric dipole moment and larger reduced mass of the complex, is evaluated. Calculations of vibrational energies, including all 5 degrees of freedom, are performed and compared with experimental data from high resolution spectroscopy. Different structural models, such as 'police-nightstick' and 'linear', together with the traditional tetrahedral ones are predicted [3],[4]. For first time results on energetics and vibrationally averaged structures of such species are presented and their comparison with recent experimental predictions attributes to evaluate the quality of the surface [5].

Further, the extension of the above model to He$_n$-ICl clusters allowed us to simulate the infra-red (IR) spectra of complexes with n up to 30 helium atoms and to study the influence on the quantum statistics of the environment [6],[7],[8]. A theoretical model based on an adiabatic quantum-chemistry-like approach is developed, using Hartree-

CP905, *Frontiers of Fundamental Physics (FFP8), Eighth International Symposium*
edited by B. G. Sidharth, A. Alfonso-Faus, and M. J. Fullana

Fock or Hartree treatments to consider both fermionic or bosonic clusters [9].

For ^{4}He boson species the simulated IR spectra exhibit well separated P and R branches, while for ^{3}He fermion ones a Q branch appears. Moreover, in the later case the quasi-degeneration of the spin-multiplets contributes to the congestion of the spectrum. These findings are in perfect agreement with experimental measurements on linear OCS molecule in He nanodroplets, where well structured spectra have been observed in ^{4}He, in contrast with the broad and unstructed profiles obtained in ^{3}He environment [10]. The above experimental observations were intepreted as a manifestation of superfluidity of bosonic helium at a microscopic scale. In this sense our current studies serve to bridge the gap between small cluster and large cluster limit.

REFERENCES

1. R. Prosmiti et al., *J. Chem. Phys.* **117**, 7017 (2002).
2. A. Valdés et al., *Mol. Phys* **102**, 2277 (2004).
3. A. Valdés et al., *J. Chem. Phys.* **122**, 044305 (2005).
4. A. Valdés et al., *J. Chem. Phys.* **125**, 014313 (2006).
5. R.A. Loomis, private communication.
6. D. López-Durán et al., *Phys. Rev. Lett.* **93**, 053401 (2004).
7. D. López-Durán et al., *J. Chem. Phys.* **121**, 2975 (2004).
8. M.P. de Lara-Castells et al., *Phys. Rev. A* **74**, 053201 (2006).
9. M.P. de Lara-Castells et al., *Phys. Rev. A* **71**, 033203 (2005).
10. S. Grebenev et al., *Science* **279**, 2083 (1998).

The vacuum energy: Casimir effect and the cosmological constant

Emilio Elizalde[1]

Instituto de Ciencias del Espacio (CSIC)
Institut d'Estudis Espacials de Catalunya (IEEC/CSIC)
Campus UAB, Facultat de Ciències, Torre C5-Parell-2a planta
E-08193 Bellaterra (Barcelona) Spain

Abstract. This is a short summary of the talk given at the meeting, wich comprised considerations on the zero point energy and some aspects of the Casimir effect. In particular, recent developments on the construction by the author and J. Haro of the first consistent proof of the dynamical Casimir effect, on the possible contribution of vacuum fluctuations to the cosmological constant, and on some new theories about it, discuseed at the meeting, are also reported.

Keywords: Vacuum energy, cosmological constant, Casimir effect, dynamical Casimir effect
PACS: 98.80.Es, 98.80.-k, 02.30.Gp, 11.10.Gh

The ordinary and the dynamical Casimir effects

The Casimir effect takes its name after the late H.B.G. Casimir, who in 1948 published a paper in the Proceedings of the Academy of Sciences of the Netherlands, in which a rather remarkable property, namely the attraction of two neutral metallic plates, was predicted theoretically [1]. This paper was the beginning of a whole branch of research, which aims nowadays at answering very profound questions about the vacuum structure of Quantum Field Theory. The interest of the subject, which is reflected by the impressive number of papers which are dealing with it in the last years [2]. It is fair to say, however, that the 1948 paper by Casimir attracted comparatively small attention during the following two or three decades. Another paper by Casimir and Polder [3], which was published in Physical Review also in 1948, got by far much more citations from experimental and theoretical colleagues. Fluctuations of the quantum vacuum appear everywhere, the key issue being to ascertain if these contributions are relevant or not for the process in question. They appear nowadays to be irrelevant for sonoluminiscence, but quite on the contrary, they are very important for accurate calculations in laser cavities, wetting of alcali compounds by Helium3 phenomena, sticking of small plates in MEM and NEM devices, and so on [2].

When the plates move quickly, as with a high-frequency vibration, one has the so called Dynamical Casimir effect, a phenomenon studied for the first time by S. Fulling and P. Davies in 1976 [4]. In fact, moving mirrors further modify the structure of the

---

[1] Presently on leave at Department of Physics & Astronomy, Dartmouth College, 6127 Wilder Laboratory, Hanover, NH 03755, USA. E-mail: elizalde@ieec.fcr.es, elizalde@math.mit.edu

CP905, *Frontiers of Fundamental Physics (FFP8), Eighth International Symposium*
edited by B. G. Sidharth, A. Alfonso-Faus, and M. J. Fullana
© 2007 American Institute of Physics 978-0-7354-0412-0/07/$23.00

quantum vacuum, what manifests in the creation and annihilation of particles. Once the mirrors return to rest, a number of the produced particles may still remain which can be interpreted as radiated particles. This flux has been calculated in several situations by using different methods, as averaging over fast oscillations [5, 6], by multiple scale analysis, with the rotating wave approximation [7], with numerical techniques [8], and others. One is interested in the production of the particles and their possible energy all the time while the mirrors are moving. In the case of a single, perfectly reflecting mirror, the number of produced particles as well as their energy diverge while the mirror moves. Several renormalization prescriptions have been used in order to obtain a well-defined energy, however, for some trajectories this finite energy is not a positive quantity and cannot be identified with the energy of the produced particles (see e.g. [4]).

Our approach [9] relies on two basic ingredients: proper use of a Hamiltonian method and the consideration of partially transmitting mirrors, which become transparent to very high frequencies. We prove, in this way, both that the number of created particles is finite and also that their energy is always positive for the whole trajectory corresponding to the mirrors' displacement. We also calculate the radiation-reaction force that acts on the mirrors owing to the emission and absorption of particles, and which is related with the field's energy through the energy conservation law, so that the energy of the field at any time t is equal, with opposite sign, to the work performed by the reaction force up to time t. Such force is usually split into two parts: a dissipative force whose work equals minus the energy of the particles that remain, and a reactive force vanishing when the mirrors return to rest. We show that the radiation-reaction force calculated from the Hamiltonian approach for partially transmitting mirrors satisfies, at all time, the energy conservation law and can naturally account for the creation of positive energy particles. Also, the dissipative part we obtain agrees with the one calculated by other methods, as using the Heisenberg picture or other effective Hamiltonians. Note that those methods have problems with the reactive part, which in general yields a non-positive energy that cannot be considered as that of the particles created at any time t.

Quantum vacuum fluctuations and the cosmological constant

A fact, apparently unrelated with the considerations above, is that our universe seems to be spatially flat and to possess a non-vanishing cc. Thus, Einstein's 'great mistake' may turn out ultimately to be a great discovery, a necessary ingredient in order to explain the acceleration of the universe. In any case, for elementary particle physicists it constitutes a great embarrassment, calculations being off by the famous 120 orders of magnitude. First, physicists tried to find a way to get rid of it (Coleman, Weinberg, Polchinski, ...), in the hope it could be proven to be zero, what was hard enough. Now it turns out that it is non-vanishing, albeit very small, indeed a peculiar quantity. The cc has to do with cosmology, through Einstein's equations and the FRW universe obtained from them, and also with the local structure of elementary particle physics, as the stress-energy density μ of the vacuum $L_{cc} = \int d^4x \sqrt{-g}\,\mu^4 = \frac{1}{8\pi G} \int d^4x \sqrt{-g}\,\lambda$. In other words: two contributions appear, on the same footing, namely $\frac{\Lambda c^2}{8\pi G} + \frac{1}{\text{Vol}} \frac{\hbar c}{2} \sum_i \omega_i$.

The issue of the cc has got renewed thrust from the observational evidence of an acceleration in the expansion of our Universe, initially reported by two different groups [10]. First, there was some controversy on the reliability of the results but after new data has been gathered, there is presently reasonable consensus, supported by new analysis. But it is even more difficult to explain why the cc is so small but non-zero [11], than to build theoretical models where it exactly vanishes.

Rather than trying to understand the fine-tuned cancelation of such enormous values at this *local* level, in this section we will elaborate on a quite simple idea (but, for the same reason, of potentially far reaching consequences), related with the *global* topology of the universe [12] and in connection with the possibility that a scalar field of small mass pervading the universe exists. Fields of this kind are ubiquitous in inflationary models, quintessence theories, etc. We do not aim at solving the old cc problem, but just to investigate some models which may show the right order of magnitude for (some contributions to) ρ_V, in the precise range of astrophysical observations [13, 10], e.g. $\rho_V = (2.14 \pm 0.13 \times 10^{-3} \text{ eV})^4 \sim 4.32 \times 10^{-9}$ erg/cm^3. We address here, so to say, the 'perturbative part' of the *new* cc problem [14]: we assume the existence of a scalar field background extending through the universe and calculate the contribution to the cc of the Casimir energy density for this field with some typical BCs. Ultraviolet contributions will be safely set to zero invoking some mechanism of a fundamental theory.

Another hypothesis will be the existence of both large and small dimensions (the total number of large spatial coordinates being always three), some of which may be compactified, so that the global topology of the universe will play an important role. There is a quite extensive literature both in the subject of what is the global topology of spatial sections of the universe and also on the issue of the possible contribution of the Casimir effect as a source of some sort of cosmic energy, as in the case of the creation of a neutron star. There are arguments that favor different topologies, as a compact hyperbolic manifold for the spatial section, what would have clear observational consequences. Other interesting work along these lines was reported in and related ideas have been discussed very recently in. However, we differ from all those in that emphasis is put now in obtaining the right order of magnitude for the effect. At the present stage it has no sense to consider the whole amount of possibilities concerning the nature of the field, the different models for the topology of the universe, and the different BCs possible, with its effect on the sign of the force too. This is left to a second, more detailed analysis.

From previous experience [15], we know that the range of orders of magnitude of the vacuum energy density for the most common possibilities is not so widespread, and may only differ by at most a couple of digits. This will allow us, both for the sake of simplicity and universality, to deal with simplified situations, as a scalar field with periodic BCs or spherically compactified dimensions.

Basic cosmological models which display the observed accelerated expansion

Consider a universe with a space-time of one of the following types: $\mathbb{R}^{d+1} \times \mathbb{T}^p \times \mathbb{T}^q$, $\mathbb{R}^{d+1} \times \mathbb{T}^p \times \mathbb{S}^q, \ldots$, which are actually plausible models for the space-time topology. Here, $d \geq 0$ stands for a possible number of non-compactified dimensions. Recall the physical contribution to the vacuum or zero-point energy $< 0|H|0 >$ is obtained by subtracting the vacuum energy corresponding to the situation with the only change that compactification is absent (in practice this is done by conveniently sending the compactification radii to infinity). As well known, both of these vacuum energies are in fact infinite, but it is its *difference* $E_C = < 0|H|0 >|_R - < 0|H|0 >|_{R \to \infty}$ (R a typical compactification length) that makes physical sense, giving rise to the finite value of the Casimir energy E_C. Renormalization must then be carried out. In fact we will discuss the Casimir (or vacuum) energy *density*, $\rho_C = E_C/V$, which can account for either a finite or an infinite volume of the spatial section of the universe (from now on we shall assume that all diagonalizations already correspond to energy densities, volume factors are replaced at the end). In terms of the spectrum $\{\lambda_n\}$ of H: $< 0|H|0 >= \frac{1}{2} \sum_n \lambda_n$, where the sum over n is over the whole spectrum, which may involve several continuum and several discrete indices. The last appear tipically when compactifying the space coordinates (much as time compactification gives rise to a finite-temperature field theory), as in the cases we are going to consider. Thus, integration over d continuous dimensions and multiple summations over $p + q$ indices appear.

The physical vacuum energy density in our case, where the contribution of a scalar field ϕ living in a partly compactified spatial section of the universe is considered, with

$$S = \frac{1}{2} \int d^4x \sqrt{-g} \left[g^{\mu\nu} \partial_\mu \phi \partial_\nu \phi + (m^2 + \xi R)\phi^2 \right], \tag{1}$$

will be denoted by ρ_ϕ (this is just the contribution to ρ_V coming from this field, there might be other, in general), $\rho_\phi = \frac{1}{2} \sum_i \lambda_i = \frac{1}{2} \sum_{\mathbf{k}} \frac{1}{\mu} \left(k^2 + M^2 \right)^{1/2}$, where the sum $\sum_{\mathbf{k}}$ is a generalized one, $M^2 = m^2 + \xi R$ is an effective mass term, and μ is the usual mass-dimensional parameter to render the eigenvalues dimensionless (the renormalization parameter; we take $\hbar = c = 1$). The mass m of the field will be here considered to be arbitrarily small but different from zero, for now, for computational reasons —as well as for physical ones, since a very tiny mass for the field can never be excluded. Our model is stationary, while the universe is expanding. A more careful calculation shows that this effect can actually be dismissed at the level of order of magnitude, since its value cannot surpass the one that we will get (as is seen from the present value of the expansion rate $\Delta R/R \sim 10^{-10}$ per year or from direct consideration of the Hubble coefficient). Recent considerations on the dynamical Casimir effect may be important in a future, more detailed analysis. For simplicity we perform a static calculation. As a consequence, the values obtained correspond to the present epoch.

To exhibit explicitly a couple of the wide family of cases considered, let us write down in detail the formulas corresponding to the two first topologies, as described above. For a (p,q)-toroidal universe, with p the number of 'large' and q of 'small' dimensions:

$$\rho_\phi = \frac{\pi^{-d/2}}{2^d \Gamma(d/2) \prod_{j=1}^{p} a_j \prod_{h=1}^{q} b_h} \int_0^\infty dk\, k^{d-1} \sum_{\mathbf{n}_p=-\infty}^{\infty} \sum_{\mathbf{m}_q=-\infty}^{\infty} \left[\sum_{j=1}^{p} \left(\frac{2\pi n_j}{a_j} \right)^2 + \sum_{h=1}^{q} \left(\frac{2\pi m_h}{b_h} \right)^2 + M^2 \right]^{1/2}$$

$$\sim \frac{1}{a^p b^q} \sum_{\mathbf{n}_p, \mathbf{m}_q = -\infty}^{\infty} \left(\frac{1}{a^2} \sum_{j=1}^{p} n_j^2 + \frac{1}{b^2} \sum_{h=1}^{q} m_h^2 + M^2 \right)^{(d+1)/2+1}, \tag{2}$$

where the last formula corresponds to the case when all large (resp. all small) compactification scales are the same. In this last expression the squared mass of the field should
be divided by $4\pi^2\mu^2$, but we have renamed it again M^2 to simplify the ensuing formulas.
We also will not take care for the moment of the mass-dim factor μ in other places since
formulas would get unnecessarily complicated and there is no problem in recovering it
at the end of the calculation. For a (p-toroidal, q-spherical)-universe,

$$\rho_\phi = \frac{\pi^{-d/2}}{2^d \Gamma(d/2) \prod_{j=1}^{p} a_j \, b^q} \int_0^\infty dk\, k^{d-1} \sum_{\mathbf{n}_p=-\infty}^{\infty} \sum_{l=1}^{\infty} P_{q-1}(l) \left[\sum_{j=1}^{p} \left(\frac{2\pi n_j}{a_j} \right)^2 + \frac{Q_2(l)}{b^2} + M^2 \right]^{1/2}$$

$$\sim \frac{1}{a^p b^q} \sum_{\mathbf{n}_p=-\infty}^{\infty} \sum_{l=1}^{\infty} P_{q-1}(l) \left(\frac{4\pi^2}{a^2} \sum_{j=1}^{p} n_j^2 + \frac{l(l+q)}{b^2} + M^2 \right)^{(d+1)/2+1}, \qquad (3)$$

where $P_{q-1}(l)$ is a polynomial in l of degree $q-1$. On dealing with our observable
universe, we assume that $d = 3 - p$, the number of non-compactified, 'large' spatial
dimensions (thus, no d dependence will remain). All these expressions for ρ_ϕ need to be
regularized and we use the zeta function method, as previously explained.

Results are obtained for the vacuum energy density corresponding to a sample of different numbers of compactified (large and small) dimensions and for different values of
the small compactification length in terms of the Planck length. Good coincidence with
the observational value for the cc is reached for the contribution of a massless scalar
field, ρ_ϕ, for p large compactified dimensions and $q = p + 1$ small compactified dimensions, $p = 0, \ldots, 3$, and this for values of the small compactification length, b, of the order
of 100 to 1000 times the Planck length l_P (what is actually a very reasonable conclusion,
according also to string theories). The cases of pure spherical compactification and of
mixed toroidal (for small magnitudes) and spherical (for big ones) compactification can
be treated in this way and yield results in the same order of magnitude range. Both these
cases correspond to (observational) isotropic spatial geometries. Dimensionally speaking, within the global approach adopted in the present paper everything is dictated by
the two basic lengths in the problem, which are its Planck value and the radius of the observable Universe. By playing with these numbers we can show that the observed value
of ρ_V may be remarkably well fitted, under reasonable hypothesis, for some common
models of the space-time topology.

The Casimir effect is an ubiquitous phenomena. Its contribution may be small (as it
seems to be the case, yet controverted, to sonoluminiscence), or of some 10-30% (that
is, not dominant but important), as in wetting phenomena involving He in condensed
matter physics. Here we see that it may also provide a contribution of the needed order
of magnitude, corresponding to our present epoch in the accelerated expansion of the
universe. The implication that this calculation bears for the early universe and for inflation is not clear from our results, since the model should be adapted to the situation and
BCs corresponding to those primeval epochs, what cannot be done straightforwardly.
This is also left for further study.

More elaborate models: an actually repulsive vacuum energy

A very important issue in all the previous analysis is the specific *sign* of the resulting force. For scalar fields and the usual compactifications or BCs it seems impossible to get the right sign corresponding to the accelerated expansion of the universe. However, in worldbrane models and others involving supergravitons and fermion fields we are able to prove that the appropriate sign can be obtained under quite natural conditions. Specifically, in the case of the torus topology we have obtained [16] that the topological contributions to the effective potential have in fact a fixed sign, which depends on the BC one imposes. It is negative for periodic fermionic fields in a a supersymmetric theory, and positive for anti-periodic fields. Thus, topology provides a mechanism which, in a most natural way, permits to have a positive cc in a multi-supergraviton model with anti-periodic fermions [16]. Moreover, the value of the cc is regulated by the corresponding size of the torus (as we have also seen in the scalar case above). One can most naturally use in this case the minimum number, $N = 3$, of copies of bosons and fermions, and show that the order of magnitude of the observational values for the cc can be reproduced invoking quite natural assumptions.

ACKNOWLEDGMENTS

Part of this work was done in collaboration with G. Cognola, J. Haro, S.D. Odintsov, and S. Zerbini. Work supported in part by MEC (Spain), projects BFM2003-00620 and PR2006-0145, and by AGAUR (Generalitat de Catalunya), contract 2005SGR-00790.

REFERENCES

1. II.G.B. Casimir, Proc. Koninkl. Ned. Akad. Wetenshap **51**, 793 (1948).
2. *Quantum Field Theory under the Influence of External Conditions*, J Phys. A: Math. Gen. **39** (2006).
3. H.G.B. Casimir and D. Polder, Phys. Rev. **73**, 360 (1948).
4. S.A. Fulling and P.C.W. Davies, Proc. Roy. Soc. Lond. **A348**, 393 (1976).
5. V.V. Dodonov and A.B. Klimov, Phys. Rev. **A53**, 2664 (1996).
6. J.Y. Ji, H.H. Jung, J.W. Park, and K.S. Soh, Phys. Rev. **A56**, 4440 (1997).
7. R. Schützhold, G. Plunien, and G. Soff, Phys. Rev. **A65**, 043820 (2002); G. Schaller, R. Schützhold, G. Plunien, and G. Soff, Phys. Rev. **A66**, 023812 (2002).
8. M. Ruser, J. Opt. **B7**, 100 (2005).
9. J. Haro and E. Elizalde, Phys. Rev. Lett. **97**, 130401 (2006).
10. S. Perlmutter *et al.* [Supernova Cosmology Project Collaboration], Astrophys. J. **517**, 565 (1999); A.G. Riess *et al.* [Hi-Z Supernova Team Collaboration], Astron. Journ. **116**, 1009 (1998).
11. S. Weinberg, Phys. Rev. **D61** 103505 (2000).
12. V. Blanloeil and B.F. Roukema, Eds., *Cosmological Topology in Paris 1998*, astro-ph/0010170. See also the entire Vol. 15 of Classical and Quantum Gravity (1998).
13. M. Tegmark et al. [SDSS Collaboration], Phys. Rev **D69**, 103501 (2004).
14. C.P. Burgess et al., hep-th/0606020, 0510123; T. Padmanabhan, gr-qc/0606061.
15. E. Elizalde, Nuovo Cim. **104B**, 685 (1989).
16. G. Cognola, E. Elizalde and S. Zerbini, Phys. Lett. **B624**, 70 (2005); G. Cognola, E. Elizalde, S. Nojiri, S.D. Odintsov and S. Zerbini, Mod. Phys. Lett. **A19**, 1435 (2004).

Deterministic Quantum Mechanics Versus Classical Mechanical Indeterminism and Nonlinear Dynamics

M. S. EL NASCHIE

King Abdul Aziz City of Science & Technology,
Riyadh, Kingdom of Saudi Arabia.
P.O. Box 272, Cobham, Surrey KT11 2FQ, United Kingdom

Abstract. At a minimum it seems that deterministic quantum mechanics recently proposed by G. 't Hooft and classical mechanical indeterminism i.e. nonlinear dynamics and deterministic chaos are homomorphic conceptually. The present work gives an introduction to a transfinite spacetime theory based on the geometry and topology of a hierarchal infinite dimensional Cantor set in which both notions are exchangeable.

Keywords: Deterministic quantum mechanics, electromagnetic fine structure constant, holographic principle, super string theory, quantum gravity and nonlinear dynamics.
PACS: 11.25.-w/11./10./11.30.Pb

INTRODUCTION

In an article published as long ago as in 1955, "Ist die Klassische Mechanik tatsächlich deterministisch?" Max Born exclaimed in the title of his article [1,2] if classical mechanics is really deterministic and went on to show that it is not.

In the meantime and after the discovery of deterministic chaos, the formulation of KAM theorem and the deep appreciation of non-integrable systems as well as solitons and all apart from the difficulties encountered in giving an exact mathematical description of fluid turbulence we should be in a far better position to contemplate the advantage and drawbacks inherent in quantum mechanics [3-6] which Max Born foresaw many years ago [1,2].

In the present work we present a synthesis of the research result of G. Ord on the generalization of R. Feynman chessboard model [5] as well as the present author's conjugate complex time interpretation of the Schrödinger equation [8] and the relation to two joint non-conservative systems with opposite dissipation forming one larger conservative set [7-9]. Throughout the analysis we are guided by 't Hooft's ideas regarding deterministic quantum mechanics [10,11] and conclude that such quantum mechanics might essentially be a deterministically chaotic mechanics, akin to a generalization of Ord's model in the context of E-infinity theory [3,12,13] and the VAK conjecture [3].

CP905, *Frontiers of Fundamental Physics (FFP8), Eighth International Symposium*
edited by B. G. Sidharth, A. Alfonso-Faus, and M. J. Fullana
© 2007 American Institute of Physics 978-0-7354-0412-0/07/$23.00

ORD'S GENERALIZATION OF FEYNMAN'S CHESSBOARD MODEL

Ord's proposal which was initiated by Feynman's chessboard model sees quantum mechanics as a projection of a higher order intrinsically statistical model [5-9]. He starts from a minimal model in $1 + 1$ dimensions by considering a simple trajectory of a spacetime lattice. Much like a random peano space filling curve, the "particle" moving on the lattice has to make a binary choice of left or right. He then takes the binomial distribution $P(m,n)$ [5] as a realistic probability where $P(m,n)$ is taken to mean the number of distinct paths of a displacement m Λ x after n steps. Proceeding in this way Ord establishes a set of difference equations which will ultimately lead to a telegraph and a Dirac-like equation. In the course of doing that he arrives at the following block diagonal matrix of the form [5]

$$E = \begin{pmatrix} (a)/E & -bE & 0 & 0 \\ b/E & aE & & \\ 0 & 0 & a/E & bE \\ 0 & 0 & b/E & aE \end{pmatrix}$$

This indicates that the initial system evolved into a two block diagonally coupled sub systems. Subsequently one can extract two sets each consisting of two differential equations from writing the difference equations explicitly. To first order, the first set turned out to be the Kac model of the telegraph equation. The second set and after some algebra and reduction in terms of Pauli spin matrices, gives us a Dirac equation in $1 + 1$ dimensions.

This is a deceptively simple but in fact highly non-trivial and insightful result. Ord started with a geometrical spacetime picture of a random walk on a lattice and ended up with two sets of differential equations. The first gave us a classical dissipative system and the second turned out to be essentially the Dirac equation. Of course Dirac's equation and Schrödinger's equation for that matter are not a complete quantum theory because Born guidance is lacking [9]. However we do now recognize two vital facts about the roots of quantum mechanics. The first fact is that the origin of quantum mechanics equations lies in a classical statistical and irreducible spacetime geometry which forces a particle to behave erratically as shown by Ord. There is however a deep order in this random spacetime manifold which can be seen only when we delay the taking of the differential limit to the last moment possible. The continuum limit smears this beautiful order in chaos of the spacetime manifold.

Second, the equations of quantum mechanics are only one projection of the original spacetime motion on a random spacetime manifold. The second projection is a dissipative quasi classical system. In a manner of speaking, quantum field theory is some form of a combination of the two equations which remain at the end conservatively Hamiltonian but there is a highly complex and non-trivial local exchange of energy going on all the time inside quantum field theory. This complexity is the origin of chaotic behaviour as well as highly ordered solitonic behaviour in quantum field theory. From the outside everything is nicely frictionless Hamiltonian, but within, it is sizzling. Thus we conclude that since the origin of quantum mechanics is semi classical statistics caused by a random spacetime manifold, then the Born interpretation of probability as $P = |\Psi|^2$ is not that strange after all and is merely a mathematical shortcut pretty much as analytical continuation. E-infinity discovered another shortcut without a phase, namely the $\phi = (\sqrt{5} - 1)/2$, based irrational quantum probability [6].

     If the preceding interpretation of Ord's results is correct, then we can say that this is exactly in line with 't Hooft's efforts to show the hidden non-quantum mechanical roots of quantum mechanics. It would be a total misunderstanding of 't Hooft's work to think he wants a return to naïve determinism which is, in our opinion, virtually impossible. In fact it is undesirable because we have come a long way since the discoveries of E. Lorenz and M. Feigenbaum in understanding what determinism really means [3-6, 12]. Having said that, we must stress again that by choosing from the outset a classical binomial distribution, Ord missed a second possibility for a quantum probability with interference but without a phase. This is the irrational probability of E-infinity which takes $P = \frac{1}{2}$ to $P_{1,2} = 1/2 \pm k = (\phi \; ; \; \phi^2)$ where $\phi = (\sqrt{5} - 1)/2$ and $k = \phi^3 (1 - \phi^3)$. This is the probability of a transfinitely random manifold which is identical to that of E-infinity spacetime theory [3-6, 12,13].

GAUGE INVARIANCE, DISSIPATIVE QUANTUM MECHANICS AND SELF-ADJOINT SETS:
Prolongation, Self Adjointness and Gauge Invariant Fields

The following two nonconservative, non-self adjoint oscillations with positive and negative damping respectively [8,9]

$$\ddot{x}_1 - \dot{x}_1 b + \omega^2 x_1 = 0 \quad \text{and} \quad \ddot{x}_2 + \dot{x}_2 b + \omega^2 x_2 = 0$$

where x is the state variable $(\dot{\ }) = d(\ )/dt$, t is the time, b is the damping coefficient and ω is the frequency may be derived from the following bilinear Lagrangian

$$\mathcal{L} = \dot{x}_1 \, \dot{x}_2 + \frac{b}{2} (x_1 \dot{x}_2 - \dot{x}_1 x_2) - \omega^2 x_1 x_2$$

The highly interesting point is that this Lagrangian represents the classical-Newtonian limit of a Lagrangian of a gauge invariant field modeling an interaction between an electromagnetic field and a complex scaler field

$$\mathcal{L} = \varphi_\mu \; ; \phi^\mu ; - i e \, A^\mu (\varphi_i ; \phi - \phi_\mu ;) - (m^2 - e^2 \, A_\mu \, A^\mu) \, \varphi \, \phi$$

$$= (\varphi ; + i e \, A_\mu \phi)(\phi^\mu ; - i e \, A^\mu \, \phi) - m^2 \, \varphi \, \phi \, .$$

The reader may ascertain for himself that the following replacements transform any of the two Lagrangians into the other [7-9]:

$$x_1(t) \leftrightarrow \varphi(x) \qquad b/2 \leftrightarrow -ie A^\mu \qquad d(\,)/dt \leftrightarrow d(\,)/dx^\mu$$

$$x_2(t) \leftrightarrow \phi(x) \qquad \omega^2 \leftrightarrow m^2 - e^2 A_\mu A^\mu$$

Thus even in quantum field theory, we have a form of internal dissipation and non-conservation in action. However energy losses (dissipation) and energy gains (flutter) balance miraculously to give a total global conservation as discussed briefly earlier on.

E-INFINITY THEORY

The basic assumptions and characteristics of E-infinity theory are the following [4-6]

- Spacetime is an infinite dimensional Cantor set ($n_f = \infty$).

- This Cantor set is hierarchal.

- Because it is hierarchal it has two finite expectation values

    (a) $\quad <D_T> \,=\, 4$ is the topological dimension

    (b) $\quad <D_H> \,=\, 4 + (\overline{4}) = 4 + \phi^3 \quad \simeq 4.236067977$

is the Hausdorff dimension where $\phi = (\sqrt{5} - 1)/2$. Thus E-infinity theory is fixed not by a single topological dimension but by three, namely $n_f = \infty$, $<n_t> \,=\, 4$ and $<n_H> \,=\, 4 + \phi^3$.

Further development of this theory leads to the conclusion that E-infinity may be modeled by a fuzzy K3 (Kähler) manifold. We note that while the classical Kähler K3 is given by $b_2^+ = 3$, $b_2^- = 19$ and $\chi = 24$ where $b_i^\pm$ are the Betti numbers and χ is the Euler class, for the fuzzy Kähler we have

$$b_2^+ = 5 + \phi^3 \,, \quad b_2^- = 19 - \phi^6 \quad \text{and} \quad \chi = 26 + k$$

where $k = \phi^3(1 - \phi^3)$. In addition E-infinity may be fixed using two Kähler fuzzy manifolds, one for the volume and one for what we call holographic boundary [3,10,11]. For the volume the Kähler is the previous one while for the holographic boundary we have a $K(\Gamma_c(7))$ given by

$$b_2^+ = 3 - \phi^6 \,, \quad b_2^- = 13 - \phi^6 \quad \text{and} \quad \chi = 18 - 2\phi^6 .$$

DERIVATION OF E-INFINITY DIMENSION

The Hausdorff dimension $4 + \phi^3$ may be found from the topological entropy $\ell n\,(1/\phi)$ of the grammatical complexity of the symbolic sequences of our E-infinity and the fact that a quantum path is area like $D_Q = 2$. Thus from

$$<n> = 2/\ell n\,(1/\phi)$$

we can easily derive a corresponding transfinite expression by expansion and keeping linear terms only:

$$\sim<n> = (1 + \phi)/(1 - \phi) = 4 + \phi^3.$$

This expression is given by [3]

$$\sum_{n=0}^{+\infty} n\phi^n = 4 + \phi^3.$$

For the ground state on the other hand we have

$$\sum_{n=0}^{-\infty} n\phi^{-n} = -(4 + \phi^3) \neq 0.$$

Here we have used the Menger-Uryson deductive dimensional system[3-6]. A legitimate question to ask now is the following: What is the advantage of Cantorian spacetime geometry? The answer is:

   a. E-infinity is resolution dependent. That means the topological theorem on the invariance of dimension is no longer valid [4-6].

   b. It is discrete but has the cardinality of the continuum [3-6].

Therefore it can unify two topologically very different spaces, namely that of general relativity and that of quantum physics [3-6]. In conclusion of this section we should mention the intimate relation between COBE thermodynamics and E-infinity [13].

G'T HOOFT'S HOLOGRAPHIC PRINCIPLES AND DETERMINISTIC QUANTUM MECHANICS

Gerrard 't Hooft is probably one of the first mainstream scientists to question quantum mechanics along the lines of A. Einstein [10, 11] and made serious mathematical proposals to remedy the situation with regard to quantum gravity [11] as well as introducing the holographic principles.

The present author sees this principle as one of the most profound proposals in theoretical physics. Using this principle and symmetry group arguments, one could determine for instance the inverse value of the fine structure constant in less than elementary but water tight derivation.

Let us start from the obvious fact that string theory was created to describe all fundamental interactions, including gravity. The over riding importance of $E_8 \, E_8$

exceptional Lie symmetry groups in reaching the goal is well known. Thus the world, according to super strings, is described by the 496 dimensions of $E_8 E_8$ and these in turn are interpreted as massless gauge Bosons. Now following 't Hooft's holographic principles [3,4], we know from E-infinity [3-6] that the holographic boundary of the $E_8 E_8$ theory is given by $SL(2,7)$ symmetry group which has 336 dimensions (see Figs. 1 and 2). On compactifying this group as in an Escher picture, the number increases to 339 dimensions. Thus all what is left in the $E_8 E_8$ bulk (see Fig. 1) is [3-6]

$$\text{Dim } E_8 E_8 - \text{Dim } SL(2,7)_c = 496 - 339 = 157.$$

Since 339 must represents only particle physics, as per 't Hooft's holographic boundary theory (see Fig. 2), the 157 left must be related to electromagnetism and gravity. Taking the 20 independent components of gravity out, i.e. the $R^{(4)} = (4)^2 (4^2 - 1)/12 = 20$ of the Riemannian tensor, we are left with [3-6]:

$$157 - R^{(4)} = 157 - 20 = 137 = \bar{\alpha}$$

which is exactly the integer part of $\bar{\alpha}$. We could have obtained the exact value if we used the exact E-infinity formalism [3-6]. Such is the power of 't Hooft's holographic principles which seems to have been inspired by the duality between Kaluza-Klein theory and the surface entropy of a black hole.

It was also the holographic principle which was found by 't Hooft to obstruct a physically realistic model for the unification of general relativity with quantum mechanics and this led 't Hooft to his proposals for a deterministic and dissipative quantum mechanics which is connected to the fact that his Hamiltonian is not bounded from below [11].

We cannot discuss all aspects of 't Hooft's proposals in the present work and we confine ourselves to a point or two:

a. The first point is with regard to the phenomena of interference and in particular destructive interference. In this respect we stress that not only quantum mechanics displays this phenomena. In fact chaos theory has an analogous feature. This may best be illustrated by the iterated function system as well as the 3 points and 4 points game as explained by the author elsewhere.

b.   Ordinary randomness and deterministically chaotic randomness are in different universality classes and therefore we expect chaotic randomness to give realistic energy spectrum in a suitable quantum field theory [3-6, 11].

c. It is important to realize the vital role played by the VAK, i.e. the vague attractor of Kolmogorove [3-6] In Hamiltonian systems we have no dissipation and the golden mean of KAM theorem replaces this notion of frictional stability. Thus we have to expand our understanding of what we consider to be an attractor including "strange" attractors in "Hamiltonian" systems which is not a contradiction in view of R. Thom and J. Milnor's new definition of an attractor. We have of course an elementary example from classical mechanics which is a Hopf bifurcation in a Hamiltonian system.

Suppose we have two such Hopf bifurcations, then one could combine them as indicated by 't Hooft by writing [11]

$$1/P_{1,2} \approx (1/P_1) + (1/P_2)$$

Now suppose we have infinitely many such Hopf bifurcations, then our system explodes. To prevent explosion, we need a hierarchy of Hopf bifurcations. This is similar to Feigenbaum's bifurcation, only there we have a period doubling due to motion on a Möbius strip [3-6]. If we have infinitely many of these bifurcations, then we could use the Dunkerly and Southwell eigenvalue theorems to combine them using the frequency w_i . In E-infinity theory it is clear that all these requirements are satisfied by a single simple expectation formula following Planck distribution for transfinite discrete sets

$$\sim <n> = (1)\,(\phi) + (2)\,(\phi)^2 + (3)\,(\phi)^3 + \ldots = \sum_0^\infty (n)\,(\phi)^n = 4 + \phi^3$$

which is the well known Hausdorff dimension of E-infinity.

To conclude this part we give the exact theoretical $\overline{\alpha}_o$ by including the transfinite corrections

$$\overline{\alpha}_o = (496 - k^2) - [\,(336 + 16k) + 20)\,] = 137.082039325$$

The exact experimental value of $\overline{\alpha}_o$ is then found by projection [4-6]

$$\overline{\alpha}_o \text{ (Experimental)} = (\overline{\alpha}_o - k_o)/(\cos(\pi/\overline{\alpha}_o)) = 137.03598.$$

Finally we should draw attention to a recent work by Ji-Huan He and his group in which he used n dimensional polytopes for $n = 4$ to $n = 26$ to give a simple Euclidean analogue for the Bulk which helps visualize both E-infinity and Heterotic string theory. This may be regarded as an Euclidean E-infinity theory.

TRANSFINITE CONTINUATION

The previous analysis demonstrated how to proceed from an approximate mostly integer value of a certain physical quantity to the exact normally transfinite one. This procedure we call in analogy to analytical continuation, transfinite continuation. In essence, the procedure is related to the extension of the n! factorial function to the Euler-Gauss function or statistics to fractal statistics. That means we are taking into consideration fractional combinatorial corresponding to fractional symmetries. That is where quantum calculus, the golden binary system, Ord's model and E-infinity theory meet to produce transfinite physical reality with average fractal symmetries [3].

CONCLUSION

G. Ord's work shows clearly that quantum mechanics equations are a projection of a larger system in higher dimensions. Another projection gives us a dissipative non-quantum system. This has profound logical consequences.

First the larger system inhabits a higher dimensional spacetime and its randomness cannot be explained except by a random fuzzy spacetime manifold. This is, of course, the subject of E-infinity theory [4]. Second, the quantum mechanical part plus the dissipative classical part together form a system which is neither quantum nor dissipative [9]. It is therefore conceivable that a similar situation exists inside quantum field theory. This possibility may be behind the discovery of $4 + \phi^3$ inside quantum field theory by E. Goldfain in the USA. Goldfain also discovered certain Feigenbaum universalities inside certain quantum field equations. Similarly the author showed that between $\lambda_1 = 4$ and $\lambda_2 = 4 + \phi^3$, the Feigenbaum Möbius limit cycles chaos resembles E-infinity ground state [32], as is obvious from the equality of $n_t = \lambda_1 = 4$ and $\sim <n> = \lambda_2 = 4 + \phi^3$.

G't Hooft's proposal may be developed further in view of the preceding remarks and may be just the beginning of a new theory combining the three revolutionary theories, relativity, quantum mechanics and deterministic chaos.

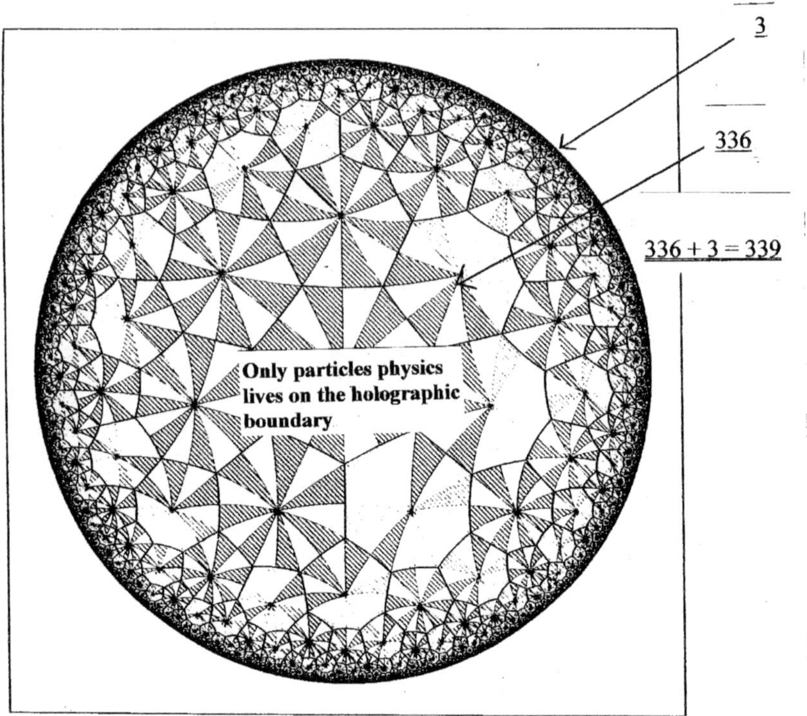

FIGURE 1. The compactified $\Gamma(7)$ modular curve is intimately connected to E-infinity spacetime as well as high energy particle physics. In fact lifting the non-super symmetric quantum gravity coupling $\overline{\alpha}_g = 42.236067977$ to the 8 dimensional super space gives exactly the dimension of $\Gamma_c(7)$, namely

$$(8)\,(\overline{\alpha}_g) \;=\; \mathrm{Dim}\ \Gamma_c(7) \;=\; |\,SL\,(2,7)\,| = 336 + 16k \cong 339.$$

Particle physics lives on the holographic boundary with 339 particle-like states.

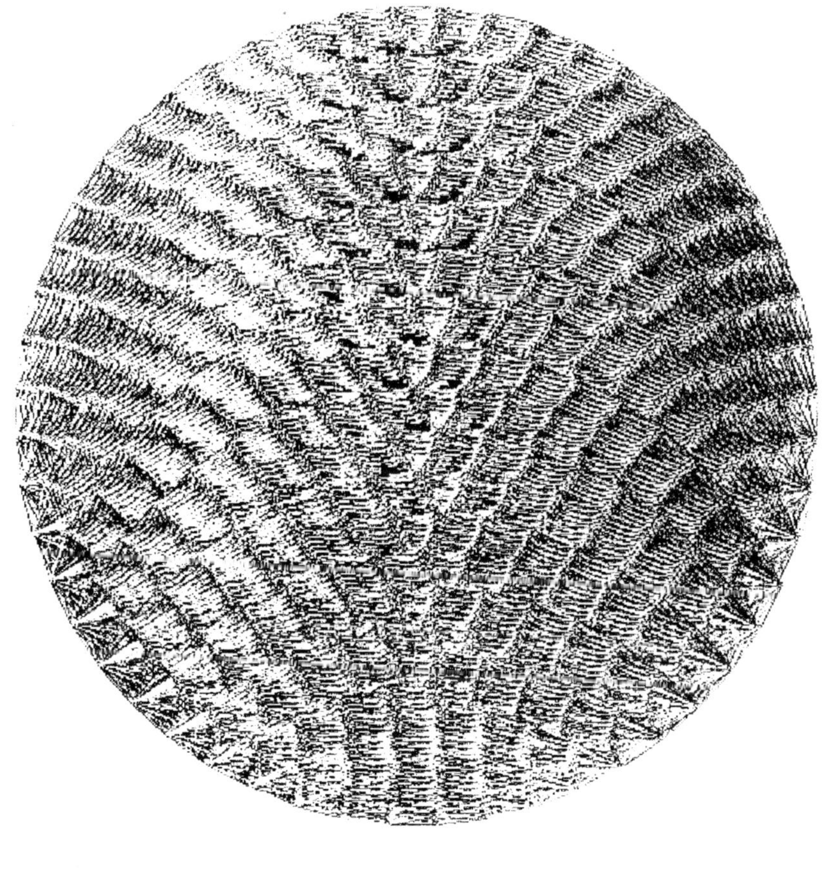

FIGURE 2. Ji-Huan He's vision of E-infinity in Euclidean but hierarchal space as a 26 dimensional cube. If $D \rightarrow \infty$, and making the geometry hyperbolic and adding fuzziness to it, we find E-infinity Bulk space with $| E_8 \, E_8 | = 496 - k^2$ and the instanton number $N = 18 - 2\phi^2$.

REFERENCES

1. M. Born Physik im Wandel meiner Zeit Vieweg – Brauenschweig 1983.
2. M. Born Ist die Klassische Mechanik tatsächlich deterministisch? Physikalishe Blätter II, 1955, p. 49.
3. M.S. El Naschie. Elementary prerequisites for E-infinity. Chaos, Solitons Fractals, 30, 2006, p. 579-605.
4. M.S. El Naschie Mohamed El Naschie answers a few questions about this month's new hot paper in the field of engineering "On a fuzzy Kähler-like manifold which is consistent with the two slit experiment. Int. J. of Nonlinear Science & Numerical Simulation, 6(2), p. 95-98 (2005)". Thomason Essential Science Indicators. http://esi-topics.com/nhp/2006/september-06-MohamedElNaschie.html
5. G. Ord Bohr, Bohm and the entwined path. In Spacetime Physics and Fractality. Festschrift in honour of Mohamed El Naschie. Edited by P. Weibel, G. Ord and O. Rössler. Springer, Vienna-New York, 2005.
6. L. Sigalotti and A. Mejias On El Naschie's conjugate complex time, fractal $E^{(\infty)}$ spacetime and faster-than-light particles. Int. J. Nonlinear Sci. & Numerical Simulation, 7(4), 2006, p. 467-472.
7. M.S. El Naschie On certain finite element-like methods for non-conservative stability sets. Solid Mechanics Archives, 4(3), August, Sijthoff & Noodhoff International Publishers, 1979.
8. M.S. El Naschie A note on quantum mechanics, diffusional interference and informions. Chaos, Solitons & Fractals, 5(5), 1995, p. 881-884.
9. M.S. El Naschie On gauge invariance, dissipative quantum mechanics and self adjoint set. Chaos, Solitons & Fractals, Vol. 32, p. 271-273. 2007, p. 271-273.
10. G. 't Hooft The mathematical basis for deterministic quantum mechanics. Arxiv: quant-ph/0604008 V2, 26 June, 2006.
11. 't Hooft Quantum gravity as a dissipative deterministic system. Classical and quantum gravity, 16, 1999, p. 3263-3279.
12. M.S. El Naschie Feigenbaum scenario for turbulence and Cantorian E-infinity theory of high energy. Chaos, Solitons & Fractals. In press. 2007.
13. Ji-Huan He Nonlinear dynamics and the Nobel prize in physics. Int. J. Nonlinear Science & Numerical Simulation, 8(1), 2007, p. 11-14.

Towards the Unity of Classical Physics.
With Some Implications for Quantum Physics

Peter Enders

Fischerinsel 2, D-10179 Berlin, Germany; Enders@dekasges.de

Abstract. Usually, classical mechanics (CM) and classical electromagnetism (CEM) are represented in textbooks as well as in monographs and, thus, are taught as would-be rather unrelated branches. This 'duality view' corresponds to the historical development, but discards the axiomatic power of CM and leaves the unity of classical physics incomplete. To overcome this, I will derive the Lorentz force and the Maxwell-Lorentz equations through purely mechanical reasoning, *viz*, exploring relationships between force and energy. This way, the first sevedn of Bopp's postulates can be actually derived. This approach continues Hertz's program of representing CM such, that other branches can be developed from it, and corresponds to Schrödinger's (1926) approach to quantum mechanics. The fields introduced develop their own live beyond CM, so that reductionism is avoided.

Keywords: Unity of classical physics, Mechanics, Electromagnetism, Lorentz force
PACS: 40

INTRODUCTION

Can quantum and gravity be unified <u>without</u> unification of CM and CEM? If yes, why takes it so long?

Are we free to exploit gauge freedom in quantum theory <u>without</u> having eliminated artificial gauge freedom in CM and CEM? If yes, why gauge can cause difficulties in QED?

The usual representation of classical physics is dualistic in that CM and CEM appear as its two pillars being connected rather loosely. This corresponds to the historical development, but the unity of physics remains on a methodological level. Newton's equation of motion has got an excellent axiomatic basis, while Maxwell's equations have not. They are introduced either axiomatically (*eg*, Jackson) or deduced from experimental results (*eg*, Mie). A third way consists in the formulation of certain postulates from which they are derived (Bopp 1962). Within my approach, the first seven of Bopp's postulates will be derived; the other ones concern the fields in matter being beyond the scope of this contribution.

To overcome this finally unsatisfying situation I will present a <u>unitary</u> view, where basic equations of CEM are derived through purely mechanical reasoning. This provides an axiomatic, <u>non</u>-reductionist foundation of them, removes not necessary assumptions, and realizes the unity of CM and CEM to a large amount.

CP905, *Frontiers of Fundamental Physics (FFP8), Eighth International Symposium*
edited by B. G. Sidharth, A. Alfonso-Faus, and M. J. Fullana
© 2007 American Institute of Physics 978-0-7354-0412-0/07/$23.00

Furthermore, this provides novel views on the vector potential; its longitudinal component is proposed to be a <u>quantum</u> field as representing an effect absent in CM.

These results rest on careful studies of the states of mechanical and non-mechanical systems as described by Newton, Euler and Helmholtz, where, to my knowledge, only Euler has given the notion of states the central role it deserves. The set of stationary states determines the set of possible motions – this fact explains a major part of the difficulties faced by purely Lagrangean approaches.

LORENTZ FORCE AS SPECIAL CASE OF LIPSCHITZ FORCE

Helmholtz asked which forces leave the total energy constant. He found the answers, (i), central forces and, (ii), time-independent gradient forces. He also asked which forces leave the kinetic energy constant. Lipschitz found $\mathbf{K}=\mathbf{v}\times\mathbf{F'}(t,\mathbf{x},\mathbf{v},\mathbf{a},\ldots)$. Requiring Newton's equation of motion, $m\mathbf{a}=\mathbf{K}$, to be equivalent to Hamilton's equations of motion restricts $\mathbf{F'}(t,\mathbf{x},\mathbf{v},\mathbf{a},\ldots)$ to $\mathbf{F'}(t,\mathbf{x})$. If $\mathbf{F'}$ is factorized into a geometric factor ('distance') and a body-dependent factor ('charge') as in Newton's theory of gravity, the magnetic Lorentz force, $q\mathbf{v}\times\mathbf{B}$, is obtained.

THE HOMOGENOUS MAXWELL EQUATIONS ARE COMPATIBILITY CONDITIONS

Consider a general force field, $q_1\mathbf{F}_1+q_2\mathbf{v}\times\mathbf{F}_2$. In the Hamilton function, $\mathbf{F}_{1,2}$ are represented by scalar, V, and vector potentials, $\mathbf{A}=\mathbf{A}_\mathrm{T}+\nabla V_A$, such that $\mathbf{F}_1=-\partial\mathbf{A}_\mathrm{T}/\partial t-\nabla(V+\partial V_A/\partial t)$, $\mathbf{F}_2=\nabla\times\mathbf{A}_\mathrm{T}$. This immediately yields $\nabla\mathbf{F}_2=0$ and $\nabla\times\mathbf{F}_1=-\partial\mathbf{F}_2/\partial t$.

THE INHOMOGENOUS MAXWELL-LORENTZ EQUATIONS. TRANSVERSE FIELDS IN THE FLUX LAW

The homogenous Maxwell-Lorentz equations determine $\nabla\mathbf{F}_2$ and $\nabla\times\mathbf{F}_1$. $\nabla\mathbf{F}_1$ and $\nabla\times\mathbf{F}_2$ can be determined by careful inspection of the CPT symmetries of $\mathbf{F}_{1,2}$. This leads to $\nabla\mathbf{F}_1=k_1\rho$ and $\nabla\times\mathbf{F}_2=k_2\mathbf{j}+k_3\partial\mathbf{F}_1/\partial t$. Removing the charge conservation (<u>longitudinal</u> fields): $k_2\mathbf{j}_\mathrm{L}+k_3\partial\nabla\mathbf{F}_{1,\mathrm{L}}/\partial t=0$, from the r.h.s. yields the 'true' flux law, $\nabla\times\mathbf{F}_2=k_2\mathbf{j}_\mathrm{T}+k_3\partial\mathbf{F}_{1,\mathrm{T}}/\partial t$ (<u>transverse</u> fields). For charge conservation is a property of charged bodies, not of the electromagnetic fields. $k_{1(2)}$ corresponds to ε_0^{-1} (μ_0) (SI units).

Consistency: $\mathbf{F}_1$ acts on static and moving charges – is created by static and moving charges; $\mathbf{F}_2$ acts on moving charges only – is created by moving charges only.

MANIFEST GAUGE-INVARIANT HAMILTONIAN

Gauge concerns only $(V+\partial V_A/\partial t)$, thus, $H = (\mathbf{p}-q\mathbf{A}_\mathrm{T})^2/2m + q(V+\partial V_A/\partial t)$. The propagating fields become transverse from the very beginning.

IS A_L A QUANTUM POTENTIAL?

In QM, gauge includes the phase of the wave function: $i\hbar\partial\psi(\varphi)/\partial t = [T(\mathbf{p}+\hbar\nabla\varphi)+V+\hbar\partial\varphi/\partial t]\psi(\varphi) \leftrightarrow i\hbar\partial\psi(0)/\partial t = [T(\mathbf{p})+V]\psi(0); \; \psi(\varphi)=\psi(0)e^{-i\varphi}$. The Aharonov-Bohm-Effekt (1959) demonstrates that $\varphi=(q/\hbar)V_A$ is a real field. In the classical limit case, φ should remain finite $\rightarrow V_A=\hbar V'_A$.

V_A represents external influences changing <u>neither</u> energy (like the Coulomb force), <u>nor</u> characteristic dimensions (like magnetic Lorentz force) of systems. The absence of such influences in CM suggests V_A to be a quantum potential: $V_A=O(\hbar)$, again.

SUMMARY AND DISCUSSION

CEM (as well as non-general-relativistic gravity) is determined through CM principles for the motion of 'charged' bodies. There is only <u>one</u> pillar of classical physics. The unity of classical physics rests not only on common methods, but on common roots in mechanics.

The electrical field strength, $\mathbf{E}=\mathbf{E}_T+\mathbf{E}_L=-\partial\mathbf{A}_T/\partial t-(\partial\mathbf{A}_L/\partial t+\nabla V)$, exhibits <u>two</u> longitudinal components: $-\partial\mathbf{A}_L/\partial t$ and ∇V. Gauging concerns only $\mathbf{E}_L$, and, thus, looks like 'maths'; what is physics behind it? Which physics can be there in view of the common statement "the results are <u>in</u>dependent of gauge"?

$\mathbf{A}_L$ is dispensable within CEM, but not within quantum mechanics. This suggest $\mathbf{A}_L$ to be a quantum potential: $\mathbf{A}_L=O(\hbar)$.

Open question: What are the transformation properties of the longitudinal and transverse fields?

The point-mechanical side can be made Lorentz-invariant through a generalization of Euler's derivation of Newton's equation of motion (Suisky & Enders 2005).

ACKNOWLEDGMENTS

I would like to thank Profs. F.-W. Hehl and O. Keller for helpful discussions.

REFERENCES

1. Y. Aharonov & D. Bohm, *Significance of Electromagnetic Potentials in the Quantum Theory*, Phys. Rev. 115 (1959) 485-491; reprint in: A. Shapere & F. Wilczek, *Geometric Phases in Physics*, Singapore: World Scientific 1989, paper [2,6].
2. F. Bopp, *Prinzipien der Elektrodynamik*, Z. Phys. 169 (1962) 45-52
3. H. Helmholtz, *Über die Erhaltung der Kraft*, ext. reprint: Leipzig: Engelmann 1889 (Ostwalds Klassiker 1).
4. H. Hertz, Die *Prinzipien der Mechanik in neuem Zusammenhange dargestellt*, Leipzig: Barth ²1910
5. J. D. Jackson, Classical Electrodynamics, New York: Wiley ³1999.
6. G. Mie, *Lehrbuch der Elektrizität und des Magnetismus*, Stuttgart: Enke ²1941.
7. E. Schrödinger, *Quantisierung als Eigenwertproblem* (4 Communs.), Ann. Phys. [4] (1926); reprint in: *Abhandlungen zur Wellenmechanik*, Leipzig: Barth 1927.
8. D. Suisky & P. Enders, *Dynamical Foundation of the Lorentz Transformation*, DPG Annual Meeting, Berlin 2005, Poster QR18.1.

Super Heavy Nuclei over Critical Fields and their Conections

Walter Greiner* and Valery Zagrebaev[†]

*Frankfurt Institute for Advanced Studies, J.W. Goethe-Universität
60325 Frankfurt am Main, Germany
[†]Flerov Laboratory of Nuclear Reaction, JINR, Dubna, 141980, Moscow region, Russia

Abstract. Low energy collisions of very heavy nuclei (^{238}U+^{238}U, ^{232}Th+^{250}Cf and ^{238}U+^{248}Cm) have been studied within the realistic dynamical model based on multi-dimensional Langevin equations. Large charge and mass transfer was found due to the "inverse quasi-fission" process leading to formation of survived superheavy long-lived neutron-rich nuclei. In many events lifetime of the composite system consisting of two touching nuclei turns out to be rather long; sufficient for spontaneous positron formation from super-strong electric field, a fundamental QED process.

Keywords: superheavy nuclei, giant quasi-atoms.
PACS: 25.70.Jj, 25.70.Lm

INTRODUCTION

Superheavy (SH) nuclei obtained in "cold" fusion reactions with Pb or Bi target [1] are along the proton drip line and very neutron-deficient with a short half-life. In fusion of actinides with ^{48}Ca more neutron-rich SH nuclei are produced [2] with much longer half-life. But they are still far from the center of the predicted "island of stability" formed by the neutron shell around N=184. In the "cold" fusion, the cross sections of SH nuclei formation decrease very fast with increasing charge of the projectile and become less than 1 pb for Z>112. Heaviest transactinide, Cf, which can be used as a target in the second method, leads to the SH nucleus with Z=118 being fused with ^{48}Ca. Using the next nearest elements instead of ^{48}Ca (e.g., ^{50}Ti, ^{54}Cr, etc.) in fusion reactions with actinides is expected less encouraging, though experiments of such kind are planned to be performed. In this connection another ways to the production of SH elements in the region of the "island of stability" should be searched for.

About twenty years ago transfer reactions of heavy ions with ^{248}Cm target have been evaluated for their usefulness in producing unknown neutron-rich actinide nuclides [3, 4, 5]. The cross sections were found to decrease very rapidly with increasing atomic number of surviving target-like fragments. However, Fm and Md neutron-rich isotopes have been produced at the level of 0.1 μb. Theoretical estimations for production of primary superheavy fragments in the damped U+U collision have been also performed those time within the semiphenomenological diffusion model [6]. In spite of obtained high probabilities for the yields of superheavy primary fragments (more than 10^{-2} mb for Z=120), the cross sections for production of heavy nuclei with low excitation energies were estimated to be very small. The authors concluded that "fluctuations and shell effects not taken into account may conciderably increase the formation probabilities".

CP905, *Frontiers of Fundamental Physics (FFP8), Eighth International Symposium*
edited by B. G. Sidharth, A. Alfonso-Faus, and M. J. Fullana
© 2007 American Institute of Physics 978-0-7354-0412-0/07/$23.00

In this paper we study the processes of low energy collisions of very heavy nuclei (such as U+Cm) within the recently proposed model of fusion-fission dynamics [7]. The purpose is to find an influence of the nearest closed shells Z=82 and N=126 (formation of ^{208}Pb as one of the primary fragments) on nucleon rearrangement between primary fragments. We try to estimate how large the charge and mass transfer could be in these reactions and what the cross sections for production of surviving neutron-rich SH nuclei are. Direct time analysis of the collision process allows us to estimate also the lifetime of the composite system consisting of two touching heavy nuclei with total charge Z>180. Such "long-lived" configurations may lead to spontaneous positron emission from super-strong electric field of quasi-atoms by a static QED process (transition from neutral to charged QED vacuum) [8, 9], see schematic Fig. 1.

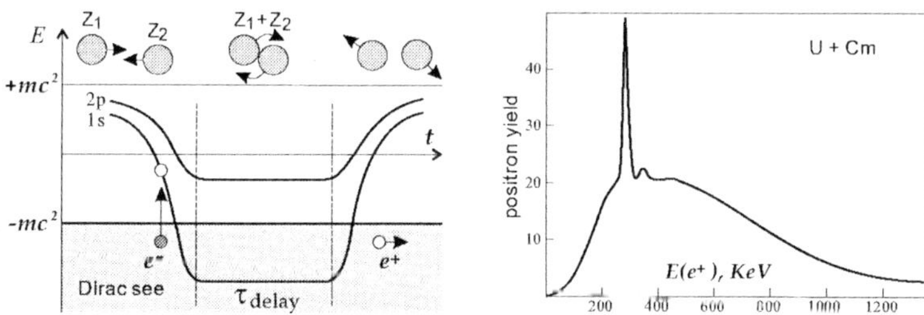

FIGURE 1. Schematic figure of spontaneous decay of the vacuum and spectrum of the positrons formed in supercritical electric field ($Z_1 + Z_2 > 173$).

ADIABATIC DYNAMICS OF HEAVY NUCLEAR SYSTEM

At incident energies around the Coulomb barrier in the entrance channel the fusion probability is about 10^{-3} for mass asymmetric reactions induced by ^{48}Ca and much less for more symmetric combinations used in the "cold synthesis". DI scattering and QF are the main reaction channels here, whereas the fusion probability [formation CN] is extremely small. To estimate such a small quantity for CN formation probability, first of all, one needs to be able to describe well the main reaction channels, namely DI and QF. Moreover, the quasi-fission processes are very often indistinguishable from the deep-inelastic scattering and from regular fission, which is the main decay channel of excited heavy compound nucleus.

To describe properly and simultaneously the strongly coupled DI, QF and fusion-fission processes of low-energy heavy-ion collisions we have to choose, first, the unified set of degrees of freedom playing the principal role both at approaching stage and at the stage of separation of reaction fragments. Second, we have to determine the unified potential energy surface (depending on all the degrees of freedom) which regulates all the processes. Finally, the corresponding equations of motion should be formulated to perform numerical analysis of the studied reactions. In contrast with other models, we take into consideration all the degrees of freedom necessary for description of all the reaction stages. Thus, we need not to split artificially the whole reaction into several stages. Moreover, in that case unambiguously defined initial conditions are easily formulated at large distance, where only the Coulomb interaction and zero-vibrations of the nuclei determine the motion. The distance between the nuclear centers R (corresponding to the elongation of a mono-nucleus), dynamic spheroidal-type surface deformations β_1 and β_2, mutual in-plane orientations of deformed nuclei φ_1 and φ_2, and mass asymmetry $\eta = \frac{A_1 - A_2}{A_1 + A_2}$ are probably the relevant degrees of freedom in fusion-fission dynamics.

The two-center shell model [10] seems to be most appropriate for calculation of the adiabatic potential energy surface. A choice of dynamic equations for the considered degrees of freedom is not so evident. The main problem here is a proper description of nucleon transfer and change of the mass asymmetry which is a discrete variable by its nature. The corresponding inertia parameter μ_η, being calculated within the Werner-Wheeler approach, becomes infinite at contact (scission) point and for separated nuclei. In Ref. [7] the inertialess Langevin type equation for the mass asymmetry has been derived from the corresponding master equation for the distribution function. Finally we use a set of 13 coupled Langevin type equations for 7 degrees of freedom (relative distance, rotation and dynamic deformations of the nuclei and mass asymmetry) which are solved numerically.

The cross sections for all the processes are calculated in a simple and natural way. A large number of events (trajectories) are tested for a given impact parameter. Those events, in which the nuclear system overcame the fission barrier from the outside and entered the region of small deformations and elongations, are treated as fusion (CN formation). The other events correspond to quasi-elastic, DI and QF processes. Subsequent decay of the excited CN ($C \rightarrow B + xn + N\gamma$) is described then within the statistical model using an explicit expression for survival probability, which directly takes into account the Maxwell-Boltzmann energy distribution of evaporated neutrons [11]. The double differential cross-sections are calculated as follows

$$\frac{d^2\sigma_\eta}{d\Omega dE}(E,\theta) = \int_0^\infty bdb \frac{\Delta N_\eta(b,E,\theta)}{N_{\text{tot}}(b)} \frac{1}{\sin(\theta)\Delta\theta\Delta E}. \tag{1}$$

Here $\Delta N_\eta(b,E,\theta)$ is the number of events at a given impact parameter b in which the system enters into the channel η (definite mass asymmetry value) with kinetic energy in the region $(E, E + \Delta E)$ and center-of-mass outgoing angle in the region $(\theta, \theta + \Delta\theta)$, $N_{\text{tot}}(b)$ is the total number of simulated events for a given value of impact parameter. In collisions of deformed nuclei an averaging over initial orientations is performed. Expression (1) describes the mass, energy and angular distributions of the *primary* fragments formed in the binary reaction (both in DI and in QF processes). Subsequent de-excitation cascades of these fragments via emission of light particles and gamma-rays in competition with fission were taken into account explicitly for each event within the statistical model leading to the *final* mass and energy distributions of the reaction fragments.

It is important that the model allows us to perform also a time analysis of the studied reactions. Each tested event is characterized by the reaction time τ_{int}, which is calculated as a difference between re-separation (scission) and contact times.

At first we applied the model to describe available experimental data on low-energy damped collision of very heavy nuclei, ^{136}Xe$+^{209}$Bi [12], where the DI process should dominate due to expected prevalence of the Coulomb repulsion over nuclear attraction.

The adiabatic potential energy surface of this nuclear system is shown on the left panel of Fig. 2 in the space of elongation and mass asymmetry at zero dynamic deformations. The colliding nuclei are very compact with almost closed shells and the potential energy has only one deep valley (just in the entrance channel, $\eta \sim 0.21$) giving rather simple mass distribution of the reaction fragments. In that case the reaction mechanism depend mainly on the nucleus-nucleus potential at contact distance, on the friction forces at this region (which determine the energy loss) and on nucleon transfer rate at contact. Note, that there is a well pronounced plateau at contact configuration in the region of zero mass asymmetry (see Fig. 2, right panel). It becomes even lower with increasing the deformations and corresponds to formation of the nuclear system consisting of strongly deformed touching fragments ^{172}Er+^{173}Tm (see Fig. 2), which means that a significant mass rearrangement may occur here leading to additional time delay of the reaction.

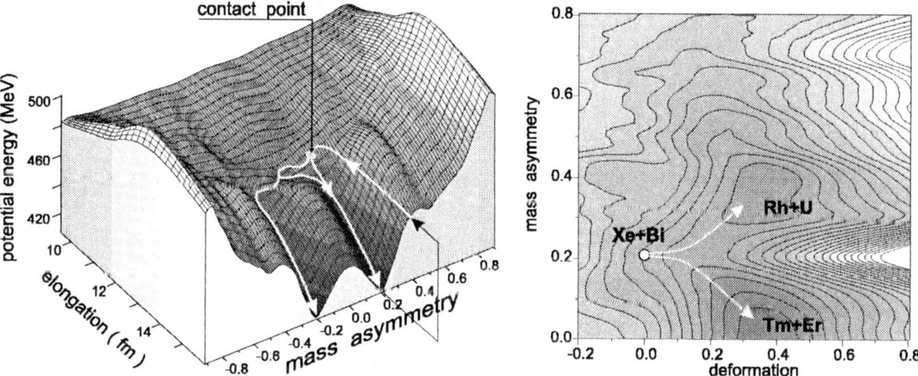

FIGURE 2. Driving potential for the nuclear system formed in ^{136}Xe+^{209}Bi collision at fixed deformations (left) and at contact configuration (right). The solid lines with arrows show schematically (without fluctuations) most probable trajectories.

On the right panel of Fig. 2 the landscape of the potential energy is shown at contact configuration depending on mass asymmetry and deformation of the fragments. As can be seen, after contact and before re-separation the nuclei aim to become more deformed. Moreover, beside a regular diffusion (caused by the fluctuations), the final mass distribution is determined also by the two well marked driving paths leading the system to more and to less symmetric configurations. They are not identical and this leads to the asymmetric mass distribution of the primary fragments, see Fig. 3(c).

In Fig. 3 the angular, energy and charge distributions of the Xe-like fragments are shown comparing with our calculations (histograms). In accordance with experimental conditions only the events with the total kinetic energy in the region of $260 < E \leq 546$ MeV and with the scattering angles in the region of $40^o \leq \theta_{c.m.} \leq 100^o$ were accumulated. The total cross section corresponding to all these events is about 2200 mb (experimental estimation is 2100 mb [12]). Due to the rather high excitation energy sequential fission of the primary heavy fragments may occur in this reaction (mainly those heavier than Bi). In the experiment the yield of the heavy fragments was found to be about 30% less comparing with Xe-like fragments. Our calculation gives 354 mb for the cross section of sequential-fission, which is quite comparable with experimental data.

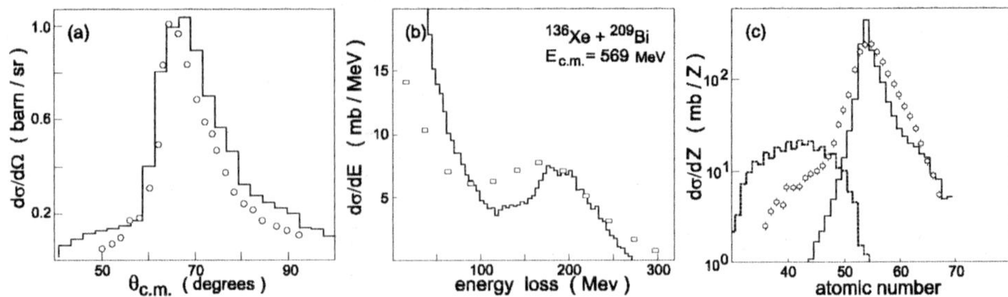

FIGURE 3. Angular (a), energy-loss (b) and charge (c) distributions of the Xe-like fragments obtained in the ^{136}Xe+^{209}Bi reaction at $E_{c.m.}$ = 568 MeV. Experimental data are taken from Ref. [12]. For other notations see the text.

Mass distribution of the fission fragments is shown in Fig. 3(c) by the dotted histogram. Note that it is a contamination with sequential fission products of heavy primary fragments leading to the bump around Z=40 in the experimental charge distribution.

LOW-ENERGY COLLISIONS OF TRANSACTINIDE NUCLEI

Reasonable agreement of our calculations with experimental data on low-energy DI and QF reactions induced by heavy ions stimulated us to study the reaction dynamics of very heavy transactinide nuclei. The purpose was to find an influence of the shell structure of the driving potential (in particular, deep valley caused by the double shell closure Z=82 and N=126) on nucleon rearrangement between primary fragments. In Fig. 4 the potential energies are shown depending on mass rearrangement at contact configuration of the nuclear systems formed in ^{48}Ca+^{248}Cm and ^{232}Th+^{250}Cf collisions. The lead valley evidently reveals itself in both cases (for ^{48}Ca+^{248}Cm system there is also a tin valley). In the first case (^{48}Ca+^{248}Cm), discharge of the system into the lead valley (normal or symmetrizing quasi-fission) is the main reaction channel, which decreases significantly the probability of CN formation. In collisions of heavy nuclei (Th+Cf, U+Cm and so on) we expect that the existence of this valley may noticeably increase the yield of surviving neutron-rich superheavy nuclei complementary to the projectile-like fragments around ^{208}Pb ("inverse" or anti-symmetrizing quasi-fission process).

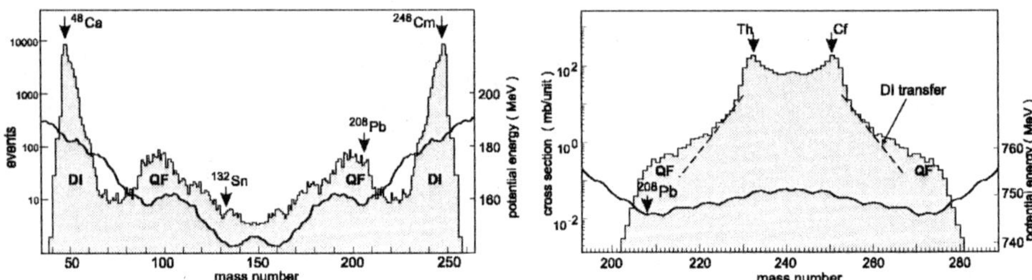

FIGURE 4. Potential energy at contact "nose-to-nose" configuration and mass distribution of primary fragments for the two nuclear systems formed in ^{48}Ca+^{248}Cm (left) and ^{232}Th+^{250}Cf (right) collisions.

Direct time analysis of the reaction dynamics allows us to estimate also the lifetime of the composite system consisting of two touching heavy nuclei with total charge Z>180. Such "long-living" configurations may lead to spontaneous positron emission from super-strong electric field of giant quasi-atoms by a static QED process (transition from neutral to charged QED vacuum) [8]. About twenty years ago an extended search for this fundamental process was carried out and narrow line structures in the positron spectra were first reported at GSI. Unfortunately these results were not confirmed later, neither at ANL, nor in the last experiments performed at GSI. These negative finding, however, were contradicted by Jack Greenberg (private communication and supervised thesis at Wright Nuclear Structure Laboratory, Yale university). Thus the situation remains unclear, while the experimental efforts in this field have ended. We hope that new experiments and new analysis, performed according to the results of our dynamical model, may shed additional light on this problem and also answer the principal question: are there some reaction features (triggers) testifying a long reaction delays? If they are, new experiments should be planned to detect the spontaneous positrons in the specific reaction channels.

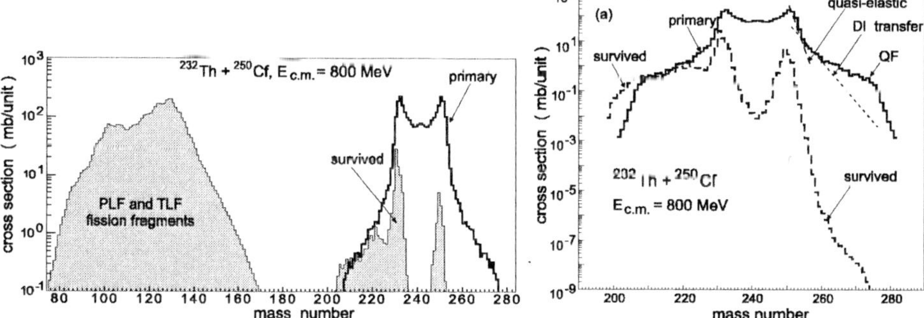

FIGURE 5. Mass distributions of primary (solid histogram), surviving and sequential fission fragments (hatched areas) in the ^{232}Th+^{250}Cf collision at 800 MeV center-of-mass energy. On the right the result of longer calculation is shown.

Using the same parameters of nuclear viscosity and nucleon transfer rate as for the system Xe+Bi we calculated the yield of primary and surviving fragments formed in the ^{232}Th+^{250}Cf collision at 800 MeV center-of mass energy. Low fission barriers of the colliding nuclei and of most of the reaction products jointly with rather high excitation energies of them in the exit channel will lead to very low yield of surviving heavy fragments. Indeed, sequential fission of the projectile-like and target-like fragments dominate in these collisions, see Fig. 5 At first sight, there is no chances to get surviving superheavy nuclei in such reactions. However, as mentioned above, the yield of the primary fragments will increase due to the QF effect (lead valley) as compared to the gradual monotonic decrease typical for damped mass transfer reactions. Secondly, with increasing neutron number the fission barriers increase on average (also there is the closed sub-shell at N=162). Thus we may expect a non-negligible yield (at the level of 1 pb) of surviving superheavy neutron rich nuclei produced in these reactions [13].

Result of much longer calculations is shown on the right panel of Fig. 5. The pronounced shoulder can be seen in the mass distribution of the primary fragments near the mass number A=208 (274). It is explained by the existence of a valley in the potential energy surface [see Fig. 4(b)], which corresponds to the formation of doubly magic nucleus ^{208}Pb ($\eta = 0.137$). The emerging of the nuclear system into this valley resembles the well-known quasi-fission process and may be called "inverse (or anti-symmetrizing) quasi-fission" (the final mass asymmetry is larger than the initial one). For $\eta > 0.137$ (one fragment becomes lighter than lead) the potential energy sharply increases and the mass distribution of the primary fragments decreases rapidly at A<208 (A>274).

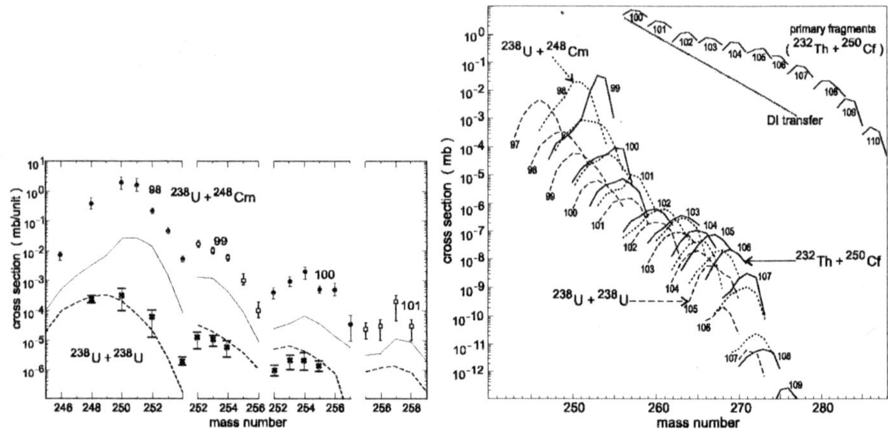

FIGURE 6.   (Left panel) Experimental and calculated yields of the elements 98÷101 in the reactions ^{238}U+^{238}U (crosses) [3] and ^{238}U+^{248}Cm (circles and squares) [4]. (Right panel) Predicted yields of superheavy nuclei in collisions of ^{238}U+^{238}U (dashed), ^{238}U+^{248}Cm (dotted) and ^{232}Th+^{250}Cf (solid lines) at 800 MeV center-of-mass energy. Solid curves in upper part show isotopic distributions of primary fragments in the Th+Cf reaction.

In Fig. 6 the available experimental data on the yield of SH nuclei in collisions of ^{238}U+^{238}U [3] and ^{238}U+^{248}Cm [4] are compared with our calculations. The estimated isotopic yields of survived SH nuclei in the ^{232}Th+^{250}Cf, ^{238}U+^{238}U and ^{238}U+^{248}Cm collisions at 800 MeV center-of-mass energy are shown on the right panel of Fig. 6. Thus, as we can see, there is a real chance for production of the long-lived neutron-rich SH nuclei in such reactions. As the first step, chemical identification and study of the nuclei up to $^{274}_{107}$Bh produced in the reaction ^{232}Th+^{250}Cf may be performed.

The time analysis of the reactions studied shows that in spite of absence of an attractive potential pocket the system consisting of two very heavy nuclei may hold in contact rather long in some cases, see Fig. 7. During this time the giant nuclear system moves over the multidimensional potential energy surface with almost zero kinetic energy (result of large nuclear viscosity). The total reaction time distribution, $\frac{d\sigma}{dlog(\tau)}$ (τ denotes the time after the contact of two nuclei), is shown in Fig. 8 for the ^{238}U+^{248}Cm collision. The dynamic deformations are mainly responsible here for the time delay of the nucleus-nucleus collision. Ignoring the dynamic deformations in the equations of motion significantly decreases the reaction time, see Fig. 8(a). With increase of the energy loss and mass transfer the reaction time becomes longer and its distribution becomes more narrow.

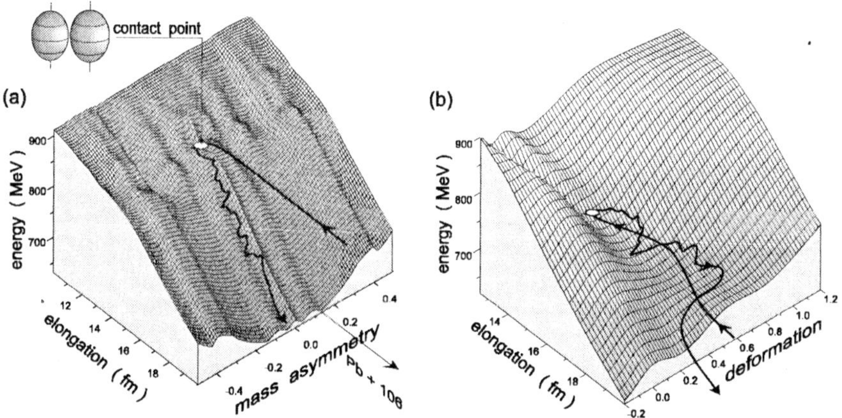

FIGURE 7. Potential energy surface for the nuclear system formed by ^{232}Th+^{250}Cf as a function of R and α ($\beta = 0.22$) (a) and as a function of R and β ($\alpha = 0.037$) (b). Typical trajectories are shown by the thick curves with arrows.

As mentioned earlier, the lifetime of a giant composite system more than 10^{-20} s is quite enough to expect positron line structure emerging on top of the dynamical positron spectrum due to spontaneous e^+e^- production from the supercritical electric fields as a fundamental QED process ("decay of the vacuum") [8]. The absolute cross section for long events is found to be maximal just at the beam energy ensuring the two nuclei to be in contact, see Fig. 8(c). The same energy is also optimal for the production of the most neutron-rich SH nuclei. Of course, there are some uncertainties in the used parameters, mostly in the value of nuclear viscosity. However we found only a linear dependence of the reaction time on the strength of nuclear viscosity, which means that the obtained reaction time distribution is rather reliable, see logarithmic scale on both axes in Fig. 8(a).

Formation of the background positrons in these reactions forces one to find some additional trigger for the longest events. Such long events correspond to the most damped collisions with formation of mostly excited primary fragments decaying by fission, see Figs. 9(a). However there is also a chance for production of the primary fragments in the region of doubly magic nucleus ^{208}Pb, which could survive against fission due to nucleon evaporation. The number of the longest events depends weakly on impact parameter up to some critical value. On the other hand, in the angular distribution of all the excited primary fragments (strongly peaked at the center-of-mass angle slightly larger than 90^0) there is the rapidly decreasing tail at small angles, see Fig. 9(b). Time distribution for the most damped events ($F_{loss} > 150$ MeV), in which a large mass transfer occurs and primary fragments scatter in forward angles ($\theta_{c.m.} < 70°$), is rather narrow and really shifted to longer time delay, see hatched areas in Fig. 8. For the considered case of ^{238}U+^{248}Cm collision at 800 MeV center-of-mass energy, the detection of the surviving nuclei in the lead region at the laboratory angles of about $25°$ and at the low-energy border of their spectrum (around 1000 MeV for Pb) could be a real trigger for longest reaction time.

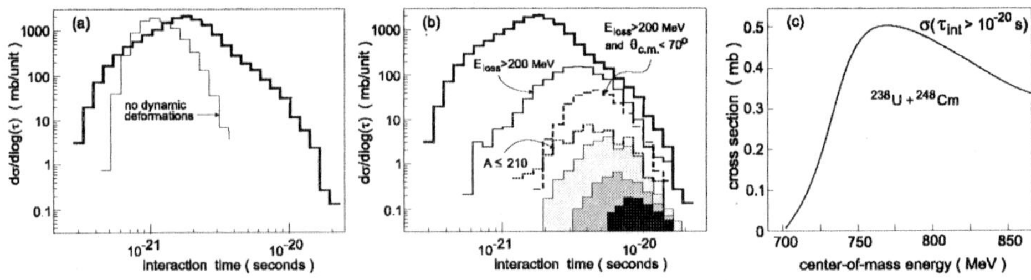

FIGURE 8. Reaction time distributions for the ^{238}U+^{248}Cm collision at 800 MeV center-of-mass energy. Thick solid histograms correspond to all events with energy loss more than 30 MeV. (a) Thin solid histogram shows the effect of switching-off dynamic deformations. (b) Thin solid, dashed and dotted histograms show reaction time distributions in the channels with formation of primary fragments with $E_{loss} > 200$ MeV, $E_{loss} > 200$ MeV and $\theta_{c.m.} < 70°$ and $A \leq 210$, correspondingly. Hatched areas show time distributions of events with formation of the primary fragments with $A \leq 220$ (light gray), $A \leq 210$ (gray), $A \leq 204$ (dark) having $E_{loss} > 200$ MeV and $\theta_{c.m.} < 70°$. (c) Cross section for events with interaction time longer than 10^{-20} s.

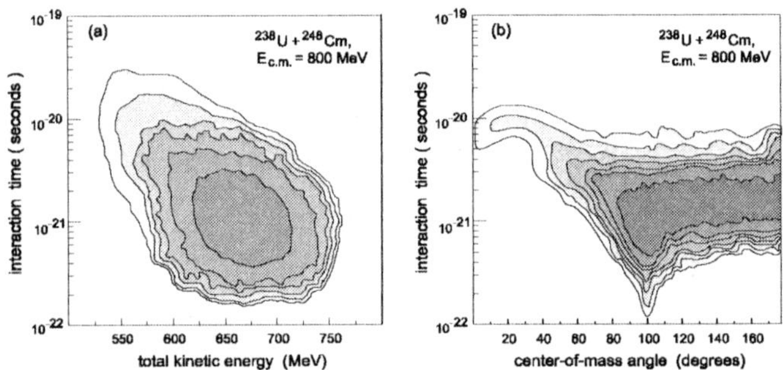

FIGURE 9. Energy-time (a) and angular-time (b) distributions of primary fragments in the ^{238}U+^{248}Cm collision at 800 MeV ($E_{loss} > 15$ MeV).

CONCLUSION

For near-barrier collisions of heavy ions it is very important to perform a combined (unified) analysis of all strongly coupled channels: deep-inelastic scattering, quasi-fission, fusion and regular fission. This ambitious goal has now become possible. A unified set of dynamic Langevin type equations is proposed for the *simultaneous* description of DI and fusion-fission processes. For the first time, the whole evolution of the heavy nuclear system can be traced starting from the approaching stage and ending in DI, QF, and/or fusion-fission channels. Good agreement of our calculations with experimental data gives us hope to obtain rather accurate predictions of the probabilities for superheavy element formation and clarify much better than before the mechanisms of quasi-fission and fusion-fission processes. The determination of such fundamental characteristics of nuclear dynamics as the nuclear viscosity and the nucleon transfer rate is now possible. The production of long-lived neutron-rich SH nuclei in the region of the "island of stability" in collisions of transuranium ions seems to be quite possible due to a large mass rearrangement in the inverse (anti-symmetrized) quasi-fission process caused by the $Z=82$ and $N=126$ nuclear shells. A search for spontaneous positron emission from a supercritical electric field of long-living giant quasi-atoms formed in these reactions is also quite promising.

REFERENCES

1. S. Hofmann and G. Münzenberg, *Rev. Mod. Phys.* **72**, 733 (2000).
2. Yu.Ts. Oganessian, V.K. Utyonkov, Yu.V. Lobanov, F.Sh. Abdullin, A.N. Polyakov, I.V. Shirokovsky, Yu.S. Tsyganov, G.G. Gulbekian, S.L. Bogomolov, B.N. Gikal, A.N. Mezentsev, S. Iliev, V.G. Subbotin, A.M. Sukhov, A.A. Voinov, G.V. Buklanov, K. Subotic, V.I. Zagrebaev, M.G. Itkis, J.B. Patin, K.J. Moody, J.F. Wild, M.A. Stoyer, N.J. Stoyer, D.A. Shaughnessy, J.M. Kenneally, P.A. Wilk, R.W. Lougheed, R.I. Il'kaev, and S.P. Vesnovskii, *Phys. Rev.* **C70**, 064609 (2004).
3. M. Schädel, J.V. Kratz, H. Ahrens, W.Brüchle, G. Franz, H. Gäggeler, I. Warnecke, G. Wirth, G. Herrmann, N. Trautmann, and M. Weis, *Phys. Rev. Lett.* **41**, 469 (1978).
4. M. Schädel, W. Brüchle, H. Gäggeler, J.V. Kratz, K. Sümmerer, G. Wirth, G. Herrmann, R. Stakemann, G. Tittel, N. Trautmann, J.M. Nitschke, E.K. Hulet, R.W. Lougheed, R.L. Hahn, and R.L. Ferguson, Phys. Rev. Lett. **48**, 852 (1982).
5. K.J. Moody, D. Lee, R.B. Welch, K.E. Gregorich, G.T. Seaborg, R.W. Lougheed, and E.K. Hulet, *Phys. Rev.* **C33**, 1315 (1986).
6. C. Riedel, W. Nörenberg, *Z. Phys.* **A290**, 385 (1979).
7. V. Zagrebaev and W. Greiner, *J. Phys.* **G31**, 825 (2005).
8. J. Reinhard, U. Müller and W. Greiner, *Z. Phys.* **A303**, 173 (1981).
9. W. Greiner (Editor), *Quantum Electrodynamics of Strong Fields*, (Plenum Press, New York and London, 1983); W. Greiner, B. Müller and J. Rafelski, *QED of Strong Fields* (Springer, Berlin and New York, 2nd edition, 1985)
10. J. Maruhn and W. Greiner, *Z. Phys.* **251**, 431 (1972).
11. V.I. Zagrebaev, Y. Aritomo, M.G. Itkis, Yu.Ts. Oganessian, M. Ohta, *Phys. Rev.* **C65**, 014607 (2002).
12. W.W. Wilcke, J.R. Birkelund, A.D. Hoover, J.R. Huizenga, W.U. Schröder, V.E. Viola, Jr., K.L. Wolf, and A.C. Mignerey, *Phys. Rev.* **C22**, 128 (1980).
13. V.I. Zagrebaev, Yu.Ts. Oganessian, M.I. Itkis and Walter Greiner, *Phys. Rev.* **C73**, 031602(R) (2006).

A Semi-Classical, Microscopic Model for Nuclear Collective Rotation Plus RPA

P. Gulshani

NUTECH Services, 3313 Fenwick Cres., Mississauga, Ontario,Canada L5L 5N1
Tel. #: 905-569-8233; matlap@rogers.com

Abstract. Collective rotation and vibration of deformed nuclei are described semiclassically but microscopically by first transforming the time-dependent Schrodinger equation to a rotating frame, while preserving time-reversal invariance, and then applying a variational method. The rotating-frame axes are chosen to coincide with the principal axes of the expectation of an arbitrary, symmetric second-rank tensor operator $\hat{\Gamma}$. It is shown that the equations derived for the rotational and vibrational motions decouple completely due to the rotational invariance of the Hamiltonian and diagonality of the expectation of $\hat{\Gamma}$ in the rotating frame. The equations describing the vibration reduce to those of the RPA. The equation describing the rotation generalizes that of the conventional cranking model (CM). The predicted rotation moment of inertia is shown to reduce to that of the CM for special types of particle interactions.

Key words: nuclear collective rotation, vibrartion, cranking model, variation methods, transformation, moment of inertia
PACS number: 21.60.Ev, 21.60.Fw, 21.60.Jz

INTRODUCTION

Certain open-shell nuclei exhibit excitation energy spectra characteristics of a quantum rotating system [1,2,3]. The moment of inertia inferred from these spectra has a value between that for rigid and irrotational flows [1,2]. To explain this observation and infer the nature of the rotation (if any), a number of microscopic and phenomenological approaches have been used over the years. In one of the microscopic approaches [4], the particle coordinates were transformed to a set of collective Euler angles θ and intrinsic coordinates ξ using the equation:

$$x = R(\theta)\, x'(\xi) \tag{1}$$

where R is an orthogonal matrix R, and x and x' denote the Cartesian coordinates of a particle in the space-fixed and rotating frames, respectively. The Euler angles were chosen to define the orientation of the principal axes of the mass quadrupole operator ($\hat{Q}$). In this transformation, the coordinates x' are constrained to make the quadrupole tensor diagonal in the rotating frame. The constraints on the coordinates x' imply that they are not all independent of each other, i.e., some of them are redundant. This redundancy introduces formidable mathematical difficulties in describing the intrinsic motion and its coupling to the rotational motion.

In this article, this redundancy problem is circumvented by adopting a semi-classical, but microscopic approach, which also includes a description of collective oscillations.

CP905, *Frontiers of Fundamental Physics (FFP8), Eighth International Symposium*
edited by B. G. Sidharth, A. Alfonso-Faus, and M. J. Fullana
© 2007 American Institute of Physics 978-0-7354-0412-0/07/$23.00

Model Derivation

Instead of Eq. (1) we use the transformation:

$$x_{ni} = R_{Ai}(t)\, x'_{nA}, \qquad\qquad n = 1,\dots\dots, N,\ i, j = 1,2,3,\ A, B = 1,2,3 \qquad (2)$$

where R is now a function of time t, with the indices i, j and A, B denoting the space-fixed and body-fixed axes, respectively, and n denoting n^{th} particle. The body-fixed axes are chosen to coincide with the principal axes of the expectation value of an arbitrary, second-rank, symmetric tensor operator ($\hat{\Gamma}$), which is a function of particle coordinates and momenta, (instead of $\hat{Q}$). Transforming the N-interacting-particle time-dependent Schrodinger equation to the rotating frame, we obtain (more detail is to appear in Ref. [5]):

$$i\hbar \frac{\partial}{\partial t}\big|F\big\rangle = (H_o - \lambda_{AB}\, L_{AB})\big|F\big\rangle \qquad (3)$$

$$\big\langle F\big|\hat{\Gamma}_{AB}\big|F\big\rangle \equiv \Gamma_{AB} = \Gamma_{AA}\,\delta_{AB} \equiv \Gamma_A\,\delta_{AB} \qquad (4)$$

$$\lambda_{AB} \equiv \sum_{j=1}^{3} \frac{dR_{Aj}}{dt} R_{Bj} = \frac{\big\langle F\big|[\hat{\Gamma}_{AB}, H_o]\big|F\big\rangle}{i\hbar\,(\Gamma_A - \Gamma_B)} \qquad (5)$$

where H_o is the nuclear Hamiltonian, L is the total angular momentum operator, and $\big|F\big\rangle$ is the nuclear wavefunction. In the transformation (2), the coordinates x' are all independent at the expense of rendering the Schrodinger equation (3) non-linear and the description of the rotation semiclassical. Eq. (3) resembles that of the CM [3] but, unlike the CM equation, it is time-reversal invariant because the angular velocity λ_{AB} in Eq (5) is a functional of the wavefunction and $\hat{\Gamma}$ [5], whereas it is so in the CM.

Eq. (3), together with Eqs. (4) and (5), is solved for rotation in two dimensions by varying $\big\langle F\big|$ in the expectation of Eq. (3), namely:

$$i\hbar\big\langle F\big|\frac{\partial}{\partial t}\big|F\big\rangle = \big\langle F\big|(H_o - \frac{1}{2}\lambda_{AB}\, L_{AB})\big|F\big\rangle \qquad (6)$$

using Hartree-Fock independent-particle determinantal wavefunction:

$$\big|F\big\rangle = e^{S_r}\, e^{S_v}\big|F_o\big\rangle \qquad (7)$$

$$S_r = \sum_{mi}(C^r_{mi}\, a^{\dagger}_m a_i - C^{r}_{mi}{}^{*}\, a^{\dagger}_i a_m) \qquad S_v = \sum_{mi}[C^v_{mi}(t)\, a^{\dagger}_m a_i - C^v_{mi}(t)^{*}\, a^{\dagger}_i a_m] \qquad (8)$$

where S_r and S_v describe rotational and vibrational motions respectively, C's are complex variational parameters, $a^{\dagger}$ and a are fermion creation and annihilation operators, and suffixes m and i refer to particle and hole states, respectively. The factor of ½ in Eq. (6) is introduced to ensure that the variation of Eq. (6) yields Eq. (3). Assuming $\hat{\Gamma}$ to be a one-body operator and for slow rotation, we expand Eq. (6), along with Eqs. (4) and (5), to second powers in S_r and S_v and vary the C's to obtain:

$$i\hbar\frac{dC^v_{nj}}{dt} = \varepsilon_{nj}\, C^v_{nj} + \sum_{mi}(C^v_{mi}\, V_{nijm} + C^{v}_{mi}{}^{*}\, V_{nmji}) + \frac{C^v_1}{2D_o}L_{nj} + \frac{C^v_2}{2D_o}T_{nj} + R_{nj} \qquad (9)$$

$$\varepsilon_{nj} \equiv \varepsilon_n - \varepsilon_j \qquad D_o \equiv \Gamma_A - \Gamma_B \qquad L_{nj} \equiv \langle n | L | j \rangle \qquad \Gamma_{nj} \equiv \langle n | \Gamma | j \rangle \qquad (10)$$

$$T_{nj} \equiv -\langle F_o | a_j^\dagger a_n \hat{t} | F_o \rangle = -\frac{1}{i\hbar}\left(-e_{nj}\Gamma_{nj} + \sum_{ml}\Gamma_{ml}^* V_{nmjl} + \sum_{ml}\Gamma_{ml}V_{nlmj}\right) \qquad (11)$$

$$C_1^v \equiv \sum_{mi}(C_{mi}^v T_{mi}^* + C_{mi}^{v\,*} T_{mi}) \qquad\qquad C_2^v = \sum_{mi}(C_{mi}^v L_{mi}^* + C_{mi}^{v\,*} L_{mi}) \qquad (12)$$

$$R_{nj} = \varepsilon_{nj} C_{nj}^r + \sum_{mi}(C_{mi}^r V_{nijm} + C_{mi}^{r\,*} V_{nmji}) + \frac{C_1^r}{2D_o}L_{nj} + \frac{C_2^r}{2D_o}T_{nj} \qquad (13)$$

$$C_1^r \equiv \sum_{mi}(C_{mi}^r T_{mi}^* + C_{mi}^{r\,*} T_{mi}) \qquad\qquad C_2^r \equiv \sum_{mi}(C_{mi}^r L_{mi}^* + C_{mi}^{r\,*} L_{mi}) \qquad (14)$$

The general solution of Eq. (9) is:

$$C_{nj}^v = X_{nj}\,e^{-i\omega t} + Y_{nj}^*\,e^{i\omega^* t} + R_{nj}\,t \qquad (15)$$

where X and Y are arbitrary complex parameters and ω is the oscillation frequency. For an oscillatory solution, we require:

$$R_{nj} = \varepsilon_{nj} C_{nj}^r + \sum_{mi}(C_{mi}^r V_{nijm} + C_{mi}^{r\,*} V_{nmji}) + \frac{C_1^r}{2D_o}L_{nj} + \frac{C_2^r}{2D_o}T_{nj} = 0 \qquad (16)$$

Eq.(8) then yields:

$$\varepsilon_{nj} X_{nj} + \sum_{mi}(X_{mi} V_{nijm} + Y_{mi} V_{nmji}) + \frac{Z_1}{2D_o}L_{nj} + \frac{Z_2}{2D_o}T_{nj} = \hbar\omega X_{nj} \qquad (17)$$

$$\varepsilon_{nj} Y_{nj}^* + \sum_{mi}(Y_{mi}^* V_{nijm} + X_{mi}^* V_{nmji}) + \frac{Z_1^*}{2D_o}L_{nj} + \frac{Z_2^*}{2D_o}T_{nj} = -\hbar\omega^* Y_{nj}^* \qquad (18)$$

$$Z_1 \equiv \sum_{mi}(X_{mi} T_{mi}^* + Y_{mi}\,T_{mi}) \qquad\qquad Z_2 \equiv \sum_{mi}(X_{mi} L_{mi}^* + Y_{mi}\,L_{mi}) \qquad (19)$$

Eqs.(17) and (18) resemble those of the RPA [6] except for the terms in Z_1 and Z_2, which couple oscillations to the rotational and intrinsic motions. These terms vanish because the expectation of the Hamiltonian and angular momentum operator are time independent (i.e., conserved, due to rotational invariance of the nuclear Hamiltonian), and because of the condition in Eq. (4), so that Eqs. (17) and (18) reduce identically to the RPA equations. Eq. (16) differs from the CM equation for rotation, namely:

$$\varepsilon_{nj} C_{nj} + \sum_{mi}(C_{mi} V_{nijm} + C_{mi}^{\,*} V_{nmji}) + \hbar\bar{\omega} L_{nj} = 0 \qquad (20)$$

in the following ways:

- The angular velocity $\lambda' = -\dfrac{C_1}{2D_o}$ (i.e., the coefficient of L_{nj}) in Eq. (16) is a dynamic variable unlike $\bar{\omega}$ in Eq. (20), which is an external parameter.
- The last term in Eq. (16) is absent from Eq. (20). This terms couples rotation to the intrinsic motion and is responsible for scaling λ' to $\bar{\omega}$. This statement is related to the observation that the transformation in Eq. (2) implicitly uses an integrable angle coordinate canonically conjugate to the angular momentum [5], whereas the CM implicitly uses a non-integrable angle whose derivative is $\bar{\omega}$.

The rotational excitation energy ΔE and the moment of inertia I_{smrm}, as defined in Eq. (21) for a given expectation of the angular momentum, are evaluated to be:

$$\Delta E = \langle F | H_o | F \rangle - \langle F_o | H_o | F_o \rangle = -\frac{C_1 C_2}{2 D_o} = \frac{\langle F | L | F \rangle^2}{2 I_{smrm}} = \frac{C_2^2}{2 I_{smrm}} \qquad (21)$$

Eq. (16) is solved in closed form for two types of residual interactions. For the first type, where the off-diagonal interaction matrix elements are assumed negligible, Eq. (16) yields:

$$C_{nj}^r = -\frac{C_1^r L_{nj} + C_2^r T_{nj}}{2 D_o (\varepsilon_{nj} + V_{njjn})} \;,\quad I_{smrm} \equiv \frac{1}{1 + \dfrac{1}{2 D_o} \sum_{mi} \dfrac{L_{nj}^* T_{mi} + L_{mi} T_{mi}^*}{e_{mi} + V_{miim}}} \sum_{nj} \frac{\left| L_{nj} \right|^2}{e_{nj} + V_{njjn}} \qquad (22)$$

and the following condition on $\hat{\Gamma}$:

$$\left(\frac{I_{smrm}}{D_o} \right)^2 = \sum_{nj} \frac{\left| L_{nj} \right|^2}{e_{nj} + V_{njjn}} \bigg/ \sum_{nj} \frac{\left| T_{nj} \right|^2}{e_{nj} + V_{njjn}} \qquad (23)$$

When the identity $\langle F | [\hat{\Gamma}_{AB} , L_{AB}] | F \rangle = i \hbar D_o$ is used, I_{smrm} in Eq. (22) becomes:

$$I_{smrm} = 2 \sum_{nj} \frac{\left| L_{nj} \right|^2}{e_{nj} + V_{njjn}} \qquad (24)$$

which is identical to the CM moment of inertia I_{cm}. Assuming a momentum-independent interaction and setting $\hat{\Gamma}$ to $\hat{Q}$, Eqs. (23) and (24) require that I_{smrm} be equal to the irrotational-flow moment of inertia. In general, this result is not valid because, in the absence of pairing, I_{cm} reduces to the irrotational-flow moment of inertia for closed-shell nuclei and to the rigid-flow value for open-shell nuclei. These results may indicate that $\hat{\Gamma}$ cannot be chosen arbitrarily. Similar results are obtained for a separable quadrupole-quadrupole effective interaction. However, the model wavefunction differs from that of the CM (the model preserves but CM violates time-reversal invariance). This implies that the values of the other parameters may not be the same in the two models. In any case, we have derived the CM-plus-RPA and CM moment of inertia from a microscopic, time-reversal invariant, non-linear theory. It is noted that the model predicts the correct nuclear mass for the centre of mass motion.

References

1. A. Bohr and B.R. Mottelson, *Nuclear Structure* Vol. II, Benjamin, N.Y., 1975.
2. D.J. Rowe, *Nuclear Collective Motion*, Methuen, London, 1970.
3. J.M. Eisenberg and W. Greiner, *Nuclear Models* Vol. 1, North Holland, Amsterdam, 1970.
4. F.M. Villars and G. Cooper, *Ann. Phys.* 56, 224 (1970).
5. P. Gulshani, *Can. J. Phys.* 84, 1-19, (2006).
6. D.J. Thouless and J.G. Valatin, *Nucl. Phys.* **31**, 211-230 (1962).

What is quantum mechanics?[*]

Gerard 't Hooft

Institute for Theoretical Physics
Utrecht University
and

Spinoza Institute
Postbox 80.195
3508 TD Utrecht, the Netherlands
e-mail: g.thooft@phys.uu.nl
internet: http://www.phys.uu.nl/~thooft/

Abstract

We discuss the arguments for suspecting that there exists a classical, *i.e.* deterministic theory underlying quantum mechanics. A difficulty is that an explanation must be found of the fact that the Hamiltonian, which is defined to be the operator that generates evolution in time, is bounded from below. The mechanism that can produce exactly such a constraint is identified in this paper. It is the fact that not all classical data are registered in the quantum description. Large sets of values of these data are assumed to be indistinguishable, forming *equivalence classes*. It is argued that this should be attributed to information loss, such as what one might suspect to happen during the formation and annihilation of virtual black holes.

The nature of the equivalence classes is further elucidated, as it follows from the positivity of the Hamiltonian. Our world is assumed to consist of a very large number of subsystems that may be regarded as approximately independent, or weakly interacting with one another. As long as two (or more) sectors of our world are treated as being independent, they all must be demanded to be restricted to positive energy states only. What follows from these considerations is a unique definition of energy in the quantum system in terms of the periodicity of the limit cycles of the deterministic model.

An example of a deterministic dissipative model producing exact quantum mechanics is provided for the case of a finite-dimensional vector space. These lecture notes have been produced partly from material published earlier, and as such contain more material than what could be presented in the talk.

---

[*]Presented at *The Eighth Symposium on Frontiers of Fundamental Physics*, Madrid, October 16—19, 2006

CP905, *Frontiers of Fundamental Physics (FFP8), Eighth International Symposium*
edited by B. G. Sidharth, A. Alfonso-Faus, and M. J. Fullana
© 2007 American Institute of Physics 978-0-7354-0412-0/07/$23.00

1. Introduction

In most theories describing the basic particles and forces of Nature, the use of quantum mechanics is unquestioned, and the framework of Hilbert space is so natural that alternatives are rarely considered. Quantum mechanics works impeccably, and there seems to be no logical reason why not to replace classical mechanics by quantum mechanics throughout. There are no difficulties with causality or other consistency requirements, and above all, all experimental observations strongly support the correctness of quantum mechanics.

Nevertheless, one may have various motivations for considering theories underlying and 'explaining' the quantum nature of our world. One of these motivations is its obscure description of reality. Should we be content with a theory that accurately predicts the outcome of any experiment, yet does not reveal what actually is going on there? When a beam of electrons, or even a single electron, is sent through two tiny slits in a screen, the probability for arriving at a particular spot behind the screen is found to display an interference pattern, as if the electron had been able to pass through the two slits at the same time. Yet we are able to formulate a causality principle for these electrons; apparently the physical events are well-orchestrated and systematic, but our description is in an obscure language.

This first objection is often wiped away by saying that it reflects the psychological needs of a physicist rather than a physical requirement. Maybe a more precise description of reality is simply not needed. But then there is a second motivation. Modern theories of matter and forces at the most fundamental level require that we subject the fabric of space and time to be subject to the rules of quantum mechanics as well. Direct attempts to do this were always bound to fail. The gravitational force is non-renormalizable, and this means that the language we use to describe the structures at the tiniest scales for time and distance must be inappropriate. (Super-) string theories and their descendant, M theory, appear to imply that there is no space-time metric at all, just stringlike objects moving in a rather poorly described continuum. This leaves us with questions concerning the causal and local nature of our physical laws. One possible alley to explore is the possibility that, at these tiniest scales, quantum mechanics is not the best language to use to describe what really is going on.

A third motivation is that theories in which the topological structure of space and time is to be included, necessarily also should deal with universes of a finite size and age. If now the dynamical laws for such a universe are phrased in a quantum mechanical setting, one has to take into account that the universe is too small to allow us to repeat experiments an infinite number of times. But if an experiment can only be carried out a few times in the entire universe, a probabilistic prediction of its outcome can never be tested accurately. In such a universe, quantum theory becomes vague and impossible to test; it must be replaced by something better.

To summarize: quantum mechanics is a theory that allows us to make *the best possible prediction* for the outcome of an experiment in any given setup. This prediction will still be probabilistic in general. In large universes having such a theory makes sense, since one can do an infinite number of experiments, and averaging procedures in that case will lead to predictions with arbitrary precision. In numerous experiments, physicists have confirmed that such quantum mechanical predictions are indeed the best one can make.

Quantum mechanics is *not* a theory that describes the actual course of events going on before an actual measurement is made. In fact, what quantum mechanics seems to be is a magnificent mathematical instrument enabling us to make accurate probabilistic predictions. One may well imagine that we have local laws of physics that in all conceivable circumstances describe myriads of physical degrees of freedom, evolving in a highly chaotic manner. Only statistical approaches can help to predict what will happen next. Such statistical approaches will be optimal if they contain two elements: one simplified deterministic law telling us how averages will evolve, and one random element reflecting fluctuations that must be ascribed to details that we cannot control. If the simplified law is identified with the Schrödinger equation, and the statistical element with the Born interpretation of the quantum amplitude, then quantum mechanics is doing exactly what one should expect from such a machine. One problem then remains: how do we identify those primordial local laws? We can phrase this as follows: Quantum mechanics appears to be an answer to a question, but which question?

Indeed the fact that quantum mechanics can be used as an instrument to solve classical problems has been observed before. The *Two dimensional Ising model* is a system of bits on an infinite square lattice where the thermodynamical free energy is given by a simple state sum. There are only interactions between nearest neighbors. Evaluating this free energy is a demanding mathematical exercise, but it can be calculated completely, using the techniques of quantum field theory[1]. Surprisingly, there was no quantum mechanics in the initial formulation of the problem; quantum mechanics simply was used as a device to solve it.

Thus, the question one would like to address is this: can one formulate local laws of physics that are themselves deterministic, but require quantum statistics to obtain meaningful solutions at large scales? How does one identify such models? Can one have a model in which events, in terms of its original degrees of freedom, are local and deterministic, while nevertheless the best one can do at large scales is to use quantum mechanics as an instrument? In that case, what then is the source of the apparently random fluctuations?

An important ingredient of our considerations is furthermore the notion of *Holography*[2]: for any system with volume $V = R^3$, the number of independent quantum degrees of freedom, at large R, can only grow linearly with the *surface*, R^2. It appears to be impossible to reconcile this with locality and causality unless one starts with other primary degrees of freedom than the quantum mechanical ones. Such a theory can be consistent with the apparent quantum mechanical nature of the world as we observe it, by assuming that these *classical* degrees of freedom are actually not preserved: there is *information loss*. This means that two different systems may evolve into one and the same final state. In quantum mechanics, this would be impossible, since it would violate unitarity and our ability to observe interference phenomena. The way to proceed would be to define *equivalence classes*: two states are equivalent if, at some finite time in the future, they evolve into the same final state. A quantum basis state of Hilbert space would then be identified with an entire equivalence class. One can imagine that the equivalence classes are so large that the number of distinct classes is controlled by the surface rather than the volume. It would mean, roughly, that two systems that are identical on the boundary, but only differ from each other deep inside the bulk, may evolve to become identical, in due time. The information in the bulk dissipates away; the information at the boundary is kept intact, as it may be refreshed continuously.

In this paper, we attempt to derive plausible assumptions from first principles. First, the formalism is displayed in Section 2. Deterministic systems are shown to be accessible by quantum mechanical procedures, although this does not turn them into acceptable quantum mechanical models just yet, because the Hamiltonian is not bounded from below. Then, we demonstrate that the most basic building blocks of any deterministic theory consists of units that would evolve with periodicity if there were no interaction (Section 3). We use the empirically known fact that the Hamiltonians are all bounded from below both before introducing the interaction and after having included the interaction. This necessitates our introduction of large equivalence classes (Section 4), such that neither the quantum mechanical nor the macroscopic observer can distinguish the elements within one equivalence class, but they can distinguish the equivalence classes. We attribute these equivalence classes to information loss, as explained above.

This procedure is necessary in particular when two systems are considered together *prior* to considering any interaction. We are led to the discovery that, besides the Hamiltonian, there must be a classical quantity ω that also corresponds to energy, and is absolutely conserved as well as positive (Section 5). It allows us to define the equivalence classes. We end up discovering a precise definition of the quantum wave function for a classical system (both amplitude and phase), and continue our procedure from there.

Physical and intuitive arguments were displayed in Ref. [3]. In that paper, it was argued that any system with information loss tends to show periodicity at small scales, and quantization of orbits. It was also argued that some lock-in mechanism was needed to relate the Hamiltonian with an ontologically observable quantity ω that is bounded from below. The lock-in mechanism was still not understood; here however we present the exact mathematical treatment and its relation to information loss. Interaction can be introduced but it is still not entirely understood what the best procedure should be.

A more satisfactory picture emerges if one realizes that energy is not directly locally observable, but determined by the periods of the limit cycles. This is explained in Section 7. We think that this interpretation is imperative, and it sheds an interesting new light on the phenomenon we call quantum mechanics. Having local determinism, information loss and limit cycles, our theory shows some resemblace with cellular automata with information loss, such as Conway's "Game of Life" [4].

2. Variables, beables and changeables

Any classical, deterministic system will contain some set of degrees of freedom $\vec{q}$ that follow some orbit $\vec{q}(t)$ in time. Time might be defined as a discrete variable or a continuous one, but this distinction is not as fundamental as one might think. If time is discrete, then the set $\vec{q}$ will have to include a clock that gives a tick at every time step $t_n = n\,\delta t$, or,

$$\frac{\mathrm{d}}{\mathrm{d}t}\, q_{\text{clock}} \;=\; 1 \;; \tag{2.1}$$

$$q_i \to q_i'(\vec{q}) \quad \text{at} \quad q_{\text{clock}} = 0 \bmod \delta t \;, \quad \forall i \neq \text{clock} \;. \tag{2.2}$$

It is not difficult to ascertain that this is just a special case of a more general equation of motion,

$$\frac{\mathrm{d}}{\mathrm{d}t}\, \vec{q} = \vec{f}(\vec{q}) \;. \tag{2.3}$$

For simplicity we therefore omit specific references to any (discretised) clock. In general, the orbit $\vec{q}(t)$ will be dictated[1] by an equation of motion of the form (2.3).

In the absence of information loss, this will correspond to a Hamiltonian

$$H = \sum_i p_i f_i(\vec{q}) + g(\vec{q}) \;, \tag{2.4}$$

where $p_i = -i\partial/\partial q_i$ is the *quantum* momentum operator. It will be clear that the quantum equations of motion generated by this Hamiltonian will exactly correspond to the *classical* equation (2.3). The function $g(\vec{q})$ is arbitrary, its imaginary part being adjusted so as to ensure hermiticity:

$$H - H^\dagger \;=\; -i\vec{\nabla} \cdot \vec{f}(\vec{q}) + 2\,i\,\text{Im}(g(\vec{q})) \;=\; 0 \;, \tag{2.5}$$

Any observable quantity $A(\vec{q})$, not depending on operators such as p_i, and therefore commuting with all q_i, will be called a *beable*. Through the time dependence of $\vec{q}$, the beables will depend on time as well. Any pair of beables, A and B, will commute with one another at all times:

$$[A(t_1),\, B(t_2)] = 0 \;, \quad \forall\, t_1, t_2 \;. \tag{2.6}$$

A *changeable* is an operator not commuting with at least one of the q_i's. Thus, the operators p_i and the Hamiltonian H are changeables. Using beables and changeables as operators[3], we can employ all standard rules of quantum mechanics to describe the classical system (2.3). The procedure described above will be referred to as *prequantization*; it is the description of a classical system using quantum Hilbert space notation.

---

[1]Thus, we do, as yet, use an absolute notion of time. Special and general relativistic transformations are left for future studies.

At this point, one is tempted to conclude that the classical systems form just a very special subset of all quantum mechanical systems. This, however, is not quite true. Quantum mechanical systems normally have a Hamiltonian that is bounded from below; the Hamiltonian (2.4) is not. At first sight, one might argue that all we have to do is *project out* all negative energy states[3][5]. We might obtain a physically more interesting Hilbert space this way, but, in general, the commutator property (2.6) between two beables is lost, if only positive energy states are used as intermediate states. As we will see, most of the beables (2.6) will not be observable in the quantum mechanical sense, a feature that they share with non-gauge-invariant operators in more conventional quantum systems with Yang-Mills fields. The projection mechanism that we need will be more delicate. As we will see, only the beables describing equivalence classes will survive as quantum observables.

We will start with the Hamiltonian (2.4), and only later project out states. Before projecting out states, we may observe that many of the standard manipulations of quantum mechanics are possible. For instance, one can introduce an integrable approximation $f_i^{(0)}(\vec{q})$ for the functions $f_i(\vec{q})$, and write

$$f_i(\vec{q}) = f_i^{(0)}(\vec{q}) + \delta f_i(\vec{q}) \ , \tag{2.7}$$

after which we do perturbation expansion with respect to the small correction terms δf_i. However, the *variation principle* in general does not work at this level, because it requires a lowest energy state, which we do not have.

3. The harmonic oscillator

We assume that a theory describing our world starts with postulating the existence of sub-systems that in some first approximation evolve independently, and then are assumed to interact. For instance, one can think of independent local degrees of freedom that are affected only by their immediate neighbors, not by what happens at a distance, baring in mind that one may have to expand the notion of immediate neighbors to include variables that are spatially separated by distances of the order of the Planck length. Alternatively, one may think of elementary particles that, in a first approximation, behave as free particles, and are then assumed to interact.

Temporarily, we switch off the interactions, even if these do not have to be small. Every sub-system then evolves independently. Imagine furthermore that some form of *information loss* takes place. Then, as was further motivated in Ref. [3], we suspect that the evolution in each domain will become *periodic*

Thus, we are led to consider the case where we have one or more independent, periodic variables $q_i(t)$. Only at a later stage, coupling between these variables will have to be introduced in order to make them observable to the outside world. Thus, the introduction of periodic variables is an essential ingredient of our theory, in addition to being just a useful exercise.

Consider a single periodic variable:

$$\frac{\partial q}{\partial t} = \omega \ , \tag{3.1}$$

while the state $\{q = 2\pi\}$ is identified with the state $\{q = 0\}$. Because of this boundary condition, the associated operator $p = -i\partial/\partial q$ is quantized:

$$p \ = \ 0, \pm 1, \pm 2, \cdots \ . \tag{3.2}$$

The inessential additive coefficient $g(q)$ of Eq. (2.4) here has to be real, because of Eq. (2.5), and as such can only contribute to the unobservable phase of the wave function, which is why we permit ourselves to omit it:

$$H = \omega\,p = \omega\,n \ ; \qquad n \ = \ 0, \pm 1, \pm 2, \cdots \ . \tag{3.3}$$

If we would find a way to dispose of the negative energy states, this would just be the Hamiltonian of the quantum harmonic oscillator with internal frequency ω (apart from an inessential constant $\frac{1}{2}\omega$).

Theorem

Consider any probability distribution $W(q)$ that is not strictly vanishing for any value of q, that is, a strictly positive, real function of q. Then a complex wave function $\psi(q)$ can be found such that

$$W(q) = \psi^*(q)\psi(q) \ , \tag{3.4}$$

and $\psi(q)$ is a convergent linear composition of eigenstates of H with non-negative eigenvalues only.

The proof is simple mathematics. Write

$$\psi(q) = \exp(\alpha(q) + i\beta(q)) \ , \qquad z = e^{iq} \ . \tag{3.5}$$

Choose $\alpha + i\beta$ to be en entire function within the unit circle of z. Then an elementary exercise in contour integration yields,

$$\alpha(q) = \tfrac{1}{2}\log(W(q)) \ ; \quad \beta(q) = \beta_0 - P \oint \frac{dq'}{2\pi}\frac{1 + \cos(q' - q)}{\sin(q' - q)}\alpha(q') \ , \tag{3.6}$$

where P stands for the principal value, and β_0 is a free common phase factor. In fact, Eq. (3.6) is not the only function obeying our theorem, because we can choose any number of zeros for $\psi(z)$ inside the unit circle and then again match (3.4). One concludes from this theorem that no generality in the function W is lost by limiting ourselves to positive energy eigenfunctions only.

In fact, we may match the function W with a wave function ψ that has a zero of an arbitrary degree at the origin of z space. This way, one can show that the lowest energy state can be postulated to be at any value of E.

In this paper, however, we shall take a different approach. We keep the negative energy states, but interpret them as representing the bra states $\langle\psi|$. These evolve with the opposite sign of the energy, since $\langle\psi(t)| = e^{+iHt}\langle\psi(0)|$. As long as we keep only one single periodic variable, it does not matter much what we do here, since energy is absolutely conserved. The kets $|n\rangle$ and bras $\langle n|$ may be given the energy values $E_n = (n + \frac{1}{2})\omega$. The time evolution of the bras goes as if $E_n = -(n + \frac{1}{2})\omega$, so that we have a sequence of energy values ranging from $-\infty$ to ∞.

Note that if we have two independent periodic systems, each with a similar Hamiltonian, our theorem does not hold. This situation is dealt with in the next section. In our bra-ket formalism, it is convenient to tune the energy of the lowest ket state at $\frac{1}{2}\omega$.

4. Two (or more) harmonic oscillators

As was explained at the beginning of Section 3, we expect that, when two periodic variables interact, again periodic motion will result. This may seem to be incorrect. If the two periods, ω_1 and ω_2 are incommensurate, an initial state will never exactly be reproduced. Well, this was before we introduced information loss. In reality, periodicity will again result. We will show how this happens, first by considering the quantum harmonic oscillators to which the system should be equivalent, according to Section 3, and then by carefully interpreting the result.

In Fig. 1, the states are listed for the two harmonic oscillators combined. Let their frequencies be ω_1 and ω_2. The kets $|n_1, n_2\rangle = |n_1\rangle|n_2\rangle$ have $n_1 \geq 0$ and $n_2 \geq 0$, so they occupy the quadrant labelled I in Fig. 1. The bra states, in view of their time dependence, occupy the quadrant labelled III. The other two quadrants contain states with mixed positive and negative energies. Those must be projected away. If we would keep those states, then any interaction between the two oscillators would result in inadmissible mixed states, in disagreement with what we know of ordinary quantum mechanics. So, although keeping the bra states is harmless because total energy is conserved anyway, the mixed states must be removed. This is very important, because now we see that the joint system cannot be regarded as a direct product. Some of the states that would be allowed classically, must be postulated to disappear. We now ask what this means in terms of the two periodic systems that we thought were underlying the two quantum harmonic oscillators.

First, we wonder whether the spectrum of combined states will still be discrete. The classical, non interacting system would only be periodic if the two frequencies have a rational ratio: $p\omega_1 - q\omega_2 = 0$, where p and q are relative primes. The smallest period would be $T = 2\pi q/\omega_1 = 2\pi p/w_2$, so that we would expect equally spaced energy levels with spacings $\omega_1/q = \omega_2/p$. Indeed, at high energies, we do get such spacings also in the quantum system, with increasing degeneracies, but at lower energies many of these levels are missing. If the frequencies have an irrational ratio, the period of the classical system is infinite, and so a continuous spectrum would have to be expected.

When two quantum harmonic oscillators are considered together, this does not happen

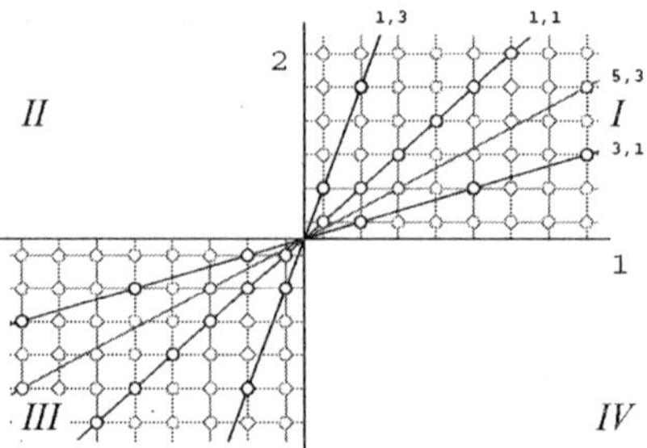

Figure 1: Combining two harmonic oscillators. Tilted lines show sequences of spectral states again associated to harmonic oscillators. For further explanation, see text.

The spectrum is always discrete. In Fig. 1, it is indicated how to avoid having missing states and variable degeneracies. We see that actually full series of equally spaced energy levels still exist:

At any given choice of a pair of odd relative primes p and q, we have a unique series of bra- and ket states with energies $\omega_{pq}(n + \frac{1}{2})$, with $\omega_{pq} = p\omega_1 + q\omega_2$.

It is easy to see that these sequences are not degenerate, that every odd relative prime pair of integers (p, q) occurs exactly once, and that all states are represented this way:

$$E_{n_1,n_2} = (n_1 + \tfrac{1}{2})\omega_1 + (n_2 + \tfrac{1}{2})\omega_2 = (n + \tfrac{1}{2})(p\omega_1 + q\omega_2) ; \tag{4.1}$$

$$\frac{2n_1 + 1}{2n_2 + 1} = \frac{p}{q} . \tag{4.2}$$

Some of these series are shown in the Figure.

We see that, in order to reproduce the quantum mechanical features, that is, to avoid the unphysical states where one energy is positive and the other negative, we have to combine two periodic systems in such a way that a new set of periodic systems arises, with frequencies ω_{pq}. Only then can one safely introduce interactions of some form. Conservation of total energy ensures that the bra and ket states cannot mix. States where one quantum oscillator would have positive energy and one has negative energy, have been projected out.

But how can such a rearrangement of the frequencies come about in a pair of classical periodic systems? Indeed, why are these frequencies so large, and why are they labelled by odd relative primes? In Fig. 2 the periodicities are displayed in configuration space, $\{q_1, q_2\}$. The combined system evolves as indicated by the arrows. The evolution might not be periodic at all. Consider now the $(5, 3)$ mode. We can explain its short period $T_{53} = 2\pi/\omega_{53}$ only by assuming that the points form equivalence classes, such that different points within one equivalence class are regarded as forming the same 'quantum' state. If all points on the lines shown in Fig. 2 (the ones slanting downwards) form one equivalence class, then this class evolves with exactly the period of the oscillator whose frequency is ω_{53}.

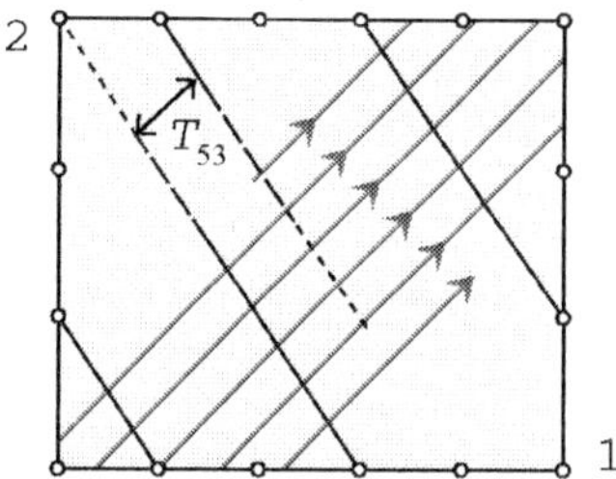

Figure 2: The equivalence class for oscillators in the case $(p, q) = (5, 3)$. Lines with arrows pointing right and up: time trajectories of individual points. Solid and broken lines going downwards: (part of) the $(5, 3)$ equivalence class at $t = 0$. For further explanation, see text.

In principle, this equivalence class can be formed in one of two ways: the information concerning the location of a point on this line is lost, either because there is an inherent information loss mechanism, implying that two different states may actually evolve to become the same state, or it could simply be that this information cannot be transmitted to macroscopically observable quantities. One could imagine a renormalization group technique that relates microscopic states to states at much larger distance scales. Not all data are being faithfully transmitted in the procedure. This latter option will later be dismissed as being impractical; it is highly revealing to assume explicit information loss.

For the time being, imagine that information loss takes place by means of processes that are random, uncontrollable or impossible to follow in detail, that cause our data point to fluctuate along the line of its equivalence class. The line itself moves with the deterministic speed of the original oscillators.

Observe in Figure 2 that, in case $(p, q) = (5, 3)$, due to these fluctuations, five points of system 1 alone now form a single equivalence class, and three points in system 2. This is because we have assigned 5 quanta of energy to system 1 for every three quanta of energy of system 2. More generally, we could represent this situation with the wave function

$$\psi_{pq} = e^{i\,(n+\frac{1}{2})\,(p\,q_1 + q\,q_2 - \omega_{pq}t)}\, e^{-\frac{1}{2}\,i(q_1+q_2)} \,, \tag{4.3}$$

where both variables $q_{1,2}$ were taken to be periodic with periods 2π. The (p, q) equivalence classes appear to be defined by the condition

$$p\,q_1 + q\,q_2 = \text{Constant}\,, \tag{4.4}$$

and this means that the n-dependent part of the wave function (4.3) has the same phase all over the entire equivalence class, if we may assume that the second term in Eq. (4.3), arising from the vacuum fluctuations $\frac{1}{2}\omega t$, may be ignored.

To describe the equivalence classes it is helpful to introduce time variables t_a for the subsystems $a = 1, 2, \cdots$ in terms of their *unperturbed* evolution law, $q_a = \omega_a t_a$. Then, writing $E_1 = p\omega_1$, $E_2 = q\omega_2$, one can characterize the equivalence classes as

$$E_1\,\delta t_1 + E_2\,\delta t_2 = 0\,, \tag{4.5}$$

which means that the reactions that induce information loss cause q_a to speed up or slow down by an amount $\pm\delta\,t_a$ obeying this equation. One can easily generalize this result for many coexisting oscillators. They must form equivalence classes such that fluctuating time differences occur that are only constrained by

$$\sum_a E_a\,\delta\,t_a = 0 \; , \tag{4.6}$$

which also are the collections of points that have the same phase in their quantum wave functions. We conclude that, in the ontological basis $\{|\vec{q}\rangle\}$, *all states $|\vec{q}\rangle$ which have the same phase in the wave function $\langle\vec{q}|\psi\rangle$ (apart from a fixed, time independent term), form one complete equivalence class.*

5. Energy and Hamiltonian

In the previous section, it was derived that the energies of the various oscillators determine the shape of the equivalence classes that are being formed. However, this would require energy to be a beable, as defined in Section 2. Of course, the Hamiltonian, being the generator of time evolution, cannot be a beable. It is important to notice here, that the parameters p and q defining the equivalence classes as in Section 4, are not exactly the energies of q_1 and q_2; the Hamiltonian eigenvalues are

$$H_1 = (n + \tfrac{1}{2})p\,\omega_1 \; ; \qquad H_2 = (n + \tfrac{1}{2})q\,\omega_2 \; , \tag{5.1}$$

with a common multiplication factor $n + \tfrac{1}{2}$. This n indeed defines the Hamiltonian of the orbit of the equivalence class. Generalizing this, the relation between the energies E_i in Eqs. (4.5) and (4.6) and the Hamiltonian H, is

$$H = (n + \tfrac{1}{2})E \; , \tag{5.2}$$

where n defines the evolution of a single clock that monitors the evolution of the entire universe.

Now that the relative primes p and q have become beables, we may allow for the fact that the periods of q_1 and q_2 depend on p and q as a consequence of some non-trivial interaction. But there is more. We read off from Fig. 2, that p points on the orbit of q_1 in fact belong to the same equivalence class. Assuming that the systems 1 and 2 that we started off with, had been obtained again by composing other systems, we must identify these points. But this forces us to redefine the original periods by dividing these by p and q, respectively, and then we end up with two redefined periodic systems that are combined in the one and only allowed way:

$$p = q = 1 \; . \tag{5.3}$$

Only a single line in Fig. 1 survives: the diagonal.

The picture that emerges is the following. We are considering a collection of variables q_a, each being periodic with different periods $T_a = 2\pi/\omega_a$. They each are associated with a positive beable E_a, such that $E_a = \omega_a$. The interactions will be such that the total energy $E = \sum_a E_a$ is conserved. Now as soon as these variables are observed together (even if they do not interact), an uncontrollable mixing mechanism takes place in such a way that the variables are sped up or slowed down by time steps δt_a obeying Eq. (4.6), so that, at any time t, all states obeying

$$\sum_a E_a t_a = \left(\sum_a E_a\right) t , \qquad (5.4)$$

form one single equivalence class.

The evolution and the mixing mechanism described here are entirely classical, yet we claim that such a system turns into an acceptable quantum mechanical theory when handled probabilistically. However, we have not yet introduced interactions.

6. Interactions

We are now in a position to formulate the problem of interacting systems. Consider two systems, labelled by an index $a = 1, 2$. System a is characterized by a variable $q_a \in [0, 2\pi)$ and a discrete index $i = 1, \cdots, N_a$, which is a label for the spectrum of states the system can be in. Without the interaction, i stays constant. Whether the interaction will change this, remains to be seen.

The frequencies are characterized by the values $E_a^i = \omega_a^i$, so that the periods are $T_a^i = 2\pi/\omega_a^i$. Originally, as in Section 4, we had $\omega_a^i = p_a\omega_a$, where $p_1 = p$ and $p_2 = q$ were relative primes (and both odd), but the periods ω_a are allowed to depend on p_a, so it makes more sense to choose a general spectrum to start with.

The non-interacting parts of the Hamiltonians of the two systems, responsible for the evolution of each, are described by

$$H_a^0|n_a, i\rangle = (n_a + \tfrac{1}{2})E_a^i|n_a, i\rangle , \qquad (6.1)$$

where the integer $n_a = -\infty, \cdots, +\infty$ is the changeable generating the motion along the circle with angular velocity ω_a. We have

$$n_a = -i\partial/\partial q_a . \qquad (6.2)$$

The total Hamiltonian describing the evolution of the combined, unperturbed, system is not $H_1^0 + H_2^0$, but

$$H_{\text{tot}}^0 = (n_{\text{tot}} + \tfrac{1}{2})(E_1^i + E_2^j) , \qquad (6.3)$$

where i, j characterize the states 1 and 2, but we have a single periodic variable $q_{\text{tot}} \in [0, 2\pi)$, and

$$n_{\text{tot}} = -i\partial/\partial q_{\text{tot}} . \qquad (6.4)$$

In view of Eq. (4.4), which here holds for $p = q = 1$, we can define

$$q_{\text{tot}} = q_1 + q_2 , \tag{6.5}$$

while $q_1 - q_2$ has become invisible. We can also say,

$$n_1 = n_2 = n_{\text{tot}} . \tag{6.6}$$

An interacting system is expected to have perturbed energy levels, so that its Hamiltonian should become

$$H^0 + H^{\text{int}} = (n_{\text{tot}} + \tfrac{1}{2})(E_1^i + E_2^j + \delta E^{ij}) , \tag{6.7}$$

where δE^{ij} are correction terms depending on both i and j. This is realized simply by demanding that the beables E_1^i and/or E_2^j get their correction terms straight from the other system. This is an existence proof for interactions in this framework, but, at first sight, it appears not to be very elegant. It means that the velocity ω_1^{ij} of one variable q_1 depends on the state j that the other variable is in, but no matrix diagonalization is required. Indeed, we still have no transitions between the different energy states i. It may seem that we have to search for a more general interaction scheme. Instead, the scheme to be discussed next differs from the one described in this section by the fact that the energies E cannot be read off directly from the state a system is in, even though they are beables. The indices i, j are locally unobservable, and this is why we usually work with superimposed states.

7. Limit cycles

Consider the evolution following from a given initial configuration at $t = t_0$, having an energy E. Let us denote the state at time t by $F(t)$. The equivalence classes $|\psi(t)\rangle$ defined in Section 4 are such that the state $F(t)$ is equivalent to the state $F(t + \Delta t)$, where

$$\Delta t = h/E , \tag{7.1}$$

in which h is Planck's constant, or:

$$|\psi(t)\rangle \stackrel{\text{def}}{=} \{F(t + n\Delta t) , \ \forall n|_{n \geq n_1}\} , \tag{7.2}$$

for some n_1. Let us now assume that the equivalence indeed is determined by information loss. That the states in Eq. (7.2) are equivalent then means that there is a smallest time t_1 such that

$$F(t_1 + n\Delta t) = F(t_1) , \ \forall n \geq 0 . \tag{7.3}$$

Thus, the system ends in a *limit cycle* with period Δt.

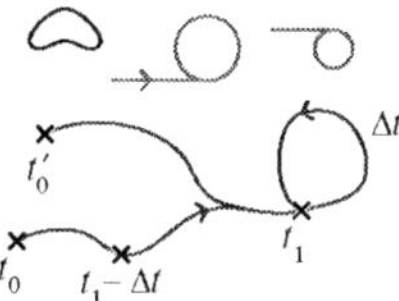

Figure 3: Configuration space showing the limit cycles of an evolving system, indicating the times t_0, t_1 and the period Δt of a limit cycle. The points at t_0 and t_0' form one equivalence class.

One now may turn this observation around. A closed system that can only be in a finite number of different states, making transitions at discrete time intervals, would necessarily evolve back into itself after a certain amount of time, thus exhibiting what is called a Poincaré cycle. If there were no information loss, these Poincaré cycles would tend to become very long, with a periodicity that would increase exponentially with the size of the system. If there is information loss, for instance in the form of some dissipation effect, a system may eventually end up in Poincaré cycles with much shorter periodicities. Indeed, time does not have to be discrete in that case, and the physical variables may form a continuum; there could be a finite set of stable orbits such that, regardless the initial configuration, any orbit is attracted towards one of these stable orbits; they are the limit cycles. The energy of a state is then simply defined to be given by Eq. (7.1), or $E \stackrel{\text{def}}{=} h/P$, where $P = \Delta t$ is the period of the limit cycle, and h is Planck's constant.

Since this period coincides with the period of the wave function, we now deduce a physical interpretation of the phase of the wave function: *The phase of a wave function (in the frame of energy eigenstates) indicates where in the limit cycle the state will be.*

In general, we will have a superposition of many possibilities, and so we add to this the interpretation of the amplitude of the wave function: the absolute value of the amplitude in the frame of energy eigenstates indicates the probability that a particular limit cycle will be reached. Thus, we have reached the exact physical meaning of a quantum wave function.

We identified any deterministic system having information loss, with a quantum mechanical system evolving with a Hamiltonian defined by Eq. (5.2). However, in order to obtain a realistic model, one has to search for a system where the energy is extrinsic. With this, we mean that the universe consists of subsystems that are weakly coupled. Uncoupled systems are described as in Section 4; weakly coupled systems must be such that the limit cycles of the combination (12) of two systems (1) and (2) must have periods P_{12} obeying

$$\frac{1}{P_{12}} \approx \frac{1}{P_1} + \frac{1}{P_2} , \tag{7.4}$$

which approaches the exact identity in the limit of large systems being weakly coupled. This is the energy conservation law.

8. The Determinant Model

In this section, we describe a simple model from which we can derive an existence theorem:

For any quantum system there exists at least one deterministic model that reproduces all its dynamics after prequantization.

To keep the argument transparent, we consider only a finite dimensional subspace of Hilbert space. Let Schrödinger's equation be

$$\frac{\mathrm{d}\psi}{\mathrm{d}t} = -iH\psi \; ; \qquad H = \begin{pmatrix} H_{11} & \cdots & H_{1N} \\ \vdots & \ddots & \vdots \\ H_{N1} & \cdots & H_{NN} \end{pmatrix} . \tag{8.1}$$

We now claim that this system, with the same Hamiltonian, is reproduced by the following deterministic system. There are two degrees of freedom, ω and φ, of which the latter is periodic: $\varphi \in [0, 2\pi)$, or $\psi(\omega, \varphi) = \psi(\omega, \varphi + 2\pi)$. Now, take as classical equations of motion:

$$\frac{\mathrm{d}\varphi(t)}{\mathrm{d}t} = \omega(t) \qquad ; \tag{8.2}$$

$$\frac{\mathrm{d}\omega(t)}{\mathrm{d}t} = -\kappa\, f(\omega)\, f'(\omega) \,, \qquad f(\omega) = \det(H - \omega) \,. \tag{8.3}$$

The function $f(\omega)$ has zeros exactly at the eigenvalues of H. Its derivative, the function $f'(\omega)$, has zeros between the zeros of f, see Fig. 4.

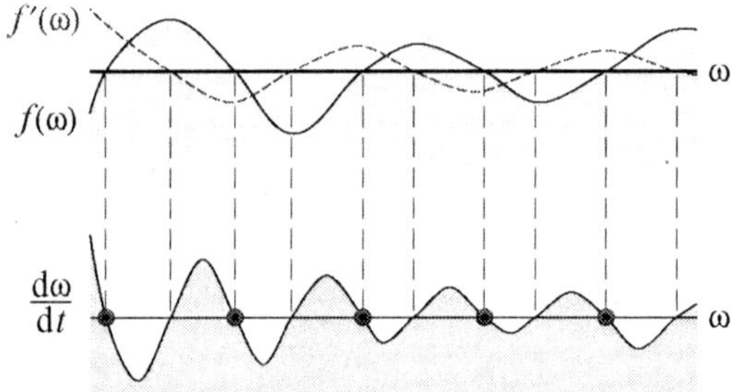

Figure 4: Above: sketches of $f(\omega)$ and of $f'(\omega)$ (dotted). Below: a sketch of the product $-\kappa f(\omega)\, f'(\omega)$. The red dots indicate its attractive zeros, which now coincide with the zeros of f.

This way, we achieve the situation that all eigenvalues ω_i, $i = 1, \cdots, N$ are attractive zeros, as can be read off from the Figure. Depending on the initial configuration, one of the eigenvalues of the matrix H is rapidly approached. If $\omega_{i+\frac{1}{2}}$ is defined to be the zero of $f'(\omega)$ between ω_i and ω_{i+1}, then the region $\omega_{i-\frac{1}{2}} < \omega < \omega_{i+\frac{1}{2}}$ is the equivalence class associated to the final state $\omega = \omega_i$. Since the convergence towards ω_i is exponential in time, one can say that for all practical purposes the limit value ω_i is soon reached, at which point the system indeed enters into a limit cycle. .At this cycle, the variable φ is periodic in time with period $T = 2\pi/\omega_i$. The pre-quantum states can be Fourier transformed in φ:

$$\psi(\omega, \varphi) = \sum_n e^{in\varphi} \psi_n(\omega) . \tag{8.4}$$

At $t \to \infty$, we have

$$\psi(\omega, \varphi, t) \to \sum_n \psi_n(\omega_i, 0)\, e^{in(\varphi - \omega_i t)} . \tag{8.5}$$

At this point, we can use the previous discussions to argue that ω_i is the ontological energy. In this particular model, we can be even more explicit. The quantum number n is absolutely conserved, so, we can use the superselection rule that the universe settles for one particular value for n, say, $n = 1$. Then, the states $\psi_1(\omega_i)$ can be exactly identified with the energy eigenstates of the original quantum system. This completes our mathematical mapping. In this model, interference is a formality: we are always free to consider probabilistic superpositions, described by superimposing wave functions ψ_i with different i, as in Section 2

9. Discussion

When we attempt to regard quantum mechanics as a deterministic system, we have to face the problem of the positivity of the Hamiltonian, as was concluded earlier in Refs [3][5][6]. There, also, the suspicion was raised that information loss is essential for the resolution of this problem. In this paper, the mathematical procedures have been worked out further, and we note that the deterministic models that we seek must have short limit cycles, obeying Eq. (7.1). Short limit cycles can easily be obtained in cellular automaton models with information loss, but the problem is to establish the addition rule (7.4), which suggests the large equivalence classes defined by Eq. (4.4).

We found that the energy eigenstates of a quantum system correspond to the limit cycles of the deterministic model. If P is the period of the limit cycle, then the energy E of this state is $E = h/P$ (see Eq. 7.1).

Comparing our models with the real universe, we note that there we have a small number of massless particle species, and many more massive ones, generating a rich spectrum of energies all very low compared to the Planck energy. If we assume that the Planck time would be Nature's natural time scale, then we observe that there must be many limit cycles whose periods are very long compared to the Planck time. Our universe appears to be built in such a way that, as soon as several of these cyclic limiting solutions are allowed to interact, new limit cycles will be reached with shorter periods, due to information loss

In cellular automaton models, one might be able to mimic such a situation best by introducing a nearly conserved, positive quantity resembling energy, which can be seen statistically to decrease slowly on the average, so that the most chaotic initial states relax into more organized states that can easily end up in a limit cycle. The more chaotic the initial state, the smaller the period of its eventual limit cycle is expected to be, but there are many special initial states with very long limit cycles: the low energy states.

States of interest, with which we might attempt to describe the universe as we know it, must be very far away from any limit cycle. They are also far away from the strictly stationary eigenstates of the Hamiltonian. This means that we *do not yet know* which of the numerous possible limit cycles our universe will land into. This is why we normally use wave functions that have a distribution of amplitudes in the basis of the Hamiltonian eigenmodes. The squares of these amplitudes indicate the probability that any particular limit cycle will be reached. Also note that, according to General Relativity, taking into account the negative energies in the gravitational potentials, the total energy of the universe should vanish, which means that the entire universe might never settle for any limit cycle, as is indeed suggested by what we know of cosmology today: the universe continues to expand. The limit cycles mentioned in this paper refer to idealized situations where small sections of the universe are isolated from the rest, so as to be able to define their energies exactly. Only when a small part of the universe is sealed off from the rest, it is destined to end up in a limit cycle.

It may be of importance to note that our definition of energy, as being the inverse of the period of the limit cycle, supplies us with an absolute scale of energy: it is not allowed to add a constant to it. Moreover, the energy E in Eq. (5.2), as opposed to the Hamiltonian H, is a beable. Thus it is allowed to couple it to gravity by imposing Einstein's equations. If indeed the vacuum has a limit cycle with a large period, it carries a very low energy, and this is why one might suspect that the true resolution of the cosmological constant problem[7] will come from deterministic quantum mechanics rather than some symmetry principle[8]. Earlier, the cosmological constant has been considered in connection with deterministic quantum mechanics by Elze[9]. The fact that the observed cosmological constant appears to be non-vanishing implies that a finite volume V of space will have a largest limit cycle with period

$$P = \frac{8\pi hG}{\Lambda V} \;, \tag{9.1}$$

which is of the order of a microsecond for a volume of a cubic micron. If Λ were negative we would have had to assume that gravity does not exactly couple to energy.

Lorentz transformations and general coordinate transformations have not been considered in this paper. Before doing that, we must find models in which the Hamiltonian is indeed extensive, that is, it can be described as the integral of an Hamiltonian density $T_{00}(\vec{x}, t)$ over 3-space, as soon as the integration volume element $\mathrm{d}^3\vec{x}$ is taken to be large compared to the 'Planck volume'. When that is achieved, we will be only one step away from generating locally deterministic quantum field theories.

What can be said from what we know presently, is that a particle with 4-momentum p_μ must represent an equivalence class that contains all translations $x^\mu \to x^\mu + \Delta x^\mu$ with $p_\mu \Delta x^\mu = n\,h$, where n is an integer. Note that a limit cycle having this large transformation group as an invariance group is hard to imagine, which probably implies that this particular limit cycle will take an infinite time to be established. Indeed, a particle in a fixed momentum state occupies the entire universe, and we already observed that the entire universe will never reach a limit cycle.

In the real world, we have only identified the observable quantum states, which we now identify with the equivalence classes of ontological states. We note that the physical states of the Standard Model in fact are also known to be gauge equivalence classes, Local gauge transformations modify our description of the dynamical variables, but not the physical observations. It is tempting to speculate that these gauge equivalence classes (partly) coincide with the equivalence classes discussed in this paper, although our equivalence classes are probably a lot larger, which may mean that many more local gauge symmetries are still to be expected.

It is even more tempting to include here the gauge equivalence classes of General Relativity: perhaps local coordinate transformations are among the dissipative transitions. In this case, the underlying deterministic theory might not be invariant under local coordinate transformations, and here also one may find novel approaches towards the cosmological constant problem and the apparent flatness of our universe.

Our reason for mentioning virtual black holes being sources of information loss might require further explanation. Indeed, the *quantum mechanical* description of a black hole is not expected to require information loss (in the form of quantum decoherence); it is the corresponding classical black hole that we might expect to play a role in the ontological theory, and that is where information loss is to be expected, since classical black holes do not emit Hawking radiation. As soon as we turn to the quantum mechanical description in accordance to the theory explained in this paper, a conventional, fully coherent quantum description of the black hole is expected.

Although we do feel that this paper is bringing forward an important new approach towards the interpretation of Quantum Mechanics, there are many questions that have not yet been answered. One urgent question is how to construct explicit models in which energy can be seen as extrinsic, that is, an integral of an energy density over space. A related problem is how to introduce weak interactions between two nearly independent systems. Next, one would like to gain more understanding of the phenomenon of (destructive) interference, a feature typical for Quantum Mechanics while absent in other statistical theories.

We think that the observations made in this paper are an important step towards the demystification of quantum mechanics.

References

[1] B. Kaufman, *Phys. Rev.* **76** (1949) 1232; B. Kaufman and L. Onsager, Phys. Rev. **76** (1949) 1244; Y. Nambu, Progr. Theor. Phys. **5** (1950) 1.

[2] G. 't Hooft, "Dimensional Reduction in Quantum Gravity", Essay dedicated to Abdus Salam, Utrecht preprint THU-93/26; gr-qc/9310026; L. Susskind, "The World

as a Hologram", J. Math. Phys. **36** (1995) 6377, hep-th/9409089; G. 't Hooft, "The Holographic Principle", Opening Lecture , in *Basics and Highlights in Fundamental Physics*, The Subnuclear series, Vol.'**37**, World Scientific, 2001 (Erice, August 1999), A. Zichichi, ed., p. 72, SPIN-2000/06, hep-th/0003004.

[3] G. 't Hooft, Class. Quant. Grav. **16** (1999) 3263, gr-qc/9903084; id., *Quantum Mechanics and determinism*, in Proceedings of the Eighth Int. Conf. on "Particles, Strings and Cosmology, Univ. of North Carolina, Chapel Hill, Apr. 10-15, 2001, P. Frampton and J. Ng, Eds., Rinton Press, Princeton, pp. 275 - 285, hep-th/0105105.

[4] M. Gardner, Scientific American, **223** (4), 120; (5), 118; (6), 114 (1970).

[5] H. Th. Elze, *Deterministic models of quantum fields*, gr-qc/0512016.

[6] M. Blasone, P. Jizba and H. Kleinert, *Annals of Physics* **320** (2005) 468, quant-ph/0504200; id., *Braz. J. Phys.* **35** (2005) 497, quant-ph/0504047.

[7] S. Nobbenhuis, *Categorizing Different Approaches to the Cosmological Constant Problem*, gr-qc/0411093, *Foundations of Physics*, to be pub.

[8] G. 't Hooft and S. Nobbenhuis, *Invariance under complex transformations, and its relevance to the cosmological constant problem*, gr-qc/0602076 , *Class. Quant. Grav.*, to be publ.

[9] H. Th. Elze, *Quantum fields, cosmological constant and symmetry doubling*, hep-th/0510267.

Neutrinos in the Electron

E.L. Koschmieder

Center for Statistical Mechanics
The University of Texas at Austin, Austin TX 78712, USA

Abstract. I will show that one half of the rest mass of the electron consists of electron neutrinos and that the other half of the rest mass of the electron consists of the mass in the energy of electric oscillations. With this composition we can explain the rest mass of the electron, its charge, its spin and its magnetic moment. We have also determined the rest masses of the muon neutrino and the electron neutrino.

Keywords: electron-mass, electron-charge, electron-spin, neutrino masses
PACS: 14.60.Cd, 14.60.Ef, 14.60.Pq

INTRODUCTION

The electron was discovered by J.J. Thomson [1] in 1897. It seems fair to say that at present, one hundred years after the discovery of the electron, we do not possess an accepted theoretical explanation of the electron. I will summarize the status of the theory of the electron as follows: One hundred years of theoretical work have made it abundantly clear that the electron is not a purely electromagnetic particle, otherwise the electron would explode through the repulsion of the elements of the charge in the electron, as was first suggested by Poincaré [2], who introduced the so-called Poincaré stresses which were meant to hold the electron together. The existence of a non-electromagnetic substance in the eletron was confirmed by Einstein, Pauli and others. They dealt only with the mass $m(c)$ of the electron and its charge e^-. The positron, the spin, the magnetic moment, and the neutrinos were not known at that time.

The electron-electron scattering experiments of the 50's and later invalidate the old theories. The measured radius of the electron is not the classical electron radius e^2/mc^2 $= 2.8 \cdot 10^{-13}$ cm, but is $\leq 10^{-16}$ cm. It is equally clear that the "something else" that must be in the electron but the charge must be non-interacting, otherwise the electron radius could not be $\leq 10^{-16}$ cm. The only non-interacting matter we know of are the neutrinos. So it seems to be natural to ask whether neutrinos are not part of the electron.

THE DECAY OF THE PIONS LEADS TO THE ELECTRON

The decay sequence of the $\pi^{\pm}$ mesons is, in the case of π^-,

$$\pi^- \rightarrow \mu^- + \bar{\nu}_\mu$$

$$\mu^- \rightarrow e^- + \nu_\mu + \bar{\nu}_e$$

CP905, *Frontiers of Fundamental Physics (FFP8), Eighth International Symposium*
edited by B. G. Sidharth, A. Alfonso-Faus, and M. J. Fullana
© 2007 American Institute of Physics 978-0-7354-0412-0/07/$23.00

In this sequence *a free electron is emitted* accompanied by the emission of a muon neutrino v_μ, an anti-muon neutrino $\bar{v}_\mu$ and an anti-electron neutrino $\bar{v}_e$, *but without the emission of an electron neutrino v_e.* I will show in the following that the missing electron neutrino v_e is emitted as part of the electron emitted in the muon decay.

If we want to learn something about the electron from the decay sequence of the pions we have first to clarify the composition of the pions. The standard model of the particles explains the $\pi^\pm$ mesons as consisting of up and down quarks and their antiparticles and of fractional electric charges. Nobody has proven the existence of the quarks nor of the fractional charges, not for lack of trying as I might add. *Nor, in particular, has the mass of the $\pi^\pm$ mesons or of the quarks been determined this way.* We also note that the standard model does not explain the mass and the charge of the electron. One can, however, explain the $\pi^\pm$ mesons differently, as we have done in [3].

The pions decay through weak decays, so it seems to be natural to assume that

* *the weak force holds the $\pi^\pm$ mesons together,*

the basic assumption made in this investigation. *From now on we deal with the consequences of this assumption.*

Since the range of the weak force is $O(10^{-16})$ cm and the measured size of the pions is $O(10^{-13})$ cm it follows, as a consequence of our assumption, that

* the *weak force can hold the $\pi^\pm$ mesons together only*

    if the $\pi^\pm$ mesons have a lattice structure,

as in a crystal, e.g. the salt crystal or a copper crystal.

Since the weak force is transmitted by neutrinos

the $\pi^\pm$ mesons must have a neutrino lattice.

The neutrinos are separated by the range of the weak force, which is $O(10^{-16})$ cm. It follows from the measured size of the $\pi^\pm$ mesons, which is $0.881 \cdot (10^{-13})$ cm, and from the range of the weak force that there must be

$$N = 2.854 \cdot 10^9 \tag{1}$$

neutrinos in the pion lattice.

Conservation of neutrino numbers during the creation of a pion, in e.g. $\gamma + p \rightarrow \pi^+ + \pi^- + p$, requires that there are equal numbers of neutrinos and antineutrinos in the lattice of the pions. The decay sequence of the pions also tells that there must be equal numbers of muon neutrinos v_μ and anti-muon neutrinos $\bar{v}_\mu$, as well as electron neutrinos v_e and anti-electron neutrinos $\bar{v}_e$ in the pions. We conclude that the lattice of the pions consists of N/4 electron neutrinos v_e, N/4 anti-electron neutrinos $\bar{v}_e$, N/4 muon neutrinos v_μ, and N/4 anti-muon neutrinos $\bar{v}_\mu$. As a consequence of the absence of electron neutrinos v_e in the π^- decay sequence and from the presence of N/4 electron neutrinos v_e in the lattice of the pions follows that *N/4 electron neutrinos must go with the electron emitted in the decay of the μ^- meson.*

THE OSCILLATION ENERGY IN THE PIONS

Fourier analysis dictates that after the $\pi^{\pm}$ mesons have been created in 10^{-23} sec in the process

$$\gamma + p \rightarrow \pi^{+} + \pi^{-} + p$$

that there must be *a continuum of frequencies in the pions*. These frequencies are picked up by the continuum of frequencies of the lattice oscillations. Elementary lattice theory, according to Born and v.Karman, says that the energy in the lattice oscillations of the $\pi^{\pm}$ mesons is given by

$$E_{v}(\pi^{\pm}) = \frac{h\nu_0 N}{2\pi(e^{h\nu/kT} - 1)} \int_{-\pi}^{\pi} \phi \, d\phi \,, \tag{2}$$

with $\qquad \nu_0 = c/2\pi a \qquad$ and $\qquad \phi = 2\pi a/\lambda \qquad$ and the lattice constant a. Eq.(2) means that the oscillation energy is the number N of the oscillations times the average of the energy of the oscillations.

The $\pi^{\pm}$ mesons are like black bodies filled with oscillating neutrinos.

It is

$$E_{v}(\pi^{+}) - 67.82\,\text{MeV} = 0.486\,\text{m}(\pi^{\pm})c^2/2 \cong 1/2 \cdot \text{m}(\pi^{\pm})c^2 \,. \tag{3}$$

1/2 of the energy in the rest mass of the $\pi^{\pm}$ mesons

is made up of the oscillation energy of the neutrinos.

OUR EXPLANATION OF THE PIONS

We have learned that the $\pi^{\pm}$ mesons can be described by a

$$\nu_e,\ \bar{\nu}_e,\ \nu_\mu,\ \bar{\nu}_\mu \quad \text{neutrino lattice,}$$

plus one elementary electrical charge. 1/2 of the energy in the rest mass of the pions is in the oscillation energy of the neutrinos in the pions. The other half of the energy in the pions is, as we will see, in the rest masses of the neutrinos. We have

$$\text{m}(\pi^{\pm})c^2 = E_{v}(\pi^{\pm}) + \Sigma\, \text{m}(\text{neutrinos})c^2 \,. \tag{4}$$

For the explanation of the $\pi^{\pm}$ mesons we use only neutrinos, whose existence is unquestionable, the neutrino oscillations and one elementary electrical charge.

We do not use hypothetical particles.

If the $\pi^{\pm}$ mesons have a neutrino lattice then it follows necessarily from the pion decay sequence that there must be neutrinos in the mass of the electron. Now we need to know the rest masses of the neutrinos.

THE REST MASS OF THE MUON NEUTRINO

The energy in the rest mass of the pions $m(\pi^{\pm})c^2 = 139.570\,\text{MeV}$ is the sum of the energy in the rest masses of the neutrinos plus the oscillation energies of the neutrinos in the pions. So

$$m(\pi^{\pm})c^2 - E_{\nu}(\pi^{\pm}) = \Sigma\,[m(\nu_{\mu}) + m(\bar{\nu}_{\mu}) + m(\nu_e) + m(\bar{\nu}_e)]c^2, \tag{5}$$

or with Eq.(3) $E_{\nu}(\pi^{\pm}) = 67.82\,\text{MeV}$

$$m(\pi^{\pm})c^2 - 67.82\,\text{MeV} = 71.75\,\text{MeV} \cong m(\pi^{\pm})c^2/2 = \Sigma\,m(\text{neutrinos})c^2. \tag{6}$$

*1/2 of the rest mass of the pions
is equal to the sum of the rest masses of all neutrinos in the pions.*

If

$$m(\nu_e) = m(\bar{\nu}_e) \ll m(\nu_{\mu}) = m(\bar{\nu}_{\mu}),$$

as will be justified immediately, it follows from Eq.(6) that

$$71.75\,\text{MeV} \cong 2\Sigma\,m(\nu_{\mu})c^2 = N/2 \cdot m(\nu_{\mu})c^2, \tag{7}$$

or, taking also the rest mass of ν_e in Eq.(10) into account, we find that
the rest mass of the muon neutrino is

$$m(\nu_{\mu}) = 49.91\,\text{milli eV}/c^2. \tag{8}$$

Since the same considerations as for π^- apply to π^+ it follows that

$$m(\nu_{\mu}) = m(\bar{\nu}_{\mu}). \tag{9}$$

THE REST MASS OF THE ELECTRON NEUTRINO

From the $1.29333\,\text{MeV}$ released in the β-decay of the neutron

$$n \rightarrow p + e^- + \bar{\nu}_e$$

follows, as explained in detail in [3], that *the rest mass of the electron neutrino is*

$$m(\bar{\nu}_e) = 0.365\,\text{milli eV}/c^2, \tag{10}$$

and from the β-decay of the anti-neutron follows that

$$m(\nu_e) = m(\bar{\nu}_e). \tag{11}$$

From the rest masses of $\nu_{\mu} = 49.91\,\text{milli-eV}/c^2$ and of $\nu_e = 0.365\,\text{milli-eV}/c^2$ follows then that

$$m(\nu_e) = 1.0021\,\alpha\,m(\nu_\mu) \cong \alpha\,m(\nu_\mu)\,, \tag{12}$$

with the *fine structure constant* α. It will probably be argued that Eq.(12) is a coincidence. However, the probability for this to be a coincidence is zero, considering the infinite pool of numbers on which the ratio $m(\nu_e)/m(\nu_\mu)$ could settle. Actually we will check the validity of Eq.(12) when we study the ratio $m(\pi^\pm)/m(e^\pm)$. In order to arrive at a correct value of this ratio it is necessary that Eq.(12) is valid. We have furthermore

$$N/4 \cdot m(\nu_e) = 0.5096\,m(e^\pm)\,. \tag{13}$$

i.e.

1/2 of the rest mass of the electron
is equal to the sum of the rest masses of the electron neutrinos
in the electron.

Since $m(\nu_e) = m(\bar{\nu}_e)$, (Eq.11), it follows from Eq.(13) that 1/2 of the rest mass of the positron is equal to the rest mass of $N/4$ anti-electron neutrinos.

NOW TO THE ELECTRON

We have learned that
N/4 electron neutrinos make up 1/2 of the rest mass of the electron.
That solves 1/2 of the problem at hand. The other half of the energy $m(e^\pm)c^2$ must be in the oscillations in the interior of the electron. The formula for the oscillation energy in the $\pi^\pm$ mesons (Eq.2) does, however, not apply to the electron, because it would lead to an oscillation energy of $34\,m(e^\pm)c^2$. Instead the oscillation energy in the electron must be

$$F_\nu(e^\pm) = \frac{h\nu_0 N \cdot \alpha}{2\pi(e^{h\nu/kT} - 1)} \int_{-\pi}^{\pi} \phi\,d\phi\,. \tag{14}$$

The appearance of the fine structure constant α in the formula for the oscillation energy in the electron means that *the nature of the oscillations in the electron is different from the oscillations in the $\pi^\pm$ mesons.* As is well-known the fine structure constant characterizes the strength of the electromagnetic forces. With $\alpha = e^2/\hbar c$ and $\nu_0 = c/2\pi a$ follows that

$$h\nu_0\alpha = e^2/a\,, \tag{15}$$

with the elementary electrical charge e and the lattice constant a. Eq.(15) means that

• *The oscillations in the interior of the electron are electric oscillations.*

From Eqs.(14,15) follows that the oscillation energy of the electron is

$$E_\nu(e^\pm) = \frac{N}{2} \cdot \frac{e^2}{a} \cdot \frac{f(T)}{2\pi} \int_{-\pi}^{\pi} \phi\,d\phi\,. \tag{16}$$

Since $E_\nu(e^\pm)$ depends on e^2 the oscillation energy is the same for the electron and positron, as it must be. There must be $N/2$ oscillations, either standing or circular oscillations, because only non-progressive waves can make up part of the rest mass of the electron. From the ratio of the oscillation energies of $\pi^\pm$ (Eq.2) and of $e^\pm$ (Eq.15) follows that the oscillation energy in the electron is

$$E_\nu(e^\pm) = \alpha/2 \cdot E_\nu(\pi^\pm), \qquad (17)$$

and with $\quad E_\nu(\pi^\pm) \cong 1/2 \cdot m(\pi^\pm) \quad$ follows that

$$E_\nu(e^\pm) = \frac{\alpha}{2} \cdot \frac{m(\pi^\pm)c^2}{2} = 0.996570 \, m(e^\pm)c^2/2. \qquad (18)$$

In other words:

1/2 of the rest mass of the electron
originates from the oscillation energy in the electron.

Because α and $m(\pi^\pm)c^2$ are known with great accuracy Eq.(18) sets a firm value for the oscillation energy $E_\nu(e^\pm)$.

The sum of the oscillation energy in the electron plus the sum of the energy in the rest masses of the neutrinos in the electron (Eq.13) is equal to

$$E_\nu(e^\pm) + N/4 \cdot m(\nu_e)c^2 = (0.9966/2 + 0.5096)m(e^\pm)c^2 = 1.0079 \, m(e^\pm)c^2. \qquad (19)$$

We have found, within $7.9 \cdot 10^{-3}$, the value of the rest mass of the electron and positron.

THE RATIO OF THE PION MASS TO THE ELECTRON MASS

The *validity* of our explanation of $m(\pi^\pm)$ as well as of $m(e^\pm)$ can be checked with the ratio $m(\pi^\pm)/m(e^\pm)$. As was first suggested by Nambu [4] and more accurately by Barut [5] it is empirically

$$\frac{m(\pi^\pm)}{m(e^\pm)} = \frac{2}{\alpha} - 1 = 273.07, \qquad (20)$$

whereas the actual ratio is 273.132, which is $1.00023 \cdot 273.07$. From our explanation of the pions follows that

$$m(\pi^\pm) = N[m(\nu_\mu) + m(\nu_e)], \qquad (21)$$

because the sum of the masses of the neutrinos in the pions is $N/2[\, m(\nu_\mu) + m(\nu_e)]$ and the oscillation energy equals the sum of the energy in the neutrino masses. We have also

$$m(e^\pm) = N/2 \cdot m(\nu_e), \qquad (22)$$

because the sum of the rest masses of the electron neutrinos in the electron is $N/4 \cdot m(\nu_e)$ and the oscillation energy is equal to the energy in the mass of the neutrinos. Consequently

$$\frac{m(\pi^\pm)}{m(e^\pm)} = \frac{N[m(\nu_\mu) + m(\nu_e)]}{N/2 \cdot m(\nu_e)}, \qquad (23)$$

and with
$$m(\nu_e) = \alpha \, m(\nu_\mu)$$

from Eq.(12) follows that

$$\frac{m(\pi^\pm)}{m(e^\pm)} = \frac{2}{\alpha} + 2 = 276.07, \qquad (24)$$

which is 1.0107 times the actual ratio $m(\pi^\pm)/m(e^\pm) = 273.132$. We have, however, only considered the ratio of the neutrino masses in $\pi^\pm$ and $e^\pm$, not the consequences of either the charge or the spin of the particles.

- The theoretical ratio of $m(\pi^\pm)/m(e^\pm)$ can only be correct if our explanation of $m(\pi^\pm)$ and of $m(e^\pm)$ is correct, and if the relation $m(\nu_e) = \alpha \, m(\nu_\mu)$ is valid.

A better approximation for $m(\pi^\pm)/m(e^\pm)$ leading to $2/\alpha$ - 1 can be found in [3].

THE MAGNETIC MOMENT OF THE ELECTRON

Uhlenbeck and Goudsmit [7] found that an electron with spin must have a magnetic moment with a magnitude of 1 Bohr magneton

$$\mu_B = e\hbar/2mc. \qquad (25)$$

Taking the spin s of the electron into account it must be

$$\vec{\mu}_e = g \cdot \frac{e\hbar}{2m(e^\pm) \cdot c} \cdot \vec{s}. \qquad (26)$$

with the dimensionless factor g = 2 which is, because of s = 1/2, required in order to bring the magnetic moment μ_e of the electron to its correct value (not considering the anomalous magnetic moment). For a particle at rest the g-factor has not been explained so far. If, however, 1/2 of the mass of the electron consists of neutrinos which are neutral and do not contribute to the magnetic moment of the electron, which to all that we know is true for the neutrinos, then the magnetic moment of the electron is

$$\vec{\mu}_e = g \cdot \frac{e\hbar}{2m(e^\pm)/2 \cdot c} \cdot \vec{s}. \qquad (27)$$

With s = 1/2 the g-factor must be equal to one, and we have

$$\mu_e = e\hbar/2m(e^\pm)c, \qquad (28)$$

or one Bohr magneton, as it must be.

- If exactly 1/2 of the mass of the electron consists of neutrinos,

    then the electron has the correct magnetic moment.

CONCLUSIONS

We have shown that

- the rest mass of the electron can be explained by the sum of the rest masses of N/4 electron neutrinos ν_e plus the mass in the energy of N/2 electric oscillations. The same considerations lead to the explanation of the positron.

- 1/2 of the rest mass of the electron is in the electron neutrinos, the other half of $m(e^\pm)$ originates from the electric oscillations.

We have verified the *validity* of our explanation of the electron mass by showing that

- our theoretical ratio $m(\pi^\pm)/m(e^\pm)$ is correct in the percent range, which can only be if we have explained $m(\pi^\pm)$ and $m(e^\pm)$ correctly, and if $m(\nu_e) = \alpha\, m(\nu_\mu)$,

and by showing that

- the correct magnetic moment of the electron follows automatically if 1/2 of the mass of the electron consists of neutrinos.

From the preceding explanation of the mass of the electron follows also the explanation of *the elementary electric charge* e carried by the electron, as detailed in [6], and the explanation of *the spin of the electron*, detailed in [3].
The rest mass of the muon can be explained similarly with a lattice of muon neutrinos, electron neutrinos and anti-electron neutrinos, as described in [3]. The rest mass of the muon is so much larger than the rest mass of the electron, $m(\mu^\pm) \cong 207\, m(e^\pm)$, because the mass of the muon neutrinos is $\cong 137$ times larger than the mass of the electron neutrinos.

REFERENCES

1. Thomsom, J.J. *Phil.Mag.* **44**, 293 (1897).
2. Poincaré, H. *Rend.Circ.Mat.Palermo* **21**, 129 (1906).
3. Koschmieder, E.L. http://arXiv.org/physics/0602037 (2006).
4. Nambu, Y. *Prog.Th.Phys.* **7**, 595 (1952).
5. Barut, A.O. *Phys.Lett.B* **73**, 310 (1978).
6. Koschmieder, E.L. http://arXiv.org/physics/0503206 (2005).
7. Uhlenbeck, G.E. and Goudsmit, S. *Naturwiss.* **13**, 953 (1925).

Casimir Energy Associated With Fractional Derivative Field

S.C. Lim

Faculty of Engineering, Multimedia University, Cyberjaya 63100, Selangor, Malaysia

Abstract. Casimir energy associated with fractional derivative scalar massless field at zero and positive temperature can be obtained using the regularization based on generalized Riemann zeta function of Epstein-Hurwitz type.

Keywords: Casimir Energy, Fractional Derivative Fields
PACS: 11.10.-z, 11.10.Lm, 11.10.Wx

INTRODUCTION

Recent advances in cosmology, in particular the strong evidences for the accelerated expansion of the universe [1-4], have rekindled considerable interest in Casimir effect. Casimir energy in extra space-time dimensions [5] has been proposed as one of the possible candidates for the origin of the accelerating cosmic expansion. However, in this paper, we shall not deal directly on the link between Casimir energy and the cosmology. Instead, we shall study Casimir energy associated with fractional derivative fields, which may exist in space-time with fractional dimensions.

Euclidean scalar field $\phi(x)$, $x \in \mathbb{R}^4$ with the Lagrangian $L = \frac{1}{2}\phi(x)\Lambda(\Box_E)\phi(x)$, satisfies the nonlocal field equation $\Lambda(\Box_E)\phi(x) = 0$. One gets fractional Klein-Gordon equation for $\Lambda(\Box_E) = (-\Box_E + m^2)^\alpha$, $\alpha > 0$. Quantization of fractional derivative fields in both conventional and stochastic approach have been studied [6-9].

CASIMIR ENERGY BETWEEN PARALLEL PLATES

We consider the Casimir energy associated with the fractional derivative scalar massless field inside perfectly conducting parallel plates. Let L^2 be the area of the plates and d be the separation between them. The system is assumed to be in thermal equilibrium with a heat reservoir at finite temperature T and the field satisfies the periodic condition $\phi(x,y,z,0) = \phi(x,y,z,\beta)$, with $\beta = 1/T$, and Dirichlet conditions at the boundary. If the coordinate system is chosen such that the z-axis is perpendicular to both plates at $z = 0$, and $z = d$. Eigenvalues of the Euclidean fractional Laplacian operator $-\Box_E^\alpha$ are

CP905, *Frontiers of Fundamental Physics (FFP8), Eighth International Symposium*
edited by B. G. Sidharth, A. Alfonso-Faus, and M. J. Fullana
© 2007 American Institute of Physics 978-0-7354-0412-0/07/$23.00

$$\lambda_{mn} = \left[k_x^2 + k_y^2 + \left(m\pi/d \right)^2 + \left(2n\pi/\beta \right)^2 \right]^\alpha , \quad m = 0,1,2,\ldots, \quad n = 0,\pm 1,\pm 2,\ldots \quad (1)$$

The corresponding generalized thermal zeta function is

$$\zeta_\alpha(s) = \frac{L^2}{\pi} \sum_{n=-\infty}^{\infty} \sum_{m=1}^{\infty} \int_0^\infty k\,dk \left[k^2 + \left(\frac{m\pi}{d} \right)^2 + \left(\frac{2n\pi}{\beta} \right)^2 \right]^{-\alpha s}$$

$$= \frac{\pi^{1-2\alpha s} L^2}{2} \frac{\Gamma(\alpha s - 1)}{2\Gamma(\alpha s)} \left\{ d^{2\alpha s-2}\zeta_R(2-2\alpha s) + E_2(\alpha s-1, 1/d^2, 4/\beta^2) \right\}, \quad (2)$$

where $\zeta_R(z)$ is the Riemann zeta function and $E_2(s,a_1,a_2) = [a_1 n_1^2 + a_2 n_2^2]^{-s}$ denotes the Epstein zeta function [10,11]. The analytic continuation of Epstein function to a meromophic function in complex plane gives

$$E_2(z,a_1,a_2) = -\frac{a_1^{-z}}{2}\zeta_R(2z) + \frac{1}{2}\sqrt{\frac{\pi}{a_2}}\frac{\Gamma(z-1/2)}{\Gamma(z)}E_1(z-1/2; a_1)$$

$$+ \frac{2\pi^z}{\Gamma(z)a_2^{z/2+1/4}} \sum_{m,n=1}^{\infty} \frac{m^{z-1/2}}{(a_1 n^2)^{[z-1/2]/2}} K_{1/2-z}\left(2\pi m\sqrt{\frac{a_1 n^2}{a_2}} \right), \quad (3)$$

where $K_\nu(z)$ is the Macdonald function. By carrying out the appropriate substitutions in the above equation, and by noting that for a well behaved differentiable function $F(s)$, then the derivative of $f(s) = F(s)/\Gamma(s)$ at $s = 0$ is given by $F(0)$, we obtain

$$\zeta_\alpha'(s=0) = \frac{\pi^2 L^2 \beta}{d^3}\left(\frac{\alpha}{720} \right) + \alpha L^2 \sqrt{\frac{2}{\beta}} \sum_{m,n=1}^{\infty} \left(\frac{n}{md} \right)^{3/2} K_{3/2}\left(\frac{\pi\beta mn}{d} \right). \quad (4)$$

The Helmholtz free energy per unit area is

$$\frac{F}{L^2} = -\frac{\pi^2 L^2}{d^3}\left(\frac{\alpha}{720} \right) - \frac{\sqrt{2}\alpha}{(d\beta)^{3/2}} \sum_{m,n=1}^{\infty} \left(\frac{n}{m} \right)^{3/2} K_{3/2}\left(\frac{\pi\beta mn}{d} \right). \quad (5)$$

Note that the first term in the above equation is just the Casimir energy associated with the fractional massless field at zero temperature. It can be shown [12] that in the low temperature limit,

$$\frac{F}{L^2} = -\frac{\pi^2 L^2}{d^3}\left(\frac{\alpha}{720}\right) + \sqrt{2}\alpha\left(\frac{1}{\pi\beta^3} + \frac{1}{d\beta^2}\right). \tag{6}$$

In the high temperature limit, one gets

$$\frac{F}{L^2} = -\frac{\pi^2 L^2}{d^3}\left(\frac{\alpha}{720}\right) + \sqrt{2}\alpha\left[\frac{\pi^2 d}{45\beta^4} + \frac{1}{\pi\beta^3}\zeta(3)\right] \tag{7}$$

Our result shows that the free energy density associated with the fractional field is α times that of ordinary scalar field.

CONCLUDING REMARKS

We have shown the regularization technique based on generalized zeta functions used in ordinary field theory can be applied to derive Casimir energy associated with the fractional derivative scalar field for simple geometry. Generalization to fractional electromagnetic field is quite straight forward. It will be interesting to see whether such a technique can still be applied for more complicated geometries. Finally, we remark that the Casimir energy associated with a quartic self-interacting fractional massless scalar field can be obtained, and it involves more complicated Epstein-Hurwitz zeta functions [12].

ACKNOWLEDGMENTS

The author would like to thank the Malaysian Academy of Sciences for the Scientific Advancement Fund Allocation (SAGA) P 96c.

REFERENCES

1. A.G. Riess *et al.* [Supernova Search Team Collaboration], Astron. J. **116**, 1009 (1998).
2. S. Perlmutter *et al.*[Supernova Cosmology Project Collaboration], Astrophys. J. **517**, 565 (1999).
3. J.L. Tonry *et al.* [Supernova Search Team Collaboration], Astrophys. J. **594**, 1 (2003).
4. A.G. Riess *et al.* [Supernova Search Team Collaboration], Astrophys. J. **607**, 665 (2004).
5. K.A. Milton, Grav. Cosmol. **9**, 66 (2003).
6. R.L.P.G. do Amaral and E.C. Marino, J.Phys. A **25**, 5183 (1992).
7. D.G. Barci, L.E. Oxman, and M Rocca, Int. J. Mod. Phys. **A11**, 2111 (1996).
8.. S.C. Lim and S.V. Muniandy, Phys. Lett. A **324**, 396 (2004).
9. S.C. Lim, Physica A **363**, 269 (2006).
10. E. Elizalde, S.D. Odintsov, A. Romeo, A.A. Bytsenko and S. Zerbini, *Zeta Regularization Techniques with Applications,* World Scientific, Singapore, 1994.
11. E. Elizalde, *Ten Physical Applications of Spectral Zeta Functions,* Lecture Notes in Physics m35, Springer, Berlin, 1995.
12. S.C. Lim and Chai-Hok Eab, paper in preparation.

Global dynamical structure in a 3D model for LiCN

J. C. Losada*, R. M. Benito[†] and F. Borondo**

*Grupo de Sistemas Complejos. Dpto. Tecnología de la Edificación. EU Arquitectura Técnica.
Universidad Politécnica de Madrid. Avda. Juan de Herrera, 6. 28040 Madrid.
[†]Grupo de Sistemas Complejos. Dpto. Física y Mecánica. ETSI Agrónomos. Universidad
Politécnica de Madrid. Avda. Complutense s/n. 28040 Madrid.
**Departamento de Química, C–IX. Universidad Autónoma de Madrid. Cantoblanco,
28049–Madrid (Spain).

Abstract. We use the frequency analysis method to characterize the phase space of a realistic 3D Hamiltonian model for the vibrations of the LiCN molecule.

Keywords: Classic mechanics; Nonlinear dynamics; Chaos theory; Frequency analysis
PACS: 05.45.Ac, 05.45.Tp

After the development of modern digital computers and following the pioneering work of Poincaré [1], surfaces of section (SOS) have been used as an effective way of mapping the motion of dynamical systems exhibiting a complicated behavior into suitable graphical representations. These constructions nicely translates dynamics into topology, thus allows the investigation of the phase space structure of chaotic systems by unveiling the rich structure (invariant tori, resonant chain of islands, cantori, chaotic regions, etc. [2]) embedded in it. Unfortunately, SOS are only practical for 2D Hamiltonian systems, since the simultaneous intersection (even approximate) of a trajectory with several planes is a very unlike event.

For systems with three or more dimensions [3], where new phenomena such as the Arnold diffusion [4], can take place, one must resort to other methods of study. An efficient alternative to study numerically the global dynamics of a system is the use of frequency analysis (FA). By computing the fundamental frequencies associated to trajectories, the existence of quasiperiodicity or chaos, as well as other dynamical characteristics, can be ascertain. Although these frequencies are, strictly speaking, only defined and fixed on these tori, the algorithm will compute a frequency vector over a finite time span for any initial condition. On the KAM tori, this vector is a very accurate approximation to the true frequencies, whereas at resonances and in chaotic regions it provides a natural interpolation between these fixed frequencies.

For an integrable Hamiltonian system, one can use action–angle variables [5] and then write the Hamiltonian in the form: $H(\mathbf{I})$, so that the resulting equations of motion

$$\dot{I}_i = \frac{\partial H}{\partial \theta_i} = 0, \quad \dot{\theta}_i = -\frac{\partial H}{\partial I_i} = v_i(\mathbf{I}) \tag{1}$$

can be easily solved, giving

$$I_i = \text{constant}, \quad \theta_i = v_i(\mathbf{I})t + \delta_i. \tag{2}$$

CP905, *Frontiers of Fundamental Physics (FFP8), Eighth International Symposium*
edited by B. G. Sidharth, A. Alfonso-Faus, and M. J. Fullana
© 2007 American Institute of Physics 978-0-7354-0412-0/07/$23.00

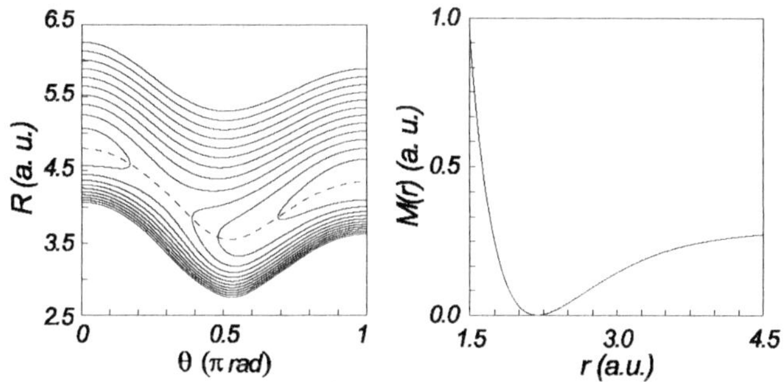

FIGURE 1. (Left) Contours plot of the potential energy surface, $v(R,\theta)$, for LiNC/LiCN at a r=2.186 a.u. The minimum energy path for the isomerization reaction has been plot superimposed in dashed line. (Right): Potential energy interaction of the C-N fragment, $M(r)$ described by a Morse function.

The function $\mathbf{I} \rightarrow v(\mathbf{I})$ gives the frequency map. If this map is invertible then the frequencies v_i can be used as coordinates, characterizing the tori, instead of the actions I_i.

For chaotic trajectories, invariant tori do not exist, and a local frequency analysis, performed by following the time evolution of the frequency ratios, provides very useful information on the way in which they evolve in phase space.

The aim of this paper is to explore the dynamic of a typical chaotic small polyatomic molecule, showing how valuable dynamical information can be extracted for multidimensional chaotic trajectories with the aid of the FA method. For this purpose we chose to study a 3D model for the LiNC/LiCN isomerizing system, described by a realistic interaction potential obtained from a quantum mechanical *ab initio* calculation [6].

Using scattering or Jacobi coordinates (R,r,θ), being R the distance from the Li atom to the center of mass of the CN fragment, r C–N distance, and θ the angle formed by these two vectors, the corresponding classical vibrational $(J-0)$ Hamiltonian is given by

$$H = \frac{P_R^2}{2\mu_{Li-CN}} + \frac{P_r^2}{2\mu_{C-N}} + \frac{1}{2}\left(\frac{1}{\mu_{Li-CN}R^2} + \frac{1}{\mu_{C-N}r^2}\right)P_\theta^2 + V(R,r,\theta). \tag{3}$$

The vibrational frequency of the C–N motion is very high and then this mode is not highly coupled with the other vibrations in the molecule. Accordingly, we will write the corresponding potential energy surface in the form:

$$V(R,r,\theta) = v(R,\theta) + M(r), \tag{4}$$

where $v(R,\theta)$ represents the potential interaction of the molecule for r fixed at its equilibrium value, $r_e = 2.186$ a.u., (see Fig. 1), and $M(r)$ describes the C–N interaction by a Morse function,

$$M(r) = D\{1 - \exp[-\beta(r-r_e)]\}^2, \tag{5}$$

The parameters D=0.29135 a.u. and β=1.4988 a.u.$^{-1}$, are directly related with the depth and width of the CN potential well, and have also been taken from the literature.

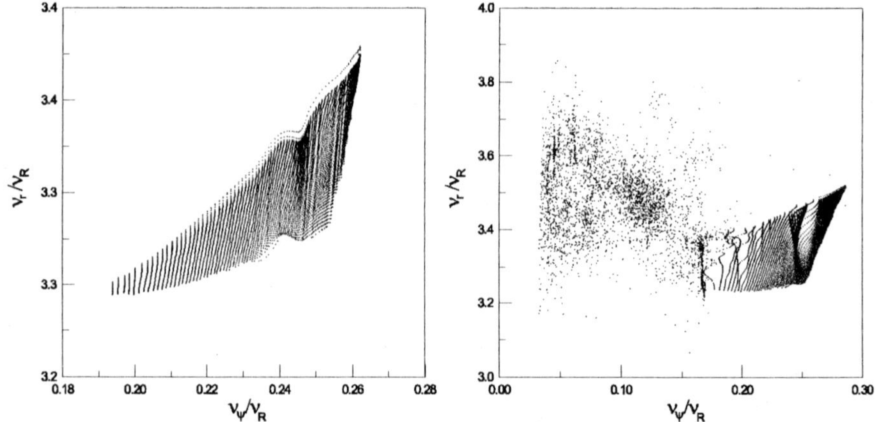

FIGURE 2. Frequency map analysis of the LiNC/LiCN system for the vibrational energies: 0.014 a.u. (left) and 0.020 a.u. (right).

The corresponding FA results are plotted in Figs. 2 and 3, for two different values of the excitation energy. At the lower energies considered ($E = 0.014$ a.u.) the majority of tori still remain. As a result, the original grid of initial conditions only gets distorted, and we observe a very regular structure of almost parallel lines, corresponding to a smooth variation of the two frequency ratios. As energy increases ($E = 0.022$ a.u.) the situation becomes more complex, and part of the regular structure is destroyed. Then, in addition to (approximately) parallel lines, three clouds of scattered points appear, indicating the destruction of tori giving rise to the chaotic regions of our system at this energy, and we can also observe the appearance of new accumulations of points; for example, those at $v_\psi/v_R = 1/16$ and $1/22$ are particularly striking.

ACKNOWLEDGMENTS

This work was supported by MEC–Spain (MTM2006–15533 and CSD2006–32).

REFERENCES

1. H. Poincaré, *Les Methodes Nouvelles de la Mecanique Celeste*, (Gauthier Villard, Paris), reprinted by Blanchard, 1987.
2. A. J. Lichtenberg and M. A. Lieberman, *Regular and Chaotic Dynamics* (Springer, New York, 1983); A. H. Nayfeh and B. Balachandran, *Applied Nonlinear Dynamics* (Wiley, New York, 1995),
3. C. Simó (ed.), *Hamiltonian Systems with Three or More Degrees of Freedom*, NATO ASI Series C, vol. 533 (Kluwer, Dordrecht, 1999).
4. V. I. Arnold, Sov. Math. Dokl. **5**, 581 (1964).
5. H. Goldstein, *Classical Mechanics* (Addison–Wesley, Reading, 1980).
6. F. Borondo and R. M. Benito, in *Non–linear Dynamics and Fundamental Interactions*, edited by F. Khanna and D. Matrasulov, NATO Science Series, vol. 213 (Springer, Dordrecht, 2006).
7. J. C. Losada, J. M. Estebaranz, R. M. Benito, and F. Borondo, J. Chem. Phys. **108**, 63 (1998).

Fundamental Differences Between Application of Basic Principles of Quantum Mechanics on Atomic and Higher Levels

Alexey Nikulov

Institute of Microelectronics Technology and High Purity Materials, Russian Academy of Sciences, 142432 Chernogolovka, Moscow District, RUSSIA

Abstract. Recent experimental result has revealed that the Bohr's quantization violates symmetry between opposite directions on the macroscopic level. This breach of symmetry is connected with challenge to the law of momentum conservation and the second law of thermodynamics.

Keywords: Foundations of quantum mechanics. Bohr's quantization, breach of symmetry, law of momentum conservation, second law of thermodynamics.
PACS: 03.65.Ta, 03.65.Ud, 05.30.Ch

INTRODUCTION

One of the fundamental problem of today is a validity of a direct extrapolation of quantum principles developed for atomic level to the macroscopic level. The most obvious doubt is connected with the contradiction between quantum mechanics and macroscopic realism [1,2]. An assumption on quantum superposition of macroscopically distinct states raises the question: "Is the flux there when nobody looks?" [1] or even "Is the moon there when nobody looks?" [3]. No direct experimental test excluding the hypothesis of macroscopic realism and giving direct evidence of a macroscopic superposition was obtained for the present. In contrast to this there is no doubt about experimental evidence of the Bohr's quantization on the macroscopic level. But an experimental result obtained recently reveals that even this principle can not be extrapolated directly on the macroscopic level without contradiction with some fundamental principles of physics.

BREACH OF SYMMETRY BETWEEN OPPOSITE DIRECTIONS

There was a logical difficulty in the Bohr's model of atom orbits until electron considered as a particle having a velocity since it was impossible to answer on the question: "What direction has this velocity?" The uncertainty relation and the wave quantum mechanics have overcame this difficulty, removing the velocity direction. Albert Einstein considered it as weakness of the quantum theory: *The weakness of the*

CP905, *Frontiers of Fundamental Physics (FFP8), Eighth International Symposium*
edited by B. G. Sidharth, A. Alfonso-Faus, and M. J. Fullana
© 2007 American Institute of Physics 978-0-7354-0412-0/07/$23.00

theory lies ... in the fact, that it leaves time and direction of the elementary process to "chance" (the citation from [4]). But thanks of this weakness the Bohr's quantization does not violate symmetry between opposite direction on the atomic level. In contrast to this there is an unambiguous experimental evidence [5] of breach of symmetry on the macroscopic level.

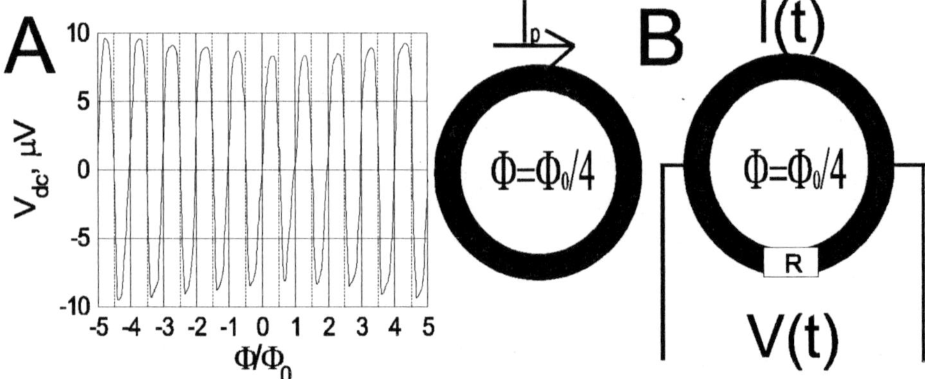

FIGURE 1. A) The quantum oscillation of the dc voltage $V_{dc}(\Phi/\Phi_0)$ observed on the semi-ring of asymmetric aluminum ring at $T = 0.97T_c$ in [5]. B) Superconducting states in a ring with closed (at left) and unclosed (at right) wave function.

The periodical change in magnetic field B of sign and value of the dc potential difference $V_{dc}(B)$, Fig.1, observed on semi-rings of an asymmetric aluminum ring below its superconducting transition $T < T_c$ [5] is unambiguous evidence of the periodical change of the direction of the dc electric field $E = -\nabla V_{dc}$ with the B value. We must ask: "Why so can the direction depend on the value?" Any possible answer should be connected with the Bohr's quantization

$$\oint dl p = \oint dl (mv + 2eA) = m\oint dl v + 2e\Phi = n2\pi\hbar \tag{1}$$

since the period of oscillations, Fig.1, corresponds to the flux quantum $\Phi_0 = \pi\hbar/e$ inside the ring with square S. The quantum oscillations $V_{dc}(BS/\Phi_0) = V_{dc}(\Phi/\Phi_0)$ are evidence of an intrinsic breach of symmetry between right and left directions [6].

CHALLENGE TO THE LAW OF MOMENTUM CONSERVATION

The dc voltage $V_{dc}(\Phi/\Phi_0)$ can be induced by switching between quantum states with different connectivity of the wave function which is not possible in atom but is observed in superconducting ring [7]. The persistent current $I_p \propto (n - \Phi/\Phi_0)$ (i.e. equilibrium direct current) is observed in the closed superconducting state [8], Fig.1, because of the Bohr's quantization (1). The potential difference $V(t) = RI(t) = RI_p \exp(-t/\tau_{RL})$ damping during a relaxation time $\tau_{RL} = L/R$ should be observed on semi-

118

rings after a transition at t = 0 of a ring segment in the normal state with a resistance R, Fig.1. The dc component of the voltage (equal V_{dc} = <V(t)> = <RI(t)> = <RI$_p$ exp(-t/τ_{RL})> $\approx$ L<I$_p$>ω_{sw} at switching with a frequency ω_{sw} << 1/τ_{RL} of the ring with a inductance L) can be observed since the pair velocity takes with overwhelming probability minimum permitted value v = (n2$\pi\hbar$ - 2eΦ)/ml = (2$\pi\hbar$/ml)(n - Φ/Φ_0) [9] and therefore the average value of the persistent current <I$_p$> = s2en$_s$<v> $\propto$ <n> - Φ/Φ_0 $\approx$ n - Φ/Φ_0 $\neq$ 0 at Φ $\neq$ nΦ_0. The induced dc voltage V_{dc} $\propto$ <I$_p$> $\propto$ <n> - Φ/Φ_0 is the periodical function V_{dc}(Φ/Φ_0) in complete agreement with the experiment, Fig.1.

The dc voltage V_{dc}(Φ/Φ_0) should be observed both on the lower semi-ring with switched segment, Fig.1, and on the upper semi-ring remaining all time in superconducting state. The latter is challenge to the law of momentum conservation, connected with the breach of the right-left symmetry, since the force F_e = 2eE of the dc electric field E = -∇V_{dc} acting on pairs 2e is not compensated by any force.

CHALLENGE TO THE SECOND LAW OF THERMODYNAMICS

The switching between superconducting states with different connectivity can be induced by both non-equilibrium and equilibrium noise [7,9]. Therefore the dc power V_{dc}^2/R observed in the V_{dc}(Φ/Φ_0) quantum phenomenon [5] challenges the second law of thermodynamics [6,10].

ACKNOWLEDGMENTS

This work has been supported by a grant "Quantum bit on base of micro- and nano-structures with metal conductivity" of the Program of ITCS department of RAS, a grant 04-02-17068 of the RFBR and a grant of the Program "Quantum Nanostructures" of the Presidium of Russian Academy of Sciences.

REFERENCES

1. A. J. Leggett and A. Garg, *Phys. Rev. Lett.* **54**, 857-860 (1985).
2. A.J.Leggett, *J.Phys.: Condens. Matter* **14**, R415-R451 (2002); *J. of Superconductivity* **12**, 683-687 (1999).
3. N.D. Mermin, *Physics Today*, **38**, 38-47 (1985).
4. A. Zeilinger, *Foundations of Physics*, **29**, 631-643 (1999).
5. S.V. Dubonos et al., *Pisma Zh.Eksp.Teor.Fiz.* **77**, 439-444 (2003) (*JETP Lett.* **77**, 371-375 (2003)); e-print arXiv: cond-mat/0303538; e-print arXiv: cond-mat/0305337.
6. A.V. Nikulov, "Brownian Motion and Intrinsic Breach of Symmetry", *the talk at the 13th General Meeting of the European Physical Society "Beyond Einstein - Physics for the 21th Century", EPS13*, Bern, Switzerland, 11-15 July 2005; e-print arXiv: cond-mat/0404717.
7. A.V. Nikulov, I.N. Zhilyaev, *J.Low Temp.Phys.* **112**, 227-236 (1998); cond-mat/9811148.
8. M.Tinkham, *Introduction to Superconductivity*, McGraw-Hill Book Company, 1975.
9. A.V. Nikulov, *Phys. Rev. B* **64**, 012505 (2001).
10. A.V. Nikulov, "About Perpetuum Mobile without Emotions" in *First International Conference on Quantum Limits to the Second Law*, edited by D. P. Sheehan, AIP Conference Proceedings 643, American Institute of Physics, Melville, NY, 2002, pp. 207-212.

Are Hadrons Shell-Structured?

Paolo Palazzi

particlez.org, PO Box 62, CH-1217 Meyrin 1, Switzerland
pp@particlez.org

Abstract. A stability analysis of the mass spectrum indicates that hadrons, like atoms and nuclei, are shell-structured. The mesonic shells mass series, combined with the results of a mass quantization analysis, reveals striking similarities with the nuclear shells. In addition, the mesonic mass patterns suggest solid-phase partonic bound states on an fcc lattice, compatible with a model by A. O. Barut with stable leptons as constituents, bound by magnetism. Baryonic shells grow with a lower population density, and only start at shell 3 with the nucleon.

Keywords: hadron, shell model; mass, quantization; statistical analysis.
PACS: 12.40.Yx, 12.90.+b, 14.40.-n.

SHELLS

Quarks and their mixing are the cornerstone of particle physics, defining an accurate CKM chemistry and a plausible but incomplete taxonomy. The organization of the mass spectrum, however, is far from satisfactory: no comprehensive mass rules are known, and attempts to derive the masses from the quarks and their binding have not succeeded in more than two decades of attempts. Paradoxically, the theory explains a number of experimental results, but cannot compute the most fundamental quantity, energy. The quark model is a set of combinatorial rules for the quantum numbers, but does not offer any structural insight the way atomic or nuclear models do.

In atoms and in nuclei, stability is organized with shells, and to probe the relevance of shells for hadrons, stability is expressed with alignments of the 1/3 power of the total number of constituents of the most stable configurations, the closed shells [1]. To represent the atomic shells, in Fig. 1a the cube root of Z for the inert gases is charted vs. the shell number. Nuclear stability is related mostly to the neutronic magic number series, but with a derived sequence based on the atomic number $A = Z + N$, stability and the spin-orbit coupling effect emerge in a similar way, as shown in Fig. 1b.

Can such patterns be detected in the particle spectrum? Not directly, for lack of an established parton count, but the mass can be used as a proxy for the number of constituents (which would work for nuclei). By analyzing the distribution of particle lifetimes as a function of the mass, stability peaks are recognized for mesons and for baryons, as shown in Fig. 2a and 2b, and indeed the cube roots of their masses follow two distinct alignments, as shown in Fig. 2c [1]. This would be a strong indication that hadrons are shell-structured if only each constituent contributed a constant amount to the total mass. This assumption is unusual, but not new.

CP905, *Frontiers of Fundamental Physics (FFP8), Eighth International Symposium*
edited by B. G. Sidharth, A. Alfonso-Faus, and M. J. Fullana
© 2007 American Institute of Physics 978-0-7354-0412-0/07/$23.00

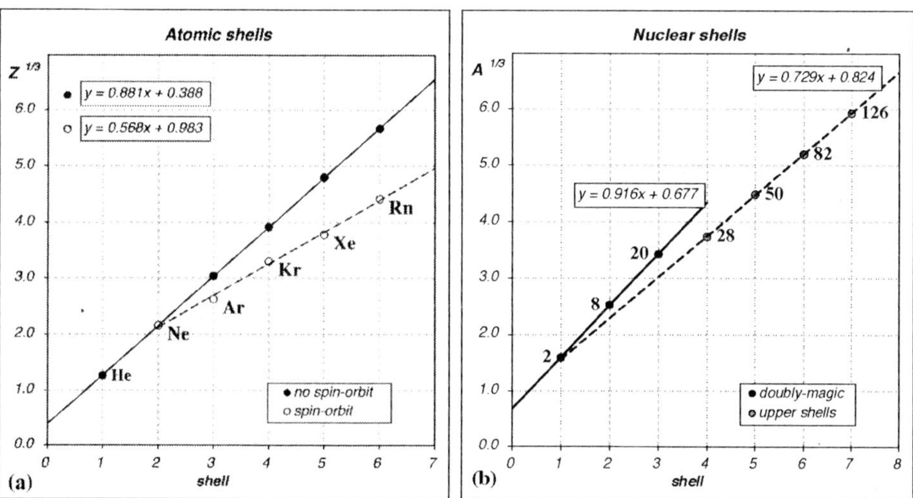

FIGURE 1. a: atomic shells, $Z^{1/3}$ vs shell number for the inert gases; the bound state equation identifies the closed shell configurations and determines the parameters of the no spin-orbit line; spin-orbit coupling is effective after shell 2, and reduces the slope. **b**: nuclear shells, $A^{1/3}$ vs shell number for the magic neutronic sequence; $A_i = Z_i + N_i$, where $N_i = 2, 8, 20, 28, \ldots$ and Z_i is looked up in the Segrè plot along N_i to correspond to maximum stability. After the first three shells, with the doubly-magic nuclei He 4, O-16 and Ca-40, spin-orbit coupling defines a reduced rate of growth (reprinted from [1]).

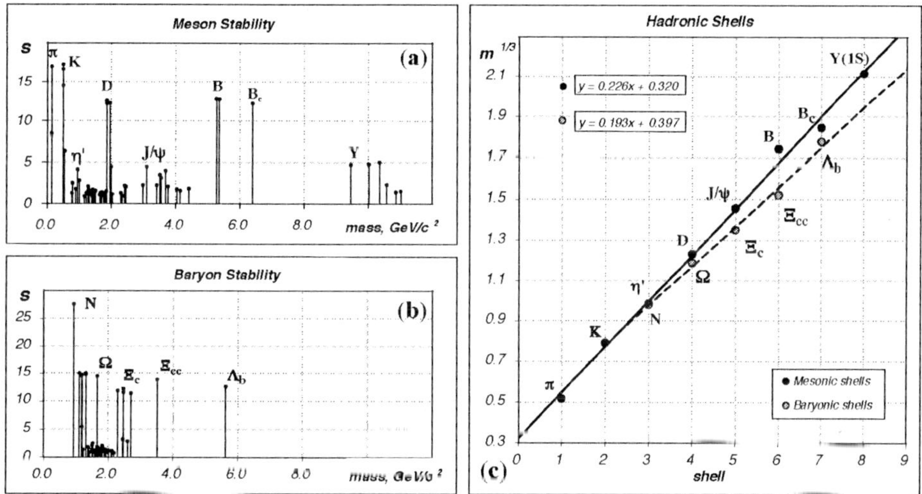

FIGURE 2. a: meson, and **b**: baryon stability plots showing $s(i) = \lg(\tau(i)/\tau(Z^0))$ versus the mass (with i running on the whole spectrum), and identification of mesonic shells 1 to 8 and baryonic shells 3 to 7. **c**: hadronic shells, expressed as $m^{1/3}$ vs. shell number (adapted from [1]). Compared to the mesons, the baryonic shells start only at shell 3, and grow with a lower slope. Deviations from the fitted line are occasionally more important than in the atomic or nuclear shells, and a rationale for this effect is outlined in Ref. [6].

MASS QUANTIZATION AND HADRONIC SHELLS

Y. Nambu observed in 1952 that meson masses are even multiples of a mass unit u of about 35 MeV/c^2, baryons (and also unstable leptons) odd multiples, so that mass differences among similar particles are quantized by 70 MeV/c^2 [2]. M. H. Mac Gregor studied this property extensively from the '70s to the present time [3] and few other authors mentioned it also. Recently this rule has been reassessed by the present author for all the mesons listed by the PDG, grouped by quark composition and J^{PC}, confirming its statistical significance [4]. In addition, the slightly different values of u for the various meson groups, all in the vicinity of 35 MeV/c^2, are also quantized on a grid of 12 intervals, and are strongly correlated with the quantum numbers. The baryon mass spectrum shows very similar regularities [5].

The hadronic shells identified by the cube root of mass plot of Fig 2c can be scaled, one constituent for each mass unit, and re-expressed with $P^{1/3}$, where P is the number of partons $P = m/u$ (plots not shown). The mesonic shells thus transformed reveal striking similarities with the nuclear shells: the shell population sequences are almost identical, doubly-magic-equivalent states are present only up to shell 3, and there are also clear indications of sub-shells [6]. The mesonic shell sequence is correlated with the quark composition up to b-bbar in shell 8, and no states are present around the mass values of the hypothetical subsequent shells. In addition, the mesonic mass quantization patterns [4] suggest solid-phase partonic bound states on an fcc lattice, with light spin-1/2 partons of charge 0, +1 and -1 with anti-parallel spin coupling and positive binding energy [6]. This interpretation strengthens the mesons-nuclei shell analogy through the fcc nuclear model of N. D. Cook [7].

A possible choice for the constituents, compatible with these results (and the only one with non-fictitious particles), are the stable leptons, as proposed by A. O. Barut in a different context, with magnetism acting as the strong interaction among partons [8].

Baryonic shells grow with a lower density compared to the mesons, and start only at shell 3 with the nucleon. The radial charge density distributions of the proton and the neutron [9], as well as the shape of the p-p elastic differential cross-section [10] also indicate that the nucleon is structured in three layers.

REFERENCES

1. P. Palazzi, "Particles and Shells", *http://cdsweb.cern.ch*, CERN-OPEN-2003-006 (2003).
2. Y. Nambu, "Empirical Mass Spectrum of Elementary Particles", *Prog. Theor. Phys.* **7**, 595 (1952).
3. M. H. Mac Gregor, "Electron generation of leptons and hadrons with reciprocal alpha-quantized lifetimes and masses", *Int. J. Mod. Phys.* **A20**, 719 and 2893 (2005).
4. P. Palazzi, "Patterns in the Meson Mass Spectrum", *http://particlez.org*, p3a-2004-001 (2004).
5. P. Palazzi, "The Baryon Mass System", in preparation.
6. P. Palazzi, "Meson Shells", *http://particlez.org*, p3a-2005-001 (2005)
7. N. D. Cook, *Models of the Atomic Nucleus*, Springer, 2006.
8. A. O. Barut, "Stable Particles as Building Blocks of Matter", *Surveys High Energ. Phys.* **1**, 113
9. R. M. Littauer, H. F. Schopper, R. R. Wilson, "Structure of the Proton and Neutron" *Phys. Rev. Letters* **7**, 144 (1961).
10. M. M. Islam, R. J. Luddy and A. V. Produkin, *Mod. Phys. Lett.*, **A18,** 743 (2003)

Vortices and chaos in the quantum fluid

D. A. Wisniacki*, E. R. Pujals† and F. Borondo**

*Departamento de Física "J.J. Giambiagi", FCEN, UBA, Pabellón 1, Ciudad Universitaria, 1428 Buenos Aires, Argentina.
†IMPA–OS, Dona Castorina 110, 22460–320, Rio de Janeiro, Brasil.
**Dep. de Química C–IX, Universidad Autónoma de Madrid, Cantoblanco, 28049–Madrid (Spain).

Abstract. The motion of a single vortex originates chaos in the quantum fluid defined in Bohm's interpretation of quantum mechanics. Here we analize this situation in a very simple case: one single vortex in a rectangular billiard.

Keywords: Quantum mechanics; Foundations of quantum mechanics; Measurement theory
PACS: 03.65.w-, 03.65.Ta

In the 1950's the physicist David Bohm presented a controversial formulation of quantum mechanics(dBB) [1] with the purpose of solving some of the fundamental interpretational difficulties existing in the standard version. This theory is based on quantum trajectories (QT) "piloted" by the de Broglie's wave, and constitutes a true theory of quantum motion [2]. On the practical side, dBB has made possible very nice interpretations of physical phenomena such as diffraction [3], or tunneling [4], and also systematic quantum dynamical calculations in systems with many degrees of freedom [5]. Recently, it has also been used to attack other fundamental problems in physics (than those for which it was originally developed). Namely, Valentini and Westman [6] and Dürr et al. [7] made some start to remove the status of postulate to Born's probability rule: $\rho = |\psi|^2$. In these works, probabilities have a dynamical origin, similar to thermal probabilities in ordinary statistical mechanics, and it was shown [6] that the standard distribution is obtained as the time evolution towards the equilibrium of initial non–equilibrium states, $\rho \neq |\psi|^2$, this taking place with a (exponential) decrease in the associated coarse–grained H–function.

Underlying the above ideas is the assumption that it exists an effective chaotic dynamics in the QTs, something that has been frequently taken for granted [8], and/or used in a poorly justified way to make claims about the important topic of "quantum chaos". The main difficulty with this kind of approaches is the fact that trajectories in the de Broglie–Bohm (dBB) theory are non–local, and the flux is more similar to a hydrodynamical flow. Taking this into account, the authors have recently shown how the vortices of the pilot wave function are at heart of complexity of QT in dBB theory, by proving [9] that the motion of a single vortex is enough to create chaos in QT (using an isotropic 2D harmonic oscillator for a numerical illustration), being this a robust effect (numerically illustrated in a rectangular billiard) [10].

In this paper, we will consider the problem of a single vortex in a 2D square billiard, and give indications on how the study can be extended progressively.

In the dBB theory the state of the system is described by a wave function, $\psi(\mathbf{r},t) =$

CP905, *Frontiers of Fundamental Physics (FFP8), Eighth International Symposium*
edited by B. G. Sidharth, A. Alfonso-Faus, and M. J. Fullana
© 2007 American Institute of Physics 978-0-7354-0412-0/07/$23.00

$R(\mathbf{r},t)\, e^{iS(\mathbf{r},t)}$ ($\hbar = 1$ throughout the paper), and the position of the particles, $\mathbf{r}$ [7]. The dynamical evolution of this two quantities (for spinless particles) is given by the time–dependent Schrödinger equation and the guidance equation, respectively:

$$\mathbf{v} = \dot{\mathbf{r}} = \frac{\nabla S}{m} = \frac{i}{2m}\frac{\psi\nabla\psi^* - \psi^*\nabla\psi}{|\psi|^2}, \tag{1}$$

where m is the mass of the particle. QTs can be obtained by numerical integration of this equation. The velocity field (1), guiding these trajectories, presents singularities (at points where the wave functions vanishes) giving rise to vortices in quantum fluid. As a model we use a particle of mass $m = 1/2$ enclosed in a 2D squared billiard with unit side length. The corresponding classical dynamics is totally regular, so that any indication of the existence of chaos is solely due to quantum effects. The corresponding eigenfunctions are: $\phi_{n_x,n_y}(x,y) = 2\sin(n_x\pi x)\sin(n_y\pi y)$, with wave numbers: $k = E_{n_x,n_y}^{1/2} = \pi(n_x^2 + n_y^2)^{1/2}$. The initial pilot wave is constructed as a linear combination of the first three eigenstates in the following way:

$$\psi(x,y) = a\phi_{1,1} + be^{i\alpha_1}\phi_{1,2} + ce^{i\alpha_2}\phi_{2,1}, \qquad a,b,c,\alpha_1,\alpha_2 \in \mathbb{R}, \tag{2}$$

with $a^2 + b^2 + c^2 = 1$ (normalization condition). Phases are chosen as: $\alpha_1 = 0, \alpha_2 = \pi/2$.

As can be easily demonstrated, this state generates a time–dependent velocity field with only one vortex, moving according to equation:

$$(x_v(t),y_v(t)) = \frac{1}{\pi}\left(\arccos\frac{a\sin(\alpha_2 - \omega)t}{2b\sin(\alpha_1 - \alpha_2)}, \arccos\frac{-a\sin(\alpha_2 - \omega)t}{2c\sin(\alpha_1 - \alpha_2)}\right), \tag{3}$$

with $\omega = E_{2,1} - E_{1,1}$. The corresponding vortex trajectories for different values of the parameters are shown in Fig. 1(a). As can be seen, they are approximately circular for the combination of lower values of a and larger $b = c$, and distort to a more squared shape as the opposite happens. Due to the periodic character of the vortex motion, QTs (moving under the influence of the corresponding velocity field) can be best monitored in phase space by computing a "stroboscopic" surface of section (SOS) at times multiple of the fundamental frequency: $t = 2/(3\pi)$. Let us recall that SOS, first described by Poincaré, are a fundamental tool in nonlinear science [11], so that much understanding of a dynamical system like our can be expected from their consideration. The corresponding results, for the same set of parameters used before, are shown in Fig. 1(b). In it, a transition to chaos takes place from regularity at $a/b = 0$ to chaos when $a/b > 0$, as a result of a pitchfork bifurcation. More interesting, the SOS shows all characteristics exhibited by ordinary generic autonomous systems. Namely, the portion or chaotic motion, or destroyed tori grows with the perturbation parameter a/b, giving the velocity of the vortex, and characteristics chains of islands are apparent for $a/b > 0$. This is not surprising, since according to Ref. [9] the dynamics of the system in the vortex vicinity corresponds to that of a non–autonomous system of 1.5D.

The next natural step in our research is to consider situations in which the number of vortices is progressively incremented. For these cases, interaction among vortices will exist, with the possibility of pair destruction maintaining the overall vorticity. The

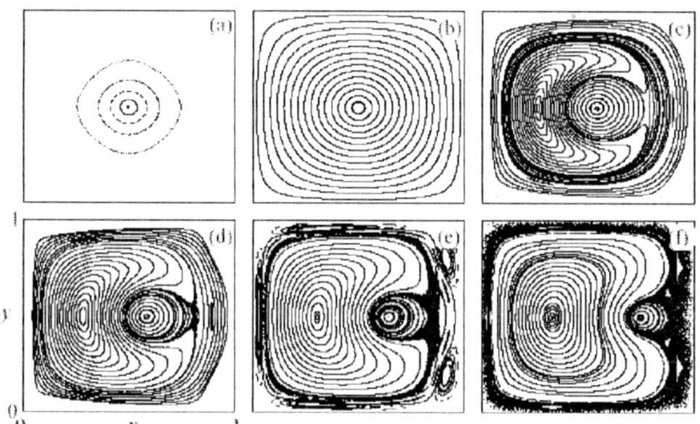

FIGURE 1. (a) Trajectories for the vortex generated by wave function (2) for $a = 0$, $b = c = 0.707106781$ (central point), $a = 0.175$, $b = c = 0.696195016$ (full line), $a = 0.35$, $b = c = 0.662382065$ (dashed), $a = 0.525$, $b = c = 0.601820156$ (dot–dashed), and $a = 0.7$, $b = c = 0.504975247$ (dotted). (b)–(f) Composite stroboscopic Poincaré surfaces of section for QT's with the same conditions as in (a).

limit of many vortices originated by pilot wave functions of increasing complexity has been already numerically explored from a statistical point of view [10]. Now, it seems interesting to examine in detail the case of two vortices, which can be done with a generalization of the same technique used here, and this will be the subject of a future publication. The main idea is to transform the original non–autonomous flux into a dynamical system of $n + 1/2$ dimensions, being n the number of fundamental frequencies existing in the pilot wave function, and examine the corresponding phase space with surrogates of the associated Poincaré surface of section.

ACKNOWLEDGMENTS

This work was supported by CONICET and UBACYT (X248) (Argentina), MEC–Spain (contracts MTM2006–15533 and CSD2006–00032), AECI–Spain (A/4574/05).

REFERENCES

1. D. Bohm, *Phys. Rev.* **85**, 166 (1952); *ibid.* **85**, 194 (1952).
2. P. R. Holland, *The Quantum Theory of Motion*, Cambridge U. Press, Cambridge, 1993.
3. C. Philippidis, C. Dewdney, and B. J. Hiley, *Nuovo Cimento* **52B**, 15 (1979).
4. C. L. Lopreore and R. E. Wyatt, *Phys. Rev. Lett.*, **82**, 5190 (1999).
5. R. E. Wyatt, *Quantum Dynamics with Trajectories: Introduction to Quantum Hydrodynamics*, Springer, New York, 2005.
6. A. Valentini and H. Westman, *Proc. R. Soc. London, Ser. A* **461**, 253 (2005).
7. D. Dürr, S. Goldstein, and N. Zanghi, *J. Stat. Phys.* **67**, 843 (1992).
8. H. Frisk, *Phys. Lett. A* **227**, 139 (1997).
9. D. A. Wisniacki and E. R. Pujals, *Europhys. Lett.* **71**, 159 (2005).
10. D. A. Wisniacki, E. R. Pujals and F. Borondo, *Europhys. Lett.* **73**, 671 (2006).
11. M. A. Lieberman and A. J. Lichtenberg, *Regular and Chaotic Dynamics*, Springer, New York, 1992.

III. FOUNDATIONS OF PHYSICS AND SCIENCE IN GENERAL

Harmonizing General Relativity with Quantum Mechanics

Antonio Alfonso-Faus[1]

E.U.I.T. Aeronáutica, Plaza Cardenal Cisneros s/n, 28040 Madrid, Spain

Abstract. Gravitation is the common underlying texture between General Relativity and Quantum Mechanics. We take gravitation as the link that can make possible the marriage between these two sciences. We use here the duality of Nature for gravitation: A continuous warped space, wave-like, and a discrete quantum gas, particle-like, both coexistent and producing an equilibrium state in the Universe. The result is a static, non expanding, spherical, unlimited and finite Universe, with no cosmological constant and no dark energy. Macht's Principle is reproduced here by the convergence of the two cosmological equations of Einstein. From this a Mass Boom concept is born given by M = t, M the mass of the Universe and t its age. Also a decreasing speed of light is the consequence of the Mass Boom, c = 1/t, which explains the Supernovae Type Ia observations without the need of expansion (nor, of course, accelerated expansion). Our Mass Boom model completely wipes out the problems and paradoxes built in the Big Bang model, like the horizon, monopole, entropy, flatness, fine tuning, etc. It also eliminates the need for inflation.

Keywords: gravitation, relativity, gravity quanta, cosmology, Supernovae, big bang, Mach's Principle

PACS: 03.65. -W, 04.60. -M, 04.62. +V, 98.80. -K

INTRODUCTION

Einstein's effort to extend the validity of classical mechanics to velocities close to the speed of light was very successful. He created Special Relativity and doing this he liberated the limits imposed on Newtonian Mechanics: it was not applicable to the case of bodies moving at a speed close to the speed of light. Later on he continued his effort to include Gravitation in his ideas on relativity and created General Relativity, a very successful effort too. By then mechanics was on its foot, valid for small as well as high velocities, as compared with the speed of light. It was also valid for normal sizes, the see and touch world around us, as well as very large sizes, as compared with the size of the see able Universe. But a dilemma was still puzzling the scientific community. Nature appeared to answer questions about its essence in accordance with the type of experiment set up for these questions: Wave nature? Particle nature? Both answers were affirmative if the questions were put in the proper way. Then a dual nature was clear for everybody: waves and particles were at the base of all Nature, both at the same time. The theory that connected these two aspects of nature was Quantum Mechanics, a wave theory and a particle theory (because Heisenberg uncertainty principle is built in this wave theory) that extended the validity of mechanics to the small world of atoms

---

[1] E-mail. aalfonsofaus@yahoo.es

CP905, *Frontiers of Fundamental Physics (FFP8), Eighth International Symposium*
edited by B. G. Sidharth, A. Alfonso-Faus, and M. J. Fullana
© 2007 American Institute of Physics 978-0-7354-0412-0/07/$23.00

and fundamental particles, the Quantum world. Again, a big puzzle was still plaguing science: General Relativity, a theory of continuous variables, had produced the two cosmological equations to deal with the Universe as a whole. And Quantum Mechanics, the wave and particle theory of Nature, appeared to explain the world satisfactorily. But, what about Gravitation? Only a theory of the wave-type, General Relativity, dared to treat it. No quantum treatment exists today that has a well developed theory to deal with the problem of quantum gravity. One way to introduce a new approach to this problem is to look at the cosmological equations and try to include the wave and particle nature into them. General Relativity is clear; it is a geometrical view of gravitation where curvature plays a central role. A curved space mechanically implies a centripetal force, which is just the force of gravitation that gives an inward "push". And a gas of gravity quanta implies an outward force, the gas pressure that in fact balances the inward push. Hence we have here the picture of gravitation with a dual nature: Wave-like inward force because of the curvature of space, and particle-like, outward force due to the gas pressure of the gravity quanta.

EINSTEIN COSMOLOGICAL EQUATIONS

The Mass-Boom concept presented elsewhere [1] implies a decreasing speed of light c, inversely proportional to cosmological time t. Then the product ct is a constant, and this can be taken as the size of the Universe, constant, with no expansion [2]. The cosmological scale factor R being constant drastically reduces the number of terms in the cosmological equations ($R' = R'' = 0$). Also the Hubble "constant" H is zero in this case, the same as the Lambda cosmological constant that equals (or is proportional to) H. Then we are left with only two terms in each of the Einstein cosmological equations, the pressure (density) term and the curvature term:

$$8\pi G\frac{p}{c^2} + \frac{Kc^2}{R^2} = 0 \tag{1}$$

$$-\frac{8\pi}{3}G\rho + \frac{Kc^2}{R^2} = 0 \tag{2}$$

Both equations are reduced to just one if, and only if, the gravity quanta gas pressure p is made equal to $-1/3\rho c^2$, where ρ is the density of the Universe:

$$p = -\frac{1}{3}\rho c^2 \tag{3}$$

$$\frac{8\pi}{3}G\rho = \frac{Kc^2}{R^2} \tag{4}$$

It is clear that this implies a model of the Universe that is spherical, $K = 1$, closed, finite and unlimited.

The density of the Universe is equal to the mass M divided by the volume $V = 2p^2R^3$. Then one has

$$\frac{8\pi}{3}G\frac{M}{2\pi^2 R^3} = \frac{c^2}{R^2} \tag{5}$$

And using $R = ct$ as the size of the Universe we arrive at

$$\frac{4}{3\pi}\frac{GM}{c^2} = R = ct \tag{6}$$

We can interpret this equation as a form of Mach's principle if we enunciate it as follows: The relativistic energy mc^2 of any mass m is of the order of its gravitational potential energy with respect to the mass M of the rest of the Universe:

$$\frac{4}{3\pi}\frac{GMm}{R} = mc^2 \tag{7}$$

The conclusion is that the two cosmological equations, which are at the base of the General Relativity approach to any model of the Universe, are just Mach's Principle. This Principle is built in the Einstein's field equations,

THE MASS-BOOM

Presented elsewhere [1] is the idea that the Action Principle may be used to derive Einstein's field equations if the factors in front of the integrals are constant:

$$\frac{G}{c^3} = \text{constant} \tag{8}$$

$$mc = \text{constant} \tag{9}$$

The second equation is just a manifestation of the constancy of momentum. Then from Mach Principle in the form in equation (6) we get:

$$M = t \tag{10}$$

in a certain system of units. This is the Mass-Boom: the equivalence of mass and time. Hence one has

$$Mc = ct = R \tag{11}$$

and one can take the size of the Universe $R = ct$ as unity. The relation

$$c = \frac{1}{t} \tag{12}$$

explains [2] the Supernovae type Ia findings without the need of postulating any accelerating expansion of the Universe, nor the existence of dark energy (in fact the cosmological constant is also zero [3], and we are left with the initial static model of Einstein that we now know is stable). In our model there is no horizon problem nor any of the numerous problems built in the Big Bang theory which motivated the construction of inflation theories.

A NEW COSMOLOGIAL MODEL WITH TIME-VARYING "CONSTANTS"

The Hubble red shift was interpreted as an expansion of the Universe. Here we have a Mass-Boom non-expanding model [4] and therefore we can interpret the red shift as an effect of a shrinking Quantum world [5], atoms and fundamental particles, immersed in a static, constant size Universe. This can be achieved by showing that Planck's "constant" h is proportional to the speed of light c.

If we impose the constancy of nuclear forces we get:

$$\frac{mc^2}{r_p} = \frac{m^2c^2}{h} = \text{constant}\frac{c}{h} = \text{constant} \tag{13}$$

Also by imposing the constancy of electrical forces we have:

$$\frac{e^2}{r_p^2} = \frac{e^2m^2c^2}{h^2} = \text{constant}\frac{\hbar c}{h^2} = \text{constant}\frac{c}{h} = \text{constant} \tag{14}$$

Then the constancy of quantum forces imposes that h must be equal to $c = 1/t$ (in a certain system of units).

For the mass of the quantum of gravity we have [6]

$$m_g = \frac{\hbar}{c^2t} = \frac{\hbar}{c}\frac{1}{ct} = 1 \tag{15}$$

Then, this mass can be taken as the unit of mass in the Universe, of the order of 10^{-65} grams. For the time we have

$$t_1 = \frac{\hbar}{Mc^2} = \frac{1}{Mc} = 1 \tag{16}$$

This unit of time is about 10^{-104} sec. The mass, time and length units in this model are

$$m_1 = m_g = 10^{-65}gr$$
$$t_1 = h/Mc^2 10^{-104}sec \tag{17}$$
$$l_1 = ct = 10^{28}cm$$

A PREDICTION

The fine structure constant α is given by

$$\alpha = \frac{e^2}{\hbar c} = \frac{e^2}{c^2} \tag{18}$$

No significant time variations have been observed for this constant, in the cosmological scale of size and time. Then we have $e = $ constant c and the Zeeman displacement d is therefore

$$d = \frac{e}{mc} = \frac{\text{constant}}{m} = \frac{\text{constant}}{t} \tag{19}$$

Hence, by measuring time variations in this displacement we may validate this theory.

ACKNOWLEDGMENTS

I am grateful to all sponsors of the Eighth Symposium on Frontiers of Fundamental Physics, Madrid 16-19 October 2006, for their support, in particular to the Madrid and València Technical Universities. I am also grateful to Prof. Fullana for his help in preparing the paper for the A.I.P. Proceedings.

REFERENCES

1. A. Alfonso-Faus, V International Symposium "Frontiers of Fundamental Physics" Hyderabad, India 2003, and arXiv:physics/0302058 v1 Ú 18 February 2003.
2. Richard Lieu and Don A. Gregory, arXiv astro-ph/0605611, May 24 2006.
3. Yves-Henri Sanejouand, arXiv:astro-ph/0509582, 20 Sep 2005.
4. Recai Erdem, arXiv,gr-qc/0611111, Nov 21 06, "Asymmetry for vanishing cosmological constant".
5. A.Alfonso-Faus, *Journal of Theoretics*, June 12, 2003, and arXiv:physics/0309108.
6. A.Alfonso-Faus, *Physics Essays*, Vol 12, nž 4, 2000 and arXiv, gr-qc/0006009v2- Aug 11, 2000.

Non invasive Measurements of Myocardial Hypertrophy in Patients with Essential Hypertension Treated with Eprosartan: Contribution of the Physics.

Ricardo Cabrera Solé, MD, PhD.

University General Hospital of Albacete. Clinical Associate Professor of Cardiology. Faculty of Medicine. Castilla La Mancha University. ZP: 02005 Albacete. Spain.

Abstract: Objective: The main objective of this study was to evaluate the effects of the treatment with eprosartan on cardiac hypertrophy in hypertensive patients using the echocardiogram to measure the hypertrophy of left ventricle. We studied 60 untreated patients diagnosed of mild to moderate hypertension which received after the diagnosis 600 mg/day of eprosartan, a novel direct angiotensin inhibitor recently introduced to treat hypertension. All patients were submitted to a standard echocardiographic study before the treatment and after 6 months of it. We evaluated by echocardiogram the following parameters: left ventricular septum and posterior wall thickness, left ventricular mass , E/A index of mitral flow considering abnormal when this index was less than 1, and left ventricular ejection fraction . Results: at the beginning we found a systolic/diastolic pressures of $165\pm9/96\pm4$ mmHg compared with the end of study of $124\pm2/79\pm3$ mmHg ($p<0.05$). Septum and posterior wall thickness were respectively at baseline 13.2 ± 2 and 12.1 ± 1.1 mmHg and at the end 11.5 ± 1.2 and 10.5 ± 1.3 mmHg ($p<0.05$ for both of them). The E/A mitral flow index was less than 1 at baseline in 45 patients compared with 19 patients after treatment ($p<0.05$). Respect to left ventricular mass we found at the beginning 232 ± 7.5 gr. , compared to 194 ± 9 gr., at the end of this study ($p<0.05$). We did not find any significant differences regarding left ventricular ejection fraction between both groups. Conclusions: we can remark that eprosartan is a very useful drug to reduce not only blood pressure but also left ventricular hypertrophy and improve left ventricular diastolic function in patients with essential hypertension according with parameters measured with non invasive methods.

Keywords: echocardiogram, essential hypertension, diastolic function, relaxation of left ventricle, eprosartan.

PACS: 87.19 Hh; 87.53Tf; 87.63Df; 87.61Pk; 81.70 Cv; 82.56 Na; 07.35K;87.16Xa.

INTRODUCTION

During centuries, medical doctors, have been looking for a method to measure cardiac function of a non-invasive way. However, only in the last decades it was possible, thanks to the important development of the physics and miniaturization of the different devices for these purposes.

In certain diseases, such as hypertension or hypertrophy cardiomyopathy, the increasing of the wall thickness (so called hypertrophy) is the gross anatomical marker and likely a determinant of many of the clinical features in most patients with these diseases (1). The hypertrophy of left ventricular mass can be due to asymmetric or symmetric (concentric) pattern which can affect to all segments of the left ventricular wall or only one segment. This is particularly important, because the symptoms of patients depend on this factor.

CP905, *Frontiers of Fundamental Physics (FFP8), Eighth International Symposium*
edited by B. G. Sidharth, A. Alfonso-Faus, and M. J. Fullana
© 2007 American Institute of Physics 978-0-7354-0412-0/07/$23.00

Also, in these patients we can find apart from the increasing ventricular wall thickness, enlargement of the atria, thickening of the mitral valve leaflets and areas of fibrosis in the ventricular wall (2), such as some authors found at necropsy of those patients affected by these diseases. However, during long time, it was impossible to study all of theses aspects at a non invasive way. Fortunately, today we have possibility to measure and to see all of these aspects of different diseases thanks to an echocardiographic analyses which permits not only to see, also to measure meanwhile the heart is functioning, those measurements that are so useful in clinical practice in present medicine. So we will show in this paper the contribution of the physics in the development of all of these progresses

From the historical perspectives, echo sounding, was first applied by human beings in the 1920s for depth recording in oceanographic studies and submarine detection. However, the use as a diagnostic tool in medicine is a relatively recent development. In 1950, Keidel (3) was the first investigator to use ultrasound to examine the heart. But it was not until the mid 1950s, when echocardiography was popularized in the United States by Holmes and co-workers (4). Initially, this technique was used for the assessment of mitral stenosis, and cardiac chamber size. A major development in the field of cardiac ultrasound was the introduction of two-dimensional echocardiography. This technique took a static image of cardiac anatomy and provided images of cardiac structure in real time (5) The next significant milestone in the evolution of cardiac ultrasound was the addition of the pulsed Doppler device, permitting to add functional information to data provided by two-dimensional echocardiography . After that, appeared the transesophageal echocardiography and a recent evolutionary development in echocardiography, the digital ultrasound, which allowed a quantitative and semi-quantitative analysis of heart function as well as tissue characterization (6).

MEASUREMENT OF CARDIAC CHAMBERS. BASIC PRINCIPLES

The ability to perform and to interpret echocardiograms and to appreciate fully the capabilities and limitations of this technique depends on an understanding of the basic principles of ultrasound. Clinical echocardiography requires ultrahigh-frequency sound (2 to 7.5 MHz) so that it can be transmitted as a narrow beam and directed along a rather well defined path through the soft tissues of the body. As the sound strikes a cardiac structure, a portion is reflected back to the receiver. The processed data from within the body can be displayed in three modes in order to present the clinical information in the most meaningful way. Recent advances in digitizing and image manipulation have led to the integration of minicomputer systems within the ultrasound instruments, such that image manipulations, storage, and transfer will be readily available within the future. It is estimated that more than 5 million echocardiograms are performed every year in the United States and that more than 2% of the general population undergoes an echocardiographic examination over the course of 1 year. Furthermore, a substantial increase in the utilization of this powerful investigative tool has been noted in recent years. Quantitative echocardiographic interpretation provides invaluable information for the practicing clinician. Nonetheless, quantitative echocardiography also plays a significant role in diagnosis, treatment and prognostication. So we consider today one of the most important tool for the different areas of the medicine for studying patients.

With the aim to determine the efficacy of echocardiogram to follow results of a treatment in hypertensive patients, we designed a study to evaluate changes in left ventricular wall thickness before and after treatment.

MATERIAL AND METHODS

Objective: the main objective of this study was to evaluate the effects of the treatment on cardiac hypertrophy in hypertensive patients using the echocardiogram to measure the hypertrophy of left ventricle .
Patients:
We studied 60 untreated patients diagnosed of mild to moderate hypertension which received after the diagnosis 600 mg/day of eprosartan, a novel direct angiotensin inhibitor recently introduced to treat hypertension. We measured patients blood pressure at the beginning, every two months and at the end of this study. All patients were submitted to a standard echocardiographic study before the treatment and after 6 month of it. We evaluated by echocardiogram the following parameters: left ventricular septum and posterior wall thickness, left ventricular mass , E/A index of mitral flow considering abnormal when this index was less than 1, and left ventricular ejection fraction .
Statistical analysis:
 Results of baseline and after 6 months data were compared (see Table I and figures 1,2 and 3) considering the average of the data and the standard deviation using Student-t to compare parametric data. We considered statistically significant results of p less that 0.05.

RESULTS

TABLE I: Parameters of patients with hypertension:

Data	Systolic pressure (mmHg)	Diastolic pressure (mmHg)	Septum thickness (mm)	Posterior Wall thickness (mm)	E/A mitral flow index <1	Left ventricular mass (grs)	Left ventricle Ejection fraction
Basal	165±9	96±4	13.2 ±2	12.1±1.1	45 patients	232±7.5	58±4
After 6 month of treatment	124±2*	79±3*	11.5±1.2*	10.5±1.3*	19 patients*	194±9 *	56.5±5+

(*)= p less than 0.05 respect to basal.   (+) = non significant.

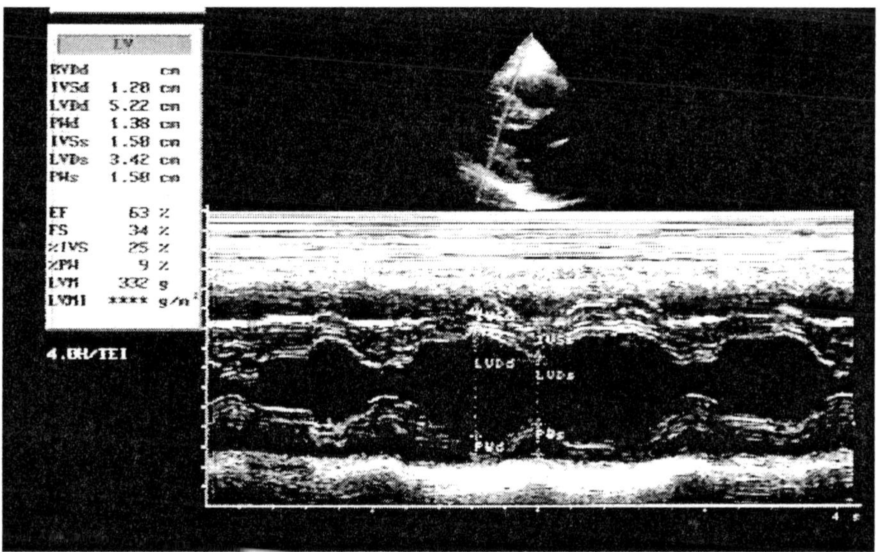

Figure 1: Image showing the different parts of the left ventricle measured by M mode echocardiogram. In the top, a two-dimensional study.

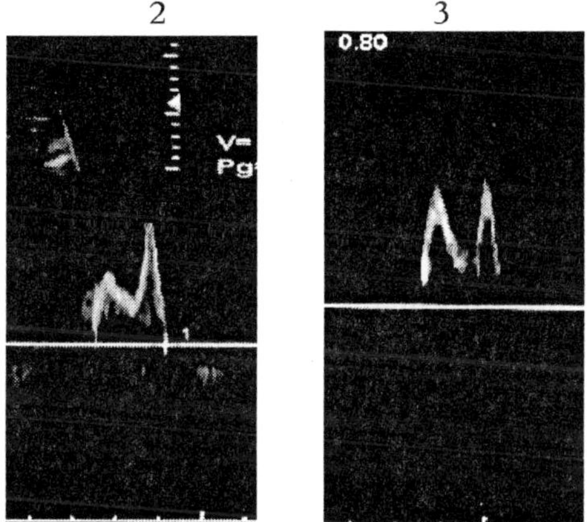

Figures 2 and 3: Observe that in figure 2 with echo Doppler study appears the second wave bigger than the first one (E/A index less than 1). However, 6 months after treatment, in figure 3 both of them appear normalized in the same patient, indicating recuperation of left ventricular diastolic function which was affected in its relaxation before treatment.

DISCUSION

We show in this study, that it is possible to reduce cardiac hypertrophy in hypertensive patients conveniently treated, and how important is to count on a non invasive method to follow patients. Echocardiography provides precise and complete information about cardiac chambers enlargement and the extent of functional impairment. So, determination of ventricular cavity dimensions, wall thickness and systolic and diastolic function allow us to a convenient evaluation of our treatment and permits to follow in time to these patients (7-8). Echocardiogram is more sensitive and specific for detecting the presence of hypertrophy and estimating left ventricular mass than electrocardiography or chest roentgenograms (9). So, it is especially useful for patients with essential hypertension, with whom it is being used to evaluate cardiac changes in population studies and assess regression of hypertrophy and mass after therapeutic intervention (10). For that, echocardiography is the non-invasive method of choice for the evaluation of this kind of patients that we have studied.

In this study we selected untreated patients to avoid the effects of previous treatment which could modify our results. It is well known that left ventricular diastolic function is one of the first things affected in hypertensive patients, however, it is also known, how difficult is to measure diastolic function in human heart. But, now we count on this method that contributes in a major way to our purposes. There is convincing evidence for an important effect of angiotensin II receptor blockers on cardiac hypertrophy. In this study, we compared the effects of eprosartan, a drug which belongs to that group, which has been recently incorporated to our therapeutic arsenal to treat hypertension. It is interesting to see in our results, that eprosartan not only reduced wall thickness of the left ventricle but also improved diastolic function. We considered that the mitral flow index E/A normally is more than 1, but in certain patients such as those with hypertension, this index is altered and inverted. In this study, we found that in more than a half of our patients (45 patients) at baseline, it was less than 1 indicating a slow relaxation of the left ventricle. When we evaluated at 6 months all of our patients again, we saw that left ventricular wall thickness (septum and posterior) have been reduced in a significant quantity. Similar results we could see respect to E/A mitral valve flow index which were altered in 45 patients at the beginning of the study improving these data in a significant way also (we saw only in 19 patients at the end of this study, p<0.05). No changes we saw respect to left ventricular ejection fraction.

CONCLUSIONS

In this study, we showed that eprosartan, is able to reduce not only hypertrophy of the wall of the left ventricle but also improves the diastolic function which was severely affected because of this disease. Also, we show the important role of the non invasive methods to measure different parameters in human medicine, that we can do thanks to devices and formula developed through physics principles. So, we can easily test our treatment results, in this case in a high prevalence disease such as hypertension which is in this moment one of the human diseases where we have an important challenge to avoid its deleterious effects on the life.

REFERENCES

1- Maron BJ,Bonow RO,Cannon RO,Leon MB,Epstein SE: Hypertrophic Cardiomyopathy: Interrelation of Clinical Manifestation, Pathophysiology, and Theraphy.N Engl J Med 1987:316:780-789.

2- Roberts CS,Roberts WC:Morphologic features . In :Zipes DP,Rowlands DJ (eds): Progress in Cardiology 2/2. Philadelphia Lea&Febiger,1989:3-22.

3- Keidel WD: Uber eine neue Methode zur Registrierung der Voluman-derungen des Herzens am Menschen. Z.Kreisl-Forsch 1950;39:257-271.

4- Holmes JH,Howry DH,Posakony GF et al: Ultrasonic Visualization of Soft Tissue Structures in Human Body. Trans Am Clin Climatol Assoc 1955;66:208-212.

5- Ebina T, Oka S,Tanaka M, et al: The Ultrasonotomography of the Heart and Great Vessels in Living Human Subjects by Means of Ultrasonic Reflection Technique. Jpn Heart J 1967;8:331-337.

6- Garcia E,Gueret P, Bennet M , et al: Digital Image Processing of two-dimensional Echocardiograms: Identification of the Endocardium. Am J Cardiol 1981;48:479-486.

7- Friedman M, Roeske WR,Sahn DJ, et al: Accuracy and Reproducibility of new M-mode Echocardiopgraphic Recommendations for Measuring Left Ventricular Dimension. Am J Cardio. 1982;49:716-723.

8- Pearson AC; Gudipati C;Nagelhort D;Sear J,Cohen JD,Labovitz AJ: Echocardiographic Evaluation of Cardiac Structure and Function in Elderly Subjects with Isolated Systolic Hypertension. J Am Coll Cardiol 1991;17:422-440.

9- Liebson PR,Savage DD: Echocardiography in Hypertension: A review: Left Ventricular Wall Mass, Standarization and Ventricular Function. Echocardiography 1986;3:181-188.

10- Liebson PR,Savage DD: Echocardiography in Hypertension: A review: II. Echocardiographic Studies of the Effects of Antihypertensive Agents on Left Ventricular Wall Mass and Function. Echocardiography 1987,4:215-249.

Avalanche-Collapse Process on Flame Spreading Over Liquid Fuels

Eugenio Degroote

Technical University of Madrid, E.U.I.T.Agricola, Dep. Ciencia y Tecnologia Aplicadas, Ciudad Universitaria, s/n, Madrid-28040, e-mail: eugenio.degroote@upm.es

Abstract. Flame spreading over liquid fuels exhibits, depending on the initial fuel surface temperature (T_∞), five different spreading regimes, separated by four critical temperatures $T_1, T_2, T_3 T_4$, being T_1 a stationnary bifurcation, T_2 a trasncritical bifurcation, T_3 a Hopf bifurcation and T_4 a homoclinic connection. For $T_4 \leq T_\infty \leq T_3$ flame spreading velocity (v_f) oscillates with a given period that diverges for $T_\infty = T_4$. The appearance of all these regimes is due to the existence of a preheating zone ahead of the flame front that modifies its spreading, and it corresponds to an avalanche-collapse process, that has been shown to possess universal characteristics. The mechanisms involved are described here and compared with the experimental results obtained in our laboratory.

Keywords: Chaos, bifurcation theory, combustion, liquid fuels, convective patterns
PACS: 82.33.Vx, 89.75.Kd

INTRODUCTION

Flame spreading over liquid fuels is still now subject of some controversy. Even though the problem has been experimentally studied, the physic mechanisms involved are not well known. Furthermore, it has been observed that solid fuels present (for a wide range of temperatures) a different behavior to the one observed in the case of liquid fuels with similar characteristics. The purpose of this work is to shed some light to the problem by doing a heat and momentum analysis that will be compared with our experimental results. Section 2 will describe the main experimental results obtained in our laboratories. Section 3 will analyze the flame spreading velocity $\left(v_f\right)$ for high temperatures (above the flash-point temperature T_2), while section 3 will describe what happens for temperatures lower than this critical temperature. Finally, a brief conclusion will be given.

EXPERIMENTAL RESULTS

Flame spreading velocity $\left(v_f\right)$ can be plotted as a function of the initial fuel surface temperature (T_∞). The result (obtained in a series of experiments conducted in our laboratories) is fully described in references [2,3]. Five different regions can be observed:

Region I For $T_\infty \geq T_1$, being T_1 a first critical value, flame spreading velocity v_f is almost constant, of order $100\,cm/s$.

Region II For $T_2 \leq T_\infty \leq T_1$, being T_2 a second critical value, flame spreading is uniform. In this region, the slope of the $T_\infty - v_f$ slope is of order $10\,cm/s \cdot K$.

CP905, *Frontiers of Fundamental Physics (FFP8), Eighth International Symposium*
edited by B. G. Sidharth, A. Alfonso-Faus, and M. J. Fullana
© 2007 American Institute of Physics 978-0-7354-0412-0/07/$23.00

Region III For $T_3 \leq T_\infty \leq T_2$, being T_3 a third critical value, flame spreading is still uniform but, in this case, the slope of the $T_\infty - v_f$ slope is of order $1 \, cm/s \cdot K$.

Region IV For For $T_4 \leq T_\infty \leq T_3$, being T_4 a third critical value, an oscillation of v_f is observed.

Region V For $T_\infty \leq T_4$ flame spreading is uniform, almost constant of order $1 \, cm/s$.

These differente regions have not been found in the case of solid fuels, where T_2, T_3, T_4 are not observed. In the solid case, flame spreading is completely uniform.

SOLID-LIKE MODEL

For values of the initial fuel surface temperature above T_2 (regions I and II) flame spreading velocity presents a similar dependence to the solid case [4]. Two different fuels (a liquid and a fuel) with similar physic paramenters (density, viscosity, etc...) would present the same behavior and the same temperature dependence in these regions. If we undertake a heat and momentum transfer between the condensed phase (either liquid or fuel) and the gas phase, the same dependence is found between T_∞ and v_f. In our case, it has been found that it should be of the following type:

$$v_f \propto \frac{1}{T_\infty \left(T_b - T_\infty\right)^2}$$

being T_b the fuel boiling temperature. This relation fits well with our experimental results. But, for $T_\infty \leq T_2$ this *solid-like* model is not appliable. For values below this critical temperature the solid-like model does not fit with our experimental data. Furthermore, it has been observed that an assistance effect appears through the liquid phase: strong temperature differences appear in the vicinity of the flame front. They are of order $T_f - T_\infty$ in the gas phase, and of order $T_b - T_\infty$ in the liquid phase. They induce strong surface tension gradients on the fuel surface, that tend to move hot liquid ahead of the flame. Therefore, an increase in the fuel vapor is observed, that accelerates flame spreading, compared to the solid case. This assistance effect through the liquid phase has been observed experimentally, and the characteristic length L of the vortex has been measured using a thermocouple technique: for $T_\infty \approx T_2$ L is of order $0 \, cm$, while for $T_\infty \approx T_3$ the approximate length of this region is $L \approx 1 \, cm$.

ASSISTED MODEL

For $T_3 \leq T_\infty \leq T_2$ (Region III) an assisted model can be proposed, that fits with our experimental data [4]. Flame spreading is assisted by a thermocapillary effect (Marangoni effect). The following dependence has been deduced from the heat and momentum transfer

$$v_f \propto \frac{1}{\left(T_b - T_\infty\right)^3}$$

Therefore, two different slopes are observed in the $T_\infty - v_f$ diagram in the vicinity of $T_\infty = T_2$. This assistance phenomena increases as we decrease temperature and, finally, it becomes unstable for $T_\infty = T_3$: for lower values (Region IV) an oscillating regime is observed. This pulsating spreading presents, for each value of T_∞, a characteristic period, that diverges finally for $T_\infty = T_4$. For even lower values (region V) no more oscillations are observed, and a slow motion regimes appears.

AVALANCHE COLLAPSE-PROCESS

When the initial fuel surface of the fuel is located in region III, a pulsating behavior occurs. Flame front velociy v_f exhibits a slow motion period, followed by a quick motion period, and so on. It corresponds to an avalanche-collapse process. In the beginning, when the flame propagates slowly $(v_f \approx 1\,cm/s)$ the preheating region in the liquid fuel at a charateristic velocity $u_s \approx 10\,cm/s$ advances flame front and, after some time, provides enough vapor fuel to increase flame speed. When this moment is attained (just a the end of the slow motion period) a quick motion period begins (with $v_f \approx 10-15\,cm/s$) until the preheating vortex is surpassed. Then flame front proceeds to spread again very slowly, and a new periodic cycle starts one more time. The characteristics of this combustion process can be extended to many fields, such as laser optics, biophysics and electronics, and a mathematical model has also been defined, that fit with our experimental data.

CONCLUSION

Tha main characteristics of flame spreading over liquid fuels have been pointed out, and a quick comparison with solid fuels has been made. This process seems to possess universal characteristics that are observed in a wide range of scientifica areas. Some of these results are now starting to be apllied to improve fire safety on fuel containers.

ACKNOWLEDGMENTS

This work has been sponsored by the Spanish Project A0304 (Technical University of Madrid - UPM, Spain).

REFERENCES

[1]K. Akita, XIV Symposium (International) on Combustion, 1973, pp.1075-1083.
[2]E.Degroote, Garcia-Ybarra, P.L., Eur. Phys. J. B, 13, 2000, pp.381-386.
[3]E. Degroote, P.L. Garcia Ybarra, J. of Th. An. and Cal., 2005, 80 (3), pp.541-548.
[4]E. Degroote, P.L. Garcia Ybarra, J. of Th. An. and Cal., 2005, 80 (3), pp.549-553.
[5]E. Degroote, P.L. Garcia Ybarra, J. of Th. An. and Cal., 2005, 80 (3), pp.554-558.

Study Of Water Hammer Effect On Weightlessness Condition

R.Elvira, A.Monzón, I.Riesgo and J.Fernández-Cabrera

Escuela Universitaria de Ingeniería Técnica Aeronáutica, Universidad Politécnica de Madrid, Plaza de Cardenal Cisneros Nº3, Madrid, Spain

Abstract. Research on fluid mechanics phenomenon called Water hammer Effect in microgravity conditions, making a comparison between several flow rates. To show how critical is in this environment and to establish a baseline of hardness in weightlessness trying to be a kick-off point in space engineering pipes development.

Keywords: Water Hammer Effect, Microgravity, Weightlessness, CFD, Parabolic Flight.
PACS:47.10.-g; 47.11.-j; 47.20.-k; 47.27.nf; 47.40.-x; 47.54.De; 47.55.dr ; 47.60.+i; 47.80.Fg; 47.85.-g

1. INTRODUCTION

Usually hydraulic installations are working on stationary conditions but there are some cases in which transient phenomenons appear by an external action. One example is a fluid moving through a pipe (Stationary conditions) and it is suddenly stopped by a quick-closing valve (Transient conditions). It causes an over-pressure in the near field of the valve that causes stresses into the pipe as well and could break and may jeopardize the performance of devices in the installation.

2. WATER HAMMER EFFECT

To explain this phenomenon is it used the elastic water column theory, that is, same assumptions than rigid water column theory with the exception that the elasticity of the pipe walls and the compressibility of the water under the action of a pressure change are also taken into account.

From the condition of dynamic equilibrium for the element of water and the condition of continuity for the element, the formula of the pressure wave is:

$$a = \sqrt{\dfrac{1}{\dfrac{\rho}{g}\left(\dfrac{1}{K}+\dfrac{Dc_1}{Ee}\right)}} \tag{1}$$

CP905, *Frontiers of Fundamental Physics (FFP8), Eighth International Symposium*
edited by B. G. Sidharth, A. Alfonso-Faus, and M. J. Fullana
© 2007 American Institute of Physics 978-0-7354-0412-0/07/$23.00

From fundamental water hammer equations (for rapid gate movements) the overpressure is this expression :

$$\Delta H = -\frac{a}{g} \Delta V \qquad (2)$$

More information of formulas see the references.

3. EXPERIMENT

Previous studies were based on CFD (Computational Fluid Dynamics), studies of Water Hammer Effect in microgravity and gravity conditions. Weakness Water Hammer Effect and symmetric path lines distribution were obtained in weightlessness condition. In Figure 1 it is shown the symmetric hypothesis, comparing the path lines.

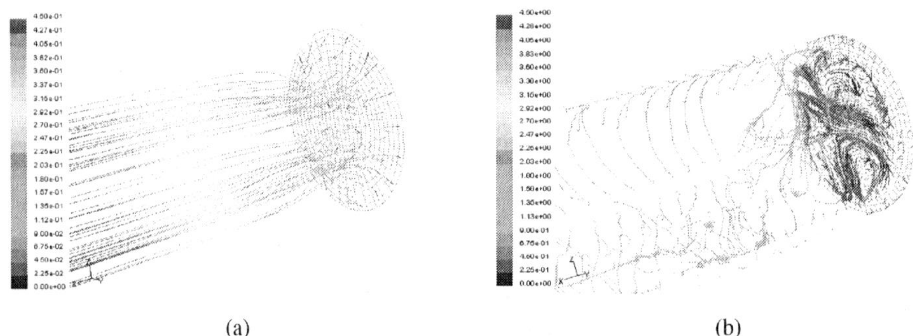

(a)   (b)

FIGURE 1. Water Hammer Effect in microgravity (a) and gravity (b) conditions. Path lines coloured by velocity magnitude in m/s.

Generally speaking the experiment is composed by a baseplate, double sealed box, close pipe circuit, pressure sensors (four), flow meter, accumulators, valves, electric pump, data acquisition system, electronic controls, computing devices, electromechanic servos and water.

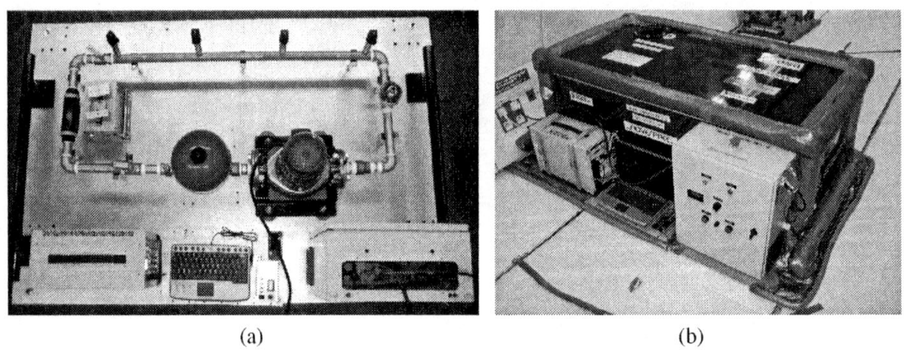

(a)   (b)

FIGURE 2. Circuit scheme (a) and experiment on-board (b)

4. RESULTS

Sensor 1 and Sensor 2 are nearest to quick-closing valve. Y-axis is static plus dynamic pressure. Figure 4 and Figure 5 show water hammer effect on microgravity for several flow rates. In terms qualitative water hammer intensity on microgravity is less than on gravity conditions.

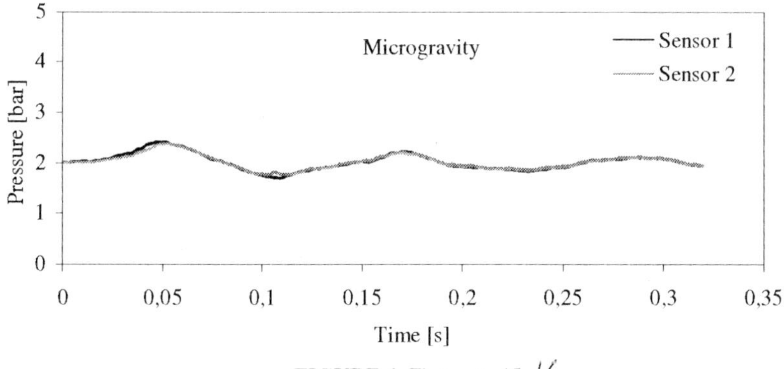

FIGURE 4. Flow rate 45 $^l/_{min}$

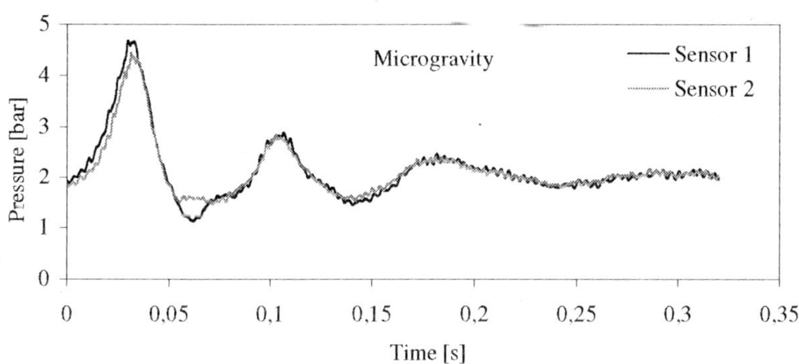

FIGURE 5. Flow rate 150 $^l/_{min}$

ACKNOWLEDGMENTS

Everybody who have believed and supported our dreams, especially our families.

REFERENCES

1. J. Parmakian, *"Water hammer analysis"*, Dover publications, 1963.
2. L. Allievi, *"Theory of Water hammer"*, translated by E. E. Halmos, printed by Riccardo Garoni, Rome, Italy, 1925.
3. L. Bergeron, *"Etude des variatioins de regime dans les conduits d'eau: solution graphique generale"*, Revue generale de l'hydraulique, Paris, Vol.1, 1935, pp.12-69.

A gravitational explanation for quantum theory - non-time-orientable manifolds

Mark J Hadley

Department of Physics, University of Warwick, Coventry CV4 7AL, UK
email: Mark.Hadley@warwick.ac.uk

Abstract. Spacetime manifolds that are not time orientable play a key role in a gravitational explanation of quantum theory. Such manifolds allow topology change, but also have fascinating additional properties such as net charge from source-free equations and spin half transformation properties. It is shown how the logical structure of propositions and the probabilities of quantum theory arise from such acausal space times.

Keywords: Topology, time-orientability, spin-half, quantum theory
PACS: 0240-k,0365.Ta,0420.Gz

Introduction

Einstein's dream was to find a grand unified theory. For him that meant not only a unification of the forces of nature, but also the unification of particles and fields. He tried repeatedly to find field theories that included gravitation and also had particle-like solutions. It is interesting to note that some early attempts were dismissed because they did not have symmetric solutions of unequal masses and opposite charge corresponding to an electron and a proton - the only know elementary particles at the time.

Although the focus of his later work was on classical field theories, Einstein also held the view that a more fundamental realist theory would eventually explain quantum theory. Einstein probed the completeness of quantum theory with his EPR experiment, but it was Bell who used the same thought experiment to show that quantum theory was fundamentally incompatible with any local hidden variable theory. This ruled out any realist explanation compatible with causality.

To model elementary particles with a classical field theory requires solutions:

1. With the right properties - such as mass, charge, spin etc
2. Interactions between them - scattering, creation annihilation etc.
3. The right behaviour - quantum theory

Although Einstein considered the first requirement in detail, the other two were largely ignored. It may have seemed expedient to get solutions first and then examine interactions and mechanics afterwards. However, the second and third requirements place severe constraints upon acceptable solutions and in so doing offer potentially helpful clues.

Interactions and topology change

If particles are modeled as topological structures of space, then interactions such as annihilation of particles, particle-antiparticle creation and some scattering experiments will require the topology of a region of spacetime to change. However there are powerful theorems due to Geroch [1] and Tipler [2] that constrain topology change in classical general relativity. The possible counterexamples can be summarised as[3]:

1. Singularities
2. Closed timelike curves
3. A lack of time orientation

The first is a breakdown of general relativity. Since General Relativity is expressed in terms of a 4D spacetime manifold, a singularity is not consistent with a description s spacetime as a manifold. Nevertheless some authors (eg Sorkin) [4] have proposed a reformulation of general relativity to alow singularities where topology change takes place.

For closed time-like curves to appear in a region of space that was previously regular requires negative energy due to Tipler's theorem. While this cannot be ruled out absolutely with current theoretical knowledge it is unwelcome as a postulate.

The third option is that some timelines from the initial surface turn around and return through the same surface. This would be a failure of time-orientability. Further work on structures that lack time orientability have proved fascinating and fruitful.

Manifolds that do not admit a time orientation can be constructed in a number of ways. The Einstein Rosen bridge is the earliest example, although its non-orientability is often overlooked or misunderstood. The common description of a path through the bridge is that the traveler goes into a black hole and comes out of a white hole. In fact both ends of the wormhole structure are blackholes, but the traveler has his time direction reversed, making all blackholes look like white holes.

Wormholes can be constructed with any combination of time and space orientability[5], and by a similar process monopole type structures of any orientation can be described[6]. The Möbius strip is non-orientable and if the central circle is considered to be a space dimension, and the other direct a time direction, then the Möbius strip models a $1 + 1$ dimensional spacetime that is space orientable but not time orientable.

Orientability and charge

Manifolds that are not time orientable have some fascinating properties. They provide counterexamples to the integral theorems linking the enclosed charge to the flux on a closed surface. When the theorems do not apply, it is possible to have the appearance of charge arising from the source free equations because there can be a net surface flux with zero enclosed charge. Sorkin [7] applied the idea using Stokes' theorem to a geon that was not space orientable and found that it could have net magnetic charge. However a 4-geon that is not time orientable does not have a consistently defined normal vector and the divergence theorem fails, allowing the appearance of net electric charge. A mathematical treatment in arbitrary dimensions is given in [6].

Working in three dimensions with the vector form of Maxwell's equations, the apparent source of charge is a net flow of flux into or out of a boundary region. A Sphere is the simplest example although all the results apply to an arbitrary compact two dimensional surface that encloses, or appears to enclose a volume of three space. The charge could be either magnetic or electric.

Considering first the magnetic charges Q_m:

$$Q_m = \oint_{S^2 = \partial V^3} \mathbf{B} . \hat{\mathbf{n}} \, dS = \int_{V^3} \mathbf{Div}.\mathbf{B} \, dV = 0 \tag{1}$$

which neatly relates the flux at a surface to the integrated charge density in the volume enclosed. It shows the apparent charge being due to the sources in the equations. The last step is a simple point by point application of Maxwell's equations. The first step is a global result that requires that the volume, dV, is compact and orientable. The integrals also require a metric to be defined. In general the equations do apply and there are no Magnetic charges as a consequence of the vanishing divergence of the magnetic vector field.

A similar treatment for the electric charge gives:

$$Q_e = \oint_{S^2 = \partial V^3} \mathbf{E} . \hat{\mathbf{n}} \, dS = \int_{V^3} \mathbf{Div}.\mathbf{E} \, dV = \int_{V^3} \rho \, dV \tag{2}$$

which relates the electric flux at a surface to the integrated charge density in the volume enclosed. As in the case above the volume has to be compact and orientable. But the volume integral also requires a consistent definition of the electric field. However the electric field depends upon the direction of time, and reverses if time is reversed. So it is not necessarily possible to give a global definition of the charge throughout the volume - it will be impossible if the field is non-zero. Consequently, on a non time orientable spacetime, it is possible to have the appearance of electric charges without any sources.

The transformation properties of $\mathbf{E}$ are evident from the Lorentz invariant description using the Faraday tensor which shows that the electric field is the space.time component of the tensor and therefore changes its value depending on the direction of the time coordinate. In the absence of a global time direction, the electric field cannot be consistently defined even when F is well defined everywhere. By contrast the magnetic field is the space.space components of the field and does not depend upon the time direction (nor even the space direction). Using the Faraday tensor, Maxwell's equations take the simple form: $\mathbf{d}\,\mathbf{F} = 0$ and $\mathbf{d} \star \mathbf{F} = \star \mathbf{J}$. The Faraday tensor is a two form and can be integrated using Stokes' theorem, which applies in any k-form, w and (k+1) dimensional volume, V:

$$\int_{S = \partial V} w = \int_V \mathbf{d}\,w \tag{3}$$

Stokes' theorem requires V to be a compact oriented manifold. (interestingly it does not require a metric).

Magnetic charge can be defined in terms of the Faraday tensor:

$$Q_m = \int_{S=\partial V} \mathbf{F} = \int_V \mathbf{d\,F} = 0 \qquad (4)$$

While a similar application to Stokes' theorem for the dual tensor gives:

$$Q_e = \int_{S=\partial V} \star\mathbf{F} = \int_V \mathbf{d} \star\mathbf{F} \qquad (5)$$

The RHS evaluates to zero in the absence of any sources ($\mathbf{J} = 0$). Although the criteria for Stokes' theorem seem to be met for a manifold that is not time orientable, the star operator is defined in terms of an oriented space-time volume element so it is not well defined if time is not orientable and so the volume integral is not well defined.

The two treatments are equivalent, as are other approaches using vector densities. The conclusion is the same, that manifolds that are not time orientable can exhibit net electric charge from the source free equations, while the divergence free magnetic field still implies that there can be no net magnetic charge.

Spin-half transformation properties

The rotational properties of a non-time orientable manifold are intriguing. Intrinsic spin is a measure of how an object transforms under a rotation. Trying to apply the concept of a rotation to a manifold is not trivial. If a particle is modeled by an asymptotically flat manifold with non trivial topology (Wheeler used the term geon) or even non-trivial causal structure (a 4-geon), then we can define a rotation as any transformation with appropriate continuity properties, that matches the usual definition, $\mathbf{R}(\theta)$, in the asymptotic region of the manifold.

$$R(\theta)M \to M \qquad (6)$$
$$R(\theta)R(\phi)x = R(\theta + \phi)x \;\; \forall x \in M \qquad (7)$$
$$R(0)x = x \;\; \forall x \in M \qquad (8)$$
$$R(\theta)x \to \mathbf{R}(\theta)x \;\; \text{as } |x| \to \infty \qquad (9)$$

The axis of rotation has been omitted, a thorough treatment requires a definition that is manifestly applicable for the full rotation group, but a restriction to a single axis is adequate for our purposes. The definition above is very general allowing a very wide range for mappings to be counted as a rotation.

The definition above may suit mathematicians, but it is not appropriate for a physical rotation, such as when a neutron is rotated in the laboratory by a magnetic field. The distinguishing feature of a physical rotation is that it is parameterised by time. The object is rotated from zero degrees to θ, as time passes from 0 to 1 say. So we refine our definition to a mapping of spacetime points: $\mathscr{R}(\theta)(x,0) \to (R(\theta)x,1)$, plus the conditions above.

The mathematical rotation defines a path through space, from the original point x to the rotated point: $x \to R(\theta)x$, $\gamma_\theta(\lambda) = \{R(\lambda\theta)x : \; x \in M, \lambda \in [0,1]\}$ Similarly, the physical rotation defines a world line starting at each point in spacetime: $(x,0) \rightarrow \mathscr{R}(\theta)(x,0) = (R(\theta)x,1)$, $\chi_\theta(t) = \{(R(t\theta),t)x : x \in M, t \in [0,1]\}$

But this construction defines a time direction throughout the manifold. It cannot apply to a spacetime that is not time orientable. A modification of the definition is required:

$$\mathscr{R}(\theta)(x,0) \rightarrow (R(\theta)x, \phi(\theta)(x)) \tag{10}$$

$$\phi(\theta)(x) \rightarrow 1 \quad \text{as } |x| \rightarrow \infty \tag{11}$$

Where ϕ is zero a time direction is not defined by the rotation. The function ϕ must be zero over a two dimensional subspace if the manifold is not time orientable. The rotations have two categories of points that do not move:

Fixed Points $(x,0) \rightarrow (x,t)$ for example the points on the axis of the rotation.

Exempt Points $(x,0) \rightarrow (x,0)$ in a sense these points do not participate in the rotation.

Consequently, a physical rotation that is 360 degrees in the asymptotic region will leave some exempt points in the manifold, and it will not be possible to find a 360 degree rotation that is the same as the identity, zero degree rotation. The topology of the rotation group is such that it is possible to find a 720 degree rotation that is an identity.

The idea has been known for centuries and is modeled by Wheeler's cube in a cube [8], Feynman's scissors trick [9] and Hartung's tethered rocks [10]. A computer animation can be seen on the web [11]. What all the demonstrations have in common is a set of fixed points that are not rotated, but all lack any explanation of how parts of an elementary particle could be anchored in free space. This model of particles as 4-geons provides an explanation and naturally describes particles with spin half.

It is notable that the requirement for non time orientable manifolds came from the desire to model elementary particles and quantum phenomena. The existence of charge and spin half arise as a natural consequence.

The Essence of Quantum Theory

Quantum theory has many strange features, but most are not absolutely unique to quantum theory. For example the uncertainty relationship between position and momentum is common to classical waves, or the wave equations themselves. While some, apparently critical, features are largely irrelevant; such as the inability to make simultaneous measurements of some observables - if this were possible (even for non-commuting observables) it would not change quantum theory.

A fundamental difference between quantum and classical physics can be found in the logical structure of the propositions. Classical physics satisfies Boolean logic while quantum theory is non-Boolean. The distributive law does not hold, instead the propositions form an orthomodular lattice [12]. This leads to the requirement to represent probabilities as subspaces of a [complex] vector space rather than as measures on a volume space. It is this fundamental distinction that leads to the dynamical and statistical features of quantum theory.

150

Although the relationship between incompatible observables has a non-Boolean characteristic, in any one experiment the propositions satisfy the normal Boolean logic. This is generally expressed by saying that quantum theory is *context dependent* - within any single context, the probabilities can be expressed in the normal way.

A consequence of the logical structure of quantum theory and the way probabilities are represented is an entirely new concept of probability. In all of classical physics probability can be ascribed to our ignorance of some variables (typically the precise initial conditions). In quantum theory this interpretation is not possible, it cannot be described in terms of local hidden variables. As shown by the violations of Bell's inequalities.

The logical structure and the new way to represent probabilities leads inevitably, and uniquely, to quantum theory and quantum field theory. The only known way to represnt probabilities of such non-Boolean logic is to use spaces and subspaces of a Hilbert space [12]. The continuity and symmetry of spacetime then give rise to the usual equations of quantum theory as is shown by [13] for the non-relativistic case, and by [14] for the relativistic case..

A formal proof that acausal spacetimes lead to the logical structure of quantum theory is given in the paper by Hadley [15]. The proof is formal and can be largely replaced by the statement that an acausal spacetime is context dependent. This can been seen simply in two ways:

With Closed timelike curves (CTCs) or a failure of time orientability, it is not in general possible to set up boundary conditions on an *initial* surface without some knowledge of *future* conditions. Future experiments can set extra boundary conditions that are not redundant. Simple models with Billiards such as those described by Carlini and Novikov [16] provide a good illustration. The standing waves on a string provide a helpful analogy; boundary conditions need to be specified at both ends before the wave can calculated.

An alternative way to look at the context dependence, with the same example, would be to consider the shape of the standing wave as a combination of forward and backward moving waves. A change of time direction at a future experiment would *send a signal back* to the start of the experiment. This can be seen as a realisation of the Cramer's transactional interpretation of quantum theory [17].

Conclusion

The results described above apply to any geometric theory of space and time that allows non trivial topology. General Relativity is the established theory of spacetime and satisfies the criteria, but most variations of general relativity would give the same result. At least in principle, General Relativity with non-trivial causal structure could explain quantum theory and much more besides. It may be the unified theory that Einstein sought for so long.

REFERENCES

1. R. P. Geroch, *Journal of Mathematical Physics* **8**, 782–786 (1967).
2. F. J. Tipler, *Annals of Physics* **108**, 1–36 (1977).
3. M. J. Hadley, *International Journal of Theoretical Physics* **38**, 1481–1492 (1999).
4. R. D. Sorkin, Forks in the road, on the way to quantum gravity, gr-qc/9706002 (1997).
5. V. Matt, *Lorentzian Wormholes: from Einstein to Hawking*, Woodbury, N.Y., 1996.
6. T. Diemer, and M. J. Hadley, *Classical and Quantum Gravity* **16**, 3567–3577 (1999).
7. R. D. Sorkin, *Journal of Physics A* **10**, 717–725 (1977).
8. C. W. Misner, K. S. Thorne, and J. A. Wheeler, *Gravitation*, W H Freedman and Sons, 1973.
9. F. R. P, and W. S, *Elementary Particles and the Laws of Physics*, Cambridge University Press, 1986, ISBN 978052165862.
10. R. W. Hartung, *American Journal of Physics* **47**, 900–910 (1979).
11. A. Norton, Mathematics of twisting (1994), URL `http://oldsite.vislab.usyd.edu.au/gallery/mathematics/diffeo/diffeo.html`.
12. E. G. Beltrametti, and G. Cassinelli, *The Logic of Quantum Mechanics*, vol. 15 of *Encyclopedia of Mathematics and its Applications*, Addison-Wesley Publishing Company, 1981.
13. L. E. Ballentine, *Quantum Mechanics*, Prentice Hall, 1989.
14. S. Weinberg, *Gravitation and Cosmology: Principles and Applications of the General Theory of Relativity*, Wiley and Sons, 1972.
15. M. J. Hadley, *Foundations of Physics Letters* **10**, 43–60 (1997).
16. A. Carlini, V. P. Frolov, M. B. Mensky, I. D. Novikov, and H. H. Soleng, *Int. J. Mod. Phys.* **D4**, 557–580 (1995).
17. J. G. Cramer, *Rev. Mod. Phys.* **58**, 647–687 (1986).

Evolution Of The Concept Of Dimension

Philippe F. Journeau

AEGEUS, 14, rue Firmin Gillot, 75015, Paris

Abstract. Concepts of time elapsing 'in' a space measuring the real emerge over the centuries. But Kant refutes absolute time and defines it, with space, as forms reacting to Newtonian mechanics. Einstein and Minkowski open a 20th century where time is a dimension, a substratum of reality 'with' space rather than 'in' it. Kaluza-Klein and String theories then develop a trend of additional spatial dimensions while de Broglie and Bohm open the possiblity that form, to begin with wave, be a reality together 'with' a space-time particle. Other recent theories, such as spin networks, causal sets and twistor theory, even head to the idea of other *"systems of dimensions."* On the basis of such progresses and recent experiments the paper then considers a background independent fourfold time-form-action-space system of dimensions.

Keywords: dimension, signature, principle, background, twistors, Kaluza-Klein, causal, fourfold
PACS. 01.55, 01.65, 01.70, 02.40, 03.65, 04.20, 04.40, 04.50, 04.60, 11.15, 14.80, 52.35, 80

INTRODUCTION

The present paper revisits the concept of dimension in physics since the Greek era, its modern times evolutions and its trends. After 19th century ether, space-time has become 20th century new background of the real and its signature a mean to differentiate space-time geometries after Minkowski.

Yet doubt is allowed after many recent remarks such as Witten [1] about a signature "irrelevant as the scattering amplitudes are holomorphic functions of the kinematic variables" and Penrose [2] idea to "regard all space-time notions as being subsidiary to those of twistor space T." Indeed two other types of dimensions used to be considered fundamental, called form and change. They are now referred to in conformal or co-homology and in action or differentiation respectively. The 'co' prefix matches a "co-necessity" principle central to this paper: while time used to be seen as a dimension 'in' another one, space, it is now understood to come 'with' in a twofold co-necessity.

First section below starts with the history of these concepts up to the idea of a 'system of dimensions' consistent with Finkelstein [3], Penrose and Bars' [4] broadenings. It then discusses recent trends and experiences supporting form and change as two extra types of dimensions 'with' rather than 'in' space-time. It concludes to a background independent fourfold system of dimensions, meaning that laws and change themselves altogether with space-time make the background and the reality itself. Conversely space-time dimensions exist with form even said 'in'-formed.

Second section states the related fourfold co-necessity principle and shows that subsequent threefold physics principles may derive from it through observer's or interaction sharing along complementary dimension or gauge connections. Last section shows impact on observer's idea abstraction and universe intelligibility issue.

CP905, *Frontiers of Fundamental Physics (FFP8), Eighth International Symposium*
edited by B. G. Sidharth, A. Alfonso-Faus, and M. J. Fullana
© 2007 American Institute of Physics 978-0-7354-0412-0/07/$23.00

1. FOUNDATION OF FOUNDATIONS

'Dimension' comes from Latin 'dimensio', derived from 'metiri', meaning measurement, an issue central to both Einstein General Relativity (GR) formal metrics and Quantum Mechanics (QM) practical quantum of action.

Roots come from 6th century BC when Parmenides poses 'being', singular continuous ontology throughout time, followed by Zeno's arrows, first vectors. Plato then sees its reflections in universals, ideas, 'forms' taking shape on our local cavern walls. Ionian Heraclites rather emphasizes dynamical 'change', fundamental a fire always renewing things, differentiating. Democritus invents atomism from the idea of a minimal length, a/tomos i.e. un/breakable. But none of these principia, that science has relied upon, may be understood without use of the others and Aristotelian co-necessity follows Plato, Democritus and Heraclites first attempts to integrate them. Many recent prominent physicists have referred to them: Niels Bohr [5] to Aristotle, Penrose to "Sheaf cohomology as a Platonic notion", others to Democritus atomism or, as Finkelstein, to change, Heraclites-like spontaneous symmetry curbing/breaking.

Fields, vectors, forms and functions, derivations and operators are physics keys. There is no dynamical equation without 'd' or 'δ' and action S sums, in $\int L\ d^4x$, such quanta of action going 'with' space-time in minimal spatial grasps or time elapses. A classical, subjective infinitesimal 'd' was before QM mingled with the reality that it depicts, until δ 'change unit', non-commutative, make it formal since "pointless" [6].

Following Newton discoveries Kant defines space and time as forms of respectively external and inner senses feeding intuition, our immanent representations. Form is therefore both internal and external, core of both GR metrics and QM measurement and Young experiment shows dual holist form vs. local particle. After Lorentz, Poincaré, Einstein and Minkowski discoveries of time 'with' rather than 'in' space, holist wave vs. particle duality is pursued by de Broglie. Yet added Bohm's "quantum potential" must be split in two light-like dimensions, form and change, within (t,φ,δ,r) time-form-action-space, each impossible to isolate, as in 'p-form'. Kaluza-Klein path rather focuses on spatial extension but GR space-time field imply 'field' is fundamental. Madore follows the idea that vector field "can be defined as derivation of the algebra C(V), linear map of C(V) into itself which satisfies the Leibniz rule [7]"

Hence space-time is derivation – Heraclites' change – co-necessary to algebra – Plato's form – as function/operator, functor/operation dual. Aspect concludes from his EPR observations: "we must get rid of the vision called 'local realist'"[8]. Bars recently states that "from the point of view of 2T-physics many aspects of 1T-physics such as Hamiltonian with interactions, time and space, are all emergent concepts that depend on the embedding" and Penrose writes: "non-locality, as exhibited by twistor functions (regarded as first co-homology elements) is tantalizing reminiscent of the non-local features of EPR effects and quanglement." A (t,φ,δ,r) fourfold co-necessity, background independent as required by Smolin [9], hence is universe basic system of dimension. Bars also proposes that "starting from a space with signature (d,2) we end up with emergent space-time with signature (d-1,1) by making various gauge choices

$$(d,2) - (1,1)^{\text{signature of extra gauge parameters}} = (d-1,1)^{\text{emergent space-time}} \text{ ,,}} \tag{1}$$

For consistency reasons it is now switched to preferred time-form-change-space order.

Starting from right side of (1) it is logical that what frames the universe doesn't change even if Bars distinguishes physical (right) from phase space (left). Therefore

$$(2,0,0,d) \circ (-1,1,1,-1) ^{\text{signature of extra gauge parameters}} = (1,1,1,d-1) ^{\text{emergent time-form-action-space}} \quad (2)$$

bears a concept of system of dimensions with gauge choices switching a combination of dimensions to another, a mechanism that will further prove useful.

From Bars' (d,2) system, retaining Riemann CP^1 or d=2 gives (1,1,1,1) signature as kept below to represent the fourfold system of dimensions drafted in this paper; modeling reality with polar coordinates rather than Cartesian helps. An example might relate to field equation with Higgs change: $(1,2,0,1) \circ (0,-2,2,0) = (1,0,2,1)$ (3)

2. FROM EXPERIENCE TO PRINCIPLES

In support of merely one spatial dimension Penrose writes (p. 978) "ordinary space-time points are represented as Riemann spheres in PN" while "points in PN are represented as light rays in space-time", echoing Aristotle's "what line is to the point, what occurs is to what already has arrived." Besides, equivalent density matrixes for any direction, shown with Bloch sphere, suggest a sole quantum spatial dimension. Above that it is our choice to use (1,0,0,3) signature or twistor-like (0,2,2,0). Completion of the zeros to obey fourfold co-necessity gives physical interactions as drafted above (3): wave-particle fits Kant scheme with legality as Locus of conformal.

Penrose [2] highlights a "conflict between the two quantum processes U and R where U is the deterministic process of Unitary evolution (as can be described by Schrödinger equation) and R is the quantum state Reduction which takes place when a measurement is performed", adding that "R represents a discontinuous change in the state vector." This dimension of 'spontaneous', change in itself Higgs 'mechanism', as Penrose 'Objective Reduction', both linked to gravity, form curvature twisting.

'Fourfold co-necessity principle' here proposed states four types of dimension altogether making reality as background. Time to and fro space formation make negentropy vs. entropy 'time arrows' as in-forming or deforming space-time through change, form symmetry breaking. From this fourfold principle, such as time-form-action-space, written (A, L, S, N) as explained below, may derive other principles:
-   Relativity: Laws of physics constant in any inertial referential, i.e. passive, commutative, without action/change S, hence (A, L, N) or (1,2,0,1) signature.
-   Space homogeneity: Laws of physics are independent from location of effect is (A, L, S) sub-principle with signature such as (1,1,2,0), as in white holes (WH)
-   Time homogeneity (in BH) with laws independent from time of effect (L, S, N)
-   Efficient or active, non-commutative, causality: cause before effect in phenomenon (A, S, N) or effects occur 'with' space-time. (1,0,2,1) reflects Reduction process, form entanglement breaking as trans-formation is caused.

Isotropy then results from N spatial dimension being no less single than the others.

The fifteen fourfold, threefold, twofold and single combinations of dimensions (that might relate to the conformal group) need further study, together with the way any background independent fourfold reality entangles to another one, such as an observer.

It requires defining three worlds, below, which may be seen as one through each of our 'I' 'Weyl' White Hole event horizon connection from continuum to quantum [10]

3. ABSTRACTION

Dimensions are at the core of ontological vs. phenomenological debate, Einstein realism versus Copenhagen interpretation: are they external, some 'outer' reality or 'our' metrics and measurement, hence about 'us'? This issue requires starting from three worlds, close to Penrose's, with following formalism and figure 2 representation:

- $K_\Sigma = (A_\Sigma, L_\Sigma, S_\Sigma, N_\Sigma)$ for physical world that non-commutative QM depicts as a discrete matrix universe. L_Σ Legality encompasses all levels of real complexity up to widest entangled wave function. Σ expresses considerable yet numerable finite degrees of complexity, assumed 'outer' experimental nature.
- $K_i = (A_i, L_i, S_i, N_i)$ represents 'i' indexed Knowers' phenomenal worlds with: a) A_i Authoring time as "form of inner sense", b) L_i is rational knowledge phenomenon, c) S_i facts, measuring and signifying, d) N_i space 'outer' senses.
- $K_0 = (A_0, L_0, S_0, N_0)$ is continuous math world. It is a space-time-form-action continuum, i.e. R^4 or C^4 (observer-observed up to i event horizon) math reality.

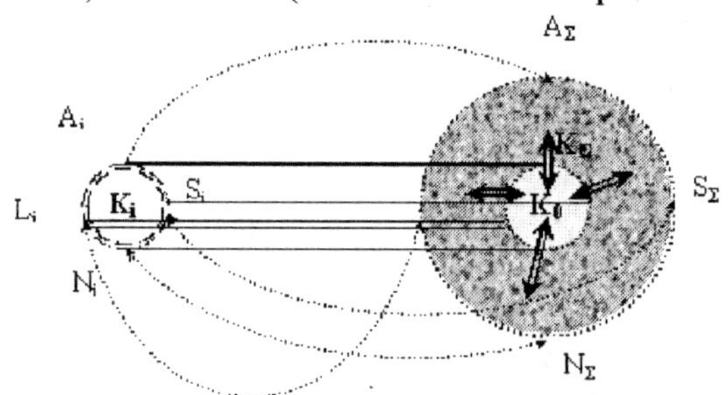

FIGURE 1. K_0-K_Σ unity between continuum and discrete worlds: K_0 space-time-form-action here shows N_0 dense below minimum N_Σ length. Others infinitely denser. Double line arrow is understanding reconciled to solve Kantian 'principle of continuity' issue [10] in a fourfold physical process: a) dT vs. δT, b) L abstraction, c) S 'd' to 'δ' creation-annihilation, d) dR to δR. Quantum vs. full fields join through (re)normalizing K_0->K_i->K_Σ or (abs)traction K_Σ->K_i realities.

QM uncovers K_Σ<->K_i measurement phenomenon, real-observer's entanglement. Discrete versus continuum starts from there. Sidharth explains how "infinities dogged physics for many decades [11]." Through phenomenal entanglement, discrete physical form crosses inner K_0<->K_i math formulation. To Kantian 'N' external vs. 'A' inner sense is substituted fourfold outer vs. inner realities between observable K_Σ discrete world and spatially tiniest, formally over-standing K_0 continuum.

REFERENCES

1. E. Witten, *Perturbative Gauge theory as a string theory in twistor space*, Princeton, 2003, p. 2
2. R. Penrose, *The Road to Reality*, Alfred A. Knopf, New York, 2004, p. 972, 978, 982, 992
3. D. R. Finkelstein, H. Saller, Z. Tang, *Beneath Gauge*, ar Xiv:quant-ph/9608023v1-1996
4. I. Bars, *Twistors and 2T-physics*, Department of Physics and Astronomy, U.S.C.L.A., 2005, p. 1
5. N. Bohr, *Atomphysik and human knowledge*, 1958, Gallimard, Paris, 1961-1991, p. 261
6. J. Madore & J. Mourad, *Non-commutative Kaluza-Klein Theory*, arXiv: hep-th/9601169, 1995a
7. J. Madore, *Introduction to Non Commutative Geometry and Physical Applications*, C.U.P. 1999b
8. A. Aspect, *Bell's theorem: the naïve view of an experimentalist*, Quantum [Un]speakables, R.A. Bertlmann and A. Zeilinger edition, Springer 2002 and in Pour la Science, 12/2004, p. 123
9. L. Smolin, *How far are we from the quantum theory of gravity?* University of Waterloo, CA, 2003
10. P. Journeau, *Kant 200 years after*, U. Navarra, R. Filosoficas 2004, in 'Connaissance', Aegeus 2005
11. B. Sidharth, *Reconciliation of Electromagnetism and Gravitation*, A.F.L.deBroglie 27/2, 2002 p.334

Experimental Test on the Applicability of the Standard Retardation Condition to Bound Magnetic Fields

A.L. Kholmetskii[*], O.V. Missevitch[†], R. Smirnov-Rueda[⋆] and R. Ivanov[‡]
and A.E. Chubykalo[‡]

[*]Faculty of Physics, Belarus State University, Minsk, Belarus
[†]Institute for Nuclear Problems, Minsk, Belarus
[⋆]Faculty of Mathematics, Complutense University, Madrid, Spain
[‡]Faculty of Physics, University of Zacatecas, Mexico

Abstract. Modern view on the fundamental structure of the whole EM field implies the existence of two components essentially different by nature. We found that it has some historical explanation. Experimental data do not support the validity of the standard retardation constraint ($v = c$) generally accepted in respect to bound fields. Besides, non-local characteristics of classical bound fields may shed a promising new light on a close relationship between quantum mechanics and classical EM theory.

Keywords: Scientific theories, fundamental physics
PACS: 01.70.+w, 43.10.Mq

INTRODUCTION

Modern view on the fundamental structure of the whole EM field implies the existence of two components essentially different by nature. More specifically, on the one hand, EM field can be decomposed on velocity-dependent (or bound) components and, on the other hand, on acceleration dependent (or radiation) components. Modern causal status of EM field means that retardation rates of bound and radiation components are equal to the velocity of light. Obviously, acceleration term disappears or takes no place for static and quasi-static phenomena and therefore, only bound components remain in this case. Thus, according to the modern viewpoint static and quasi-static EM phenomena (associated with bound fields) also have retardation rates equal to the velocity of light. Nevertheless, it leads to a rather strange situation: no one has undertaken explicit empirical analysis of causal properties of static and quasi-static fields. We found that it has some historical explanation.

To highlight to this paradoxical situation (when causal properties of bound fields are taken for granted without any explicit empirical analysis) we need to turn our attention to the peculiar historical background of the classical EM theory. Hertz's experiments on observation of EM waves (1888) provided the basis for acceptance of Maxwell's theory. However, only Lorentz's modification of Maxwell's theory (1892) provided Lienard (1898) and Wiechert (1901) with inhomogeneous wave equations. The retarded Lienard-Wiechert solutions were found under the retardation constraint which, as it is generally

CP905, *Frontiers of Fundamental Physics (FFP8), Eighth International Symposium*
edited by B. G. Sidharth, A. Alfonso-Faus, and M. J. Fullana
© 2007 American Institute of Physics 978-0-7354-0412-0/07/$23.00

perceived, was just experimentally verified by Hertz for the whole EM field. It gave a rise to the acceptance of fundamental viewpoint on equal rates of propagation for bound and radiation field components.

Thus, Hertz's experiments were not especially thought-out to test causal behavior of bound fields since the separation of the EM field on two essentially different parts was discovered later by Lienard and Wiechert. A natural question arises in this respect: should be there an additional experimental test on the applicability of the standard retardation condition to bound fields? In fact, we think that it should have been recommended and justified nearly one hundred years ago standing only on methodological grounds since the intensity of bound fields fall off much faster than the strength of radiation components. Nevertheless, since the time of Hertz's experiments no one has claimed the need of additional tests directed on verification of the causal characteristics of bound fields.

METODOLOGY AND REALIZATION OF A CONSISTENT EXPERIMENTAL APPROACH

It must be clear now that the most rigorous and methodologically consistent empirical test of the causal behavior of the whole EM field should be based on individual tests in respect to each component (radiation and bound) taken individually. We start with analysis of EM field in the plane of loop antenna. It implies the analysis of the relative contribution of bound and radiation components as functions on a distance. In the most general case we know how to calculate magnetic fields generated by the loop antenna [1],[2],[3]:

$$\mathbf{B} = \mathbf{B_u} + \mathbf{B_a} = \frac{1}{4\pi\varepsilon_0 c^2} \oint_\Gamma \left\{ \frac{[I]_c}{R^2} + \frac{[\dot{I}]_c}{cR} \right\} \mathbf{k} \times \mathbf{n} dl \tag{1}$$

where $\mathbf{n} = \mathbf{R}/R$; I is the conduction current; $\mathbf{k}$ is the unit vector in the direction of $\mathbf{I}$, i.e. $\mathbf{I} = I\mathbf{k}$ and dl is an infinitesimal element of the loop line Γ. The first and the second terms are bound and radiation components, respectively.

The emitting and receiving antennas are considered as small: (a) radius of the loop is considerably smaller than the distance R from the center of the antenna to the point of observation. (b) wave-length of EM radiation greatly exceeds the perimeter of the loop. It will be fulfilled in the practical realization of the experimental set-up.

Thus, the value of the induced e.m.f. have two terms that fall as R^{-2} and R^{-1}. At relatively large distances the radiation R^{-1}-term already predominates and it can be defined as the reference signal. In order to make verifiable predictions we specified the relative positions of zero-crossing points (a point where the signal crosses a zero line for the first time) of the whole and the R^{-1}-reference signals, respectively. The distance on the time axis between both points we denoted as $\Delta t(R)$. Starting with this analysis we calculated the value of $\Delta t(R)$ as a function of a distance between EA and RA (we used *Mathcad Professional 2000 software* for numerical evaluations). Under the assumption of the applicability of the standard retardation condition to bound fields $\Delta t(R)$ is a non-negative function in the whole range of variations of the variable R (see

[4]). Nevertheless, the numerical predictions showed that the time shift $\Delta t(R)$ reaches negative values if the retardation of bound components is not equal to the velocity of light.

To avoid interference with reflected EM radiation, practical realization was based on the excitation of one very short quasi-harmonic signal (~ 1 ns). The small antenna approximation and the maximum ratio of bound-to-radiation field strength could be easily implemented for finite size multi-section antennas.

In order to compare zero-crossing points of the whole and the reference signal, we applied the following procedure. Keeping the orientation of antennas unchanged, we varied the distance between them in the range of $R = 40 \div 280$ cm. At each position we succeeded in producing detectable signals and stored them in a digital form. We measured the instant of time when the oscillating disturbance crosses a zero line for the first time. At large distances R^{-1}-term predominates and in our case it constitutes more than 99 per cent at a distance of 300 cm. So, at this distance we can take the whole signal as the reference one and using the digital format available in modern oscilloscopes, we stored it and recovered its position on oscilloscope time scale at any other smaller distances. The resultant experimentally found $\Delta t(R)$-dependence (see [4]) for the range $R = 40 \div 280$ cm exhibits a clearly visible negative minimum which should take no place if one expects the validation of the standard retardation condition $v = c$ as the conventional EM theory requires.

CONCLUSIONS

Based on these results, one can conclude that experimental data do not support the validity of the standard retardation constraint ($v = c$) generally accepted in respect to bound fields. It breaks the fundamental causal status of modern physics and indicates on a wider fundamental framework of the classical EM theory with respect to bound fields. Experimentally observed violation of the applicability of the standard retardation condition to bound fields can be interpreted as explicit manifestation of their non-local properties. On the other hand, non-local characteristics of classical bound fields may shed a promising new light on a close relationship between quantum mechanics and classical EM theory.

REFERENCES

1. O.D. Jefimenko, *Electricity and Magnetism: An Introduction to the Theory of Electric and magnetic Fields*, 2nd Edition (Electret Scientific Co., Star City, WV, 1989)
2. W.G.V. Rosser, *Interpretation of Classical Electromagnetism*, Fundamental Theories of Physics, Vol. 78, Edt. A. Van der Merwe (Kluwer Academic Publishers, Dordrecht/Boston/London, 1997)
3. J.A. Stratton, *Electromagnetic Theory* (McGraw-Hill Book Co., New York-London, 1941)
4. A.L.Kholmetskii, O.V.Missevitch, R.Smirnov-Rueda, R.Ivanov and A.E.Chubykalo, to be published in *J. Appl. Phys.*, (2007)

Multivalued Fields

Hagen Kleinert[1]

Institut füer Theoretische Physik, Freie Universität Berlin,
Arnimallee 14 D-14195 Berlin, Germany

Abstract. The lecture discusses some of the many physical systems which can be described efficiently in terms of multivalued fields.

Keywords: Foundations of quantum mechanics
PACS: 03.65.Ta

INTRODUCTION

Multivalued fi elds play an important role in understanding a great variety of phase transitions. They arise typically in hydrodynamic descriptions of condensed matter systems, and the multivaluedness manifests itself in presence of grand-canonical ensemble of various defects. The defects can be point-like, line-like, and surface-like. The line-like defects are most interesting since they they are similar to worldlines of point particles whose properties can be studied by fi eld theory. The fi elds used to describe line-like defects are called disorder fi eld.

SUPERFLUID TRANSITION

The simplest phase transitions which can be explained by multivalued fi eld theory is the so-called λ-transition of superfluid helium. The name has its origin in the shape of the peak in the specifi c heat observed at a critical temperature $T_c \approx 2.18\,\mathrm{K}$ shown in Fig. 1.

For temperatures T much below T_c, the fluid shows no friction and possesses only massless excitations. These are the quanta of the *second sound*, called phonons. They cause the typical temperature behavior of the specifi c heat

$$C \sim T^D \tag{1}$$

in D space dimensions, as fi rst explained by Debye in his 1912 theory of specifi c heat [1].

As the temperature is raised, another type of excitations appears in the superfluid. These are the famous *rotons* whose existence Landau deduced from the thermodynamic properties of the superfluid [2, 3]. Rotons are the excitations at the minimum of the phonon dispersion curve which can be measured by neutron scattering and is displayed

---

[1] E-mail: kleinert@physik.fu-berlin.de

CP905, *Frontiers of Fundamental Physics (FFP8), Eighth International Symposium*
edited by B. G. Sidharth, A. Alfonso-Faus, and M. J. Fullana
© 2007 American Institute of Physics 978-0-7354-0412-0/07/$23.00

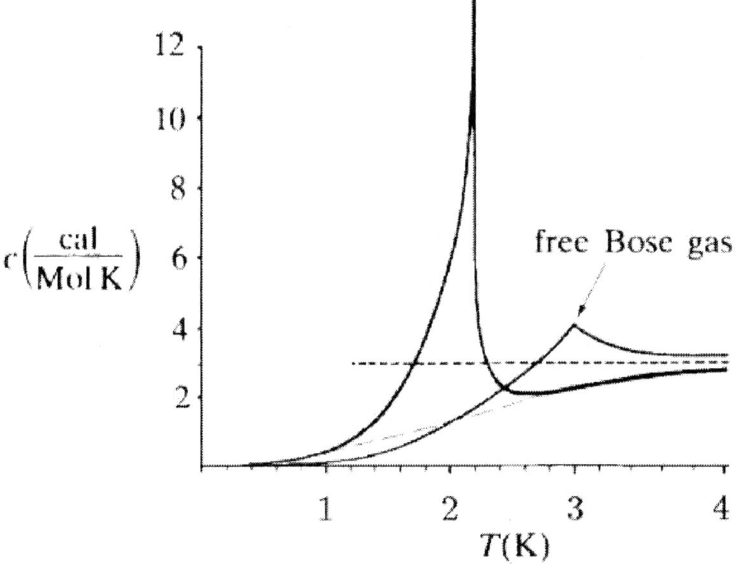

FIGURE 1. Specific heat of superfluid ^{4}He. For very small T, it shows the typical power behavior $\propto T^3$ characteristic for masseless excitations in three dimension in the Debye theory of specific heat. Here these axcitations are phonons of second sound. The peak is caused by the proliferation of vortex loops at the superfluid-normal transition.

in Fig. 2. As long as T stays suffi ciently far below T_c, the thermodynamic properties of the superfluid are dominated by phonons and rotons.

If the temperature approaches T_c, the rotons join each other and form large surfaces. The adjacent boundaries of the rotons cancel each other, so that the memory of the surfaces is lost, their shape becomes irrelevant, and only the boundaries of the surfaces are physical objects, observable as *vortex loops*. At T_c, the vortex loops become infi nitely long and proliferate. The large activation energies for creating single rotons are overcome by the high confi gurational entropy of the long vortex loops.

The inside of a vortex line is a normal fluid since the large rotation velocity destroys the superfluid. For this reason, the proliferation of the vortex loops fi lls the system with normal fluid, and the fluid looses its superfluid properties. This proliferation mechanism was suspected more than fi fty years ago by Onsager in 1949 [4], re-emphasized by Feynman in 1955 [5], and turned into a proper disorder fi eld theory in the 1980's by the author [6]. The same idea was taken up by Shockley in 1952 [7] who proposed a proliferation of defect lines in solids to be responsible for the melting transition. This led to a detailed theory of melting by the author [8].

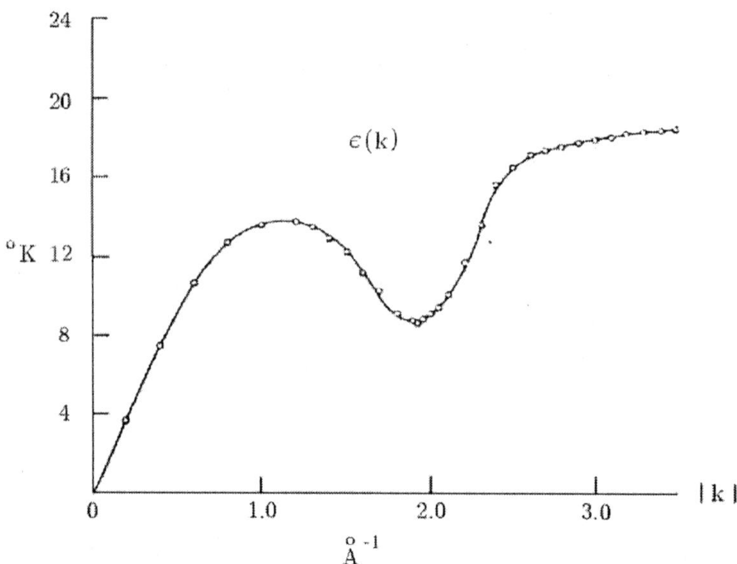

FIGURE 2. Energies of the elementary excitations in superfluid ^{4}He measured by neutron scattering showing the roton minimum near $k \approx 2$ Å.[after R.A. Cowley and A.D. Woods, Can. J. Phys **49**, 177 (1971).

ROLE OF CONFIGURATION ENTROPY

There is a simple estimate for the temperature of such a phase transition. A line-like excitation of length l with an excitation energy ε per unit length is suppressed strongly by a Boltzmann factor $e^{-\varepsilon l/T}$. This suppression is decreased by configurational entropy. The line can bend easily on a length scale ξ, which is of the order of the coherence length of the system, the line has roughly $(2D)^{l/\xi}$ possible configurations, where D is the space dimension [9]. A rough approximation for the partition function of a grand-canonical ensemble of such loops with arbitrary lengths is

$$Z \approx \oint dl\, (2D)^{l/\xi} e^{-\varepsilon l/T}. \tag{2}$$

The integral converges only below a critical temperature

$$T < T_c = \varepsilon \xi / \log(2D). \tag{3}$$

Above T_c, the integral diverges and the ensemble undergoes a phase transition in which the loops proliferate and become infinitely long. This will be called a *condensation process* of loops.

From Eq. (3) we can immediately deduce a relation between the critical temperature and the roton energy in superfluids. The circumference the rotons is bounded below by about $\pi\xi$. Their energy is $E_{\mathrm{roton}} \approx \pi\xi\varepsilon$. Inserting this into Eq. (3) we obtain that the critical temperature of a line-like system in $D = 3$ dimensions is proportional to the roton energy

$$T_c = cE_{\text{rot}} \tag{4}$$

with a proportionality constant

$$c_{\text{lines}} \approx 1/\pi \log 6 \approx 1/5.6. \tag{5}$$

ORIGIN OF MASSLESS EXCITATIONS

The massless excitations in superfluid helium are a consequence of a spontaneous breakdown of a continuous symmetry of the Hamiltonian. Such massless excitations are called *Nambu-Goldstone modes*. These arise as follows. Superfluid ^{4}He is described by a complex *order field* $\phi(\mathbf{x})$ which is the wave function of the condensate. Near the transition and for smooth spatial variations, the energy density is given by the Hamiltonian of Landau, Ginzburg, and Pitaevskii

$$H[\phi] \;=\; \frac{1}{2} \int d^3x \left\{ |\nabla \phi|^2 + \tau |\phi|^2 + \frac{\lambda}{2} |\phi|^4 \right\}, \tag{6}$$

where τ denotes the relative temperature distance from the critical temperature

$$\tau \equiv T/T_c - 1. \tag{7}$$

Below the critical temperature where $\tau < 0$, the ground state lies at

$$\phi(\mathbf{x}) = \phi_0 = \sqrt{\frac{-\tau}{\lambda}} e^{i\alpha}. \tag{8}$$

This field value is called the *order parameter* of the superfluid.

The ground state is not unique but infinitely degenerate. Only its absolute value of $|\phi_0|$ is fixed, the phase α is arbitrary. For this reason, the entropy does not go to zero at zero temperature. The degeneracy in α is is due to the fact that the Hamiltonian density (6) is invariant under constant $U(1)$ phase transformations

$$\phi(\mathbf{x}) \to e^{i\alpha} \phi(\mathbf{x}). \tag{9}$$

The Nambu-Goldstone theorem states that such a degenerate ground state possesses massless excitations, unless there is another massless excitation which prevents this by mixing with the Goldstone excitation. In order to prove this we decompose the order field $\phi(\mathbf{x})$ into size and phase variables

$$\phi(\mathbf{x}) = \rho(\mathbf{x}) e^{i\theta(\mathbf{x})}, \tag{10}$$

and rewrite (6) as

$$H[\rho, \theta] \;=\; \frac{1}{2} \int d^3x \left[(\nabla \rho)^2 + \rho^2 (\nabla \theta)^2 + \tau \rho^2 + \frac{\lambda}{2} \rho^4 \right]. \tag{11}$$

If τ is negative, the size of the order field is frozen at the minimum (8), implying $\rho_0 = \sqrt{-\tau/\lambda}$, and the Hamiltonian (6) can be approximated by its so-called *London limit* or *hydrodynamic limit*,

$$H^{\mathrm{hy}}[\theta] \;=\; \rho_0^2 \int d^3x\,(\nabla\theta)^2. \tag{12}$$

We have omitted a constant *condensation energy*

$$H_{\mathrm{c}}^{\mathrm{hy}} \;=\; -\int d^3x\,\frac{\tau^2}{2\lambda}. \tag{13}$$

The Hamiltonian density (12) shows that the energy of a plane-wave excitations of the phase grows with the square of the wave vector $\mathbf{k}$, and goes to zero for $\mathbf{k} \to 0$. These are the massless Nambu-Goldstone modes. By rewriting (12) as

$$H^{\mathrm{hy}}[\theta] \;=\; \frac{\rho_s}{2M} \int d^3x\,(\nabla\theta)^2, \tag{14}$$

we obtain the usual hydrodynamic kinetic energy, and identifies

$$\rho_s = 2M\rho_0^2 \tag{15}$$

as the *superfluid density*.

Apart from the constant field $\phi(\mathbf{x}) = \phi_0$, the Hamiltonian can be extremized by nontrivial field configurations which represent vortex lines. At the center of each line the size $\rho(\mathbf{x})$ of order field vanishes. The question arises as to what happens to these solutions in the hydrodynamic limit where $\rho(\mathbf{x})$ is constant everywhere? The alert reader may have noticed that in going from (6) to (11) we have made an important error which for the discussion of the Nambu-Goldstone mechanism was irrelevant but becomes important for the understanding of the phase transition. We have used the chain rule of differentiation to express

$$\nabla\phi(\mathbf{x}) = \{i[\nabla\theta(\mathbf{x})]\rho + \nabla\rho(\mathbf{x})\}e^{i\theta(\mathbf{x})}, \tag{16}$$

However, this equation is false. Since $\theta(\mathbf{x})$ is a phase variable, it is intrinsically multi-valued. At every point $\mathbf{x}$ it is possible to add an arbitrary integer-multiple of 2π without changing $e^{i\theta(\mathbf{x})}$.

VORTEX GAUGE FIELDS

The correct chain rule is [10]

$$\nabla\phi(\mathbf{x}) = \{i[\nabla\theta(\mathbf{x}) - 2\pi\delta(\mathbf{x};S)]\rho(\mathbf{x}) + \nabla\rho(\mathbf{x})\}e^{i\theta(\mathbf{x})} \tag{17}$$

where $\delta(\mathbf{x};S)$ is the δ-functions on the surface S across which $\theta(\mathbf{x})$ jumps by 2π. With this, we may approximate in the London limit:

$$|\nabla\phi(\mathbf{x})|^2 \;\xrightarrow[\text{London limit}]{}\; |\phi|^2\,[\nabla\theta(\mathbf{x}) - \theta^{\mathrm{v}}(\mathbf{x})]^2 \tag{18}$$

where we have introduced the field

$$\theta^v(\mathbf{x}) \equiv 2\pi\delta(\mathbf{x},S). \tag{19}$$

With this notation, the correct version of (11) reads therefore

$$H[\rho,\theta] \;=\; \frac{1}{2}\int d^3x\left[(\nabla\rho)^2 + \rho^2(\nabla\theta - \theta^v)^2 + \tau\rho^2 + \frac{\lambda}{2}\rho^4\right], \tag{20}$$

where

$$\theta^v(\mathbf{x}) \equiv 2\pi\delta(\mathbf{x},S). \tag{21}$$

In the London limit, the gradient energy density (14) must be corrected accordingly, so that the hydrodynamic limit of the Ginzburg-Landau Hamiltonian containing phonons and vortex lines reads, from now on in natural units with $\rho_s/M = 1$,

$$H_v^{\mathrm{hy}}[\theta] \;=\; \frac{1}{2}\int d^3x\,(\nabla\theta - \theta^v)^2. \tag{22}$$

This Hamiltonian density is obviously gauge invariant under deformations of the surface, under which $\theta^v(\mathbf{x})$ and $\theta(\mathbf{x})$ change by

$$\theta^v(\mathbf{x}) \to \theta^v(\mathbf{x}) + \nabla\Lambda_\delta^v(\mathbf{x}), \quad \theta(\mathbf{x}) \to \theta(\mathbf{x}) + \Lambda_\delta^v(\mathbf{x}). \tag{23}$$

with the gauge functions

$$\Lambda_\delta^v(\mathbf{x}) = 2\pi\delta(\mathbf{x};V). \tag{24}$$

The field $\theta^v(\mathbf{x})$ is called *vortex gauge field*.

VORTEX DENSITY

By Stokes' theorem we find the *vortex density*

$$\nabla\times\theta^v(\mathbf{x}) \equiv \mathbf{j}^v(\mathbf{x}) = 2\pi\delta(\mathbf{x};L) \tag{25}$$

For closed vortex lines L, the δ-function satisfies

$$\nabla\cdot\delta(\mathbf{x};L) = 0, \tag{26}$$

implying the conservation law

$$\nabla\cdot\mathbf{j}^v(\mathbf{x}) = 0. \tag{27}$$

This conservation law is a trivial consequence of $\mathbf{j}^v$ being the curl of θ^v. It is therefore a Bianchi identity associated with the vortex gauge field structure.

The expression (22) is in general not the complete energy of a vortex configuration. It is possible to add a gradient energy in the vortex gauge field, which introduces an extra *core energy* to the vortex line. The extended Hamiltonian of of the hydrodynamic limit of the Ginzburg-Landau Hamiltonian containing phonons and vortex lines with an extra core energy reads

$$H_{vc}^{hy} = \int d^3x \left[\frac{1}{2} (\nabla\theta - \theta^v)^2 + \frac{\varepsilon_c}{2} (\nabla \times \theta^v)^2 \right].\qquad (28)$$

The extra core energy does not destroy the invariance under vortex gauge transformations (23).

The core energy term is proportional to the square of a δ-function which is highly singular. The singularity is a consequence of the hydrodynamic limit in which the field $\rho(\mathbf{x})$ in (11) is completely frozen at the minimum of (20). Moreover, the coherence length of the ρ-field is zero, and this is the origin of the above δ-functions. With this in mind we may regularize the δ-functions in the core energy physically by smearing them out over the actual small coherence length ξ of the superfluid, which is of the order of a few . Whatever the size of ξ, the regularized last term yields an energy proportional to the total length of the vortex lines.

UNIVERSALITY OF VORTEX GAUGE FIELDS

The same type of gauge fields appear in many physical systems.

- First in the superconductor, where they explain the existence of a tricritical point at which the order of the superconductive phase transition changes from second to first order [16].
- Second is the abelian gauge theory of quark confinement where they explain how a condensate of magnetic monopoles produces a flux tube connecting quarks and antiquarks with a linearly rising potential [17, 18, 19].
- Third in the theory of crystal melting, where they describe dislocation and disclination lines whose proliferation explains the melting transition [8].
- In the theory of gravity where the Riemann-Cartan geometry can be understood as a Euclidean space filled with defect lines, and quantum gravity arises from quantum fluctuations of the shapes of these defect lines [8].

In the lecture these applications were discussed in detail.

REFERENCES

1. P. Debye, *Zur Theorie der spezifischen Wärmen*, Annalen der Physik 39(4), 789 (1912).
2. L.D. Landau, *J. Phys. U.S.S.R.* **11**, 91 (1947) [see also *Phys. Rev.* **75**, 884 (1949)].
3. C.A. Jones and P.H. Roberts, *J. Phys. A: Math. Gen.* **15**, 2599 (1982).
4. L. Onsager, *Nuovo Cimento Suppl.* **6**, 249 (1949).
5. R.P. Feynman, in *Progress in Low Temperature Physics*, ed. by C. J. Gorter (North-Holland, Amsterdam, 1955).
6. H. Kleinert, *Gauge fields in Condensed Matter*, Vol. I: *Superflow and Vortex Lines, Disorder Fields, Phase Transitions*, World Scientific, Singapore, 1989 (kl/b1).
7. W. Shockley, in *L'Etat Solid*, Proc. of Neuvienne-Consail de Physique, Brussels, ed. R. Stoops (Inst. de Physique, Solvay, Brussels, 1952).

8. H. Kleinert, *Gauge fi elds in Condensed Matter*, Vol. II: *Stresses and Defects, Differential Geometry, Crystal Defects*, World Scientific, Singapore, 1989 (kl/b2).

9. Note that this configurational entropy cannot be properly accounted for by a model restricted only to circular vortex lines proposed by
G. Williams, *Phys. Rev. Lett.* **59**, 1926 (1987)

10. H. Kleinert, *Theory of Fluctuating Nonholonomic Fields and Applications: Statistical Mechanics of Vortices and Defects and New Physical Laws in Spaces with Curvature and Torsion*, publ. in *Proceedings of NATO Advanced Study Institute on Formation and Interaction of Topological Defects* at the University of Cambridge, Plenum Press, New York, 1995, p. 201–232 (kl/227).

11. This relation was found in the textbook [6] pp. 517, 518 (kl/bl/gifs/vl-517s.html).

12. J.D. Jackson, *Classical Electrodynamics*, sec. ed., J. Wiley & Sons, New York 1975 (Section 5.5).

13. Similar canonical representations for the defect ensembles in a variety of physical systems are given in
H. Kleinert, *J. Phys.* **44**, 353 (1983) (Paris) (kl/102).

14. H. Kleinert, *Path Integrals in Quantum Mechanics, Statistics, Polymer Physics, and Financial Markets,*
4th ed., World Scientific, Singapore 2006 (kl/b5).

15. See the textbook [6], Part 2, Sections 9.7 and 11.9, in particular Eq.(11.133); also
H. Kleinert and W. Miller, *Phys. Rev. Lett.* **56**, 11 (1986); *Phys. Rev. D* **38**, 1239(1988).

16. H. Kleinert, *Lett. Nuovo Cimento* **35**, 405 (1982) (kl/97). The tricritical value $\kappa \approx 0.8/\sqrt{2}$ derived in this paper was confirmed only recently by Monte Carlo simulations:
J. Hove, S. Mo, and A. Sudbo, *Phys. Rev. B* **66**, 64524 (2002).

17. H. Kleinert, *The Extra Gauge Symmetry of String Deformations in Electromagnetism with Charges and Dirac Monopoles, Int. J. Mod. Phys. A* **7**, 4693 (1992). (kl/203)

18. H. Kleinert, *Double-Gauge Invariance and Local Quantum Field Theory of Charges and Dirac Magnetic Monopoles, Phys. Lett. B* **246**, 127 (1990) (kl/205).

19. H. Kleinert, *Abelian Double-Gauge Invariant Continuous Quantum Field Theory of Electric Charge Confi nement, Phys. Lett. B* **293**, 168 (1992) (kl/211).

Biological and Physical Principles in Self-Organization of Brain

H. Kröger

Département de Physique, Université Laval, Québec, Québec G1K 7P4, Canada

Abstract. Complexity is a widespread phenomenon in nature. The brain is a foremost example of complexity. We address the question: How does such complexity emerge in brain? Which are the biological, chemical or physical principles at work which organize the formation of complexity? In particular we address the topics (i) adaptive learning, (ii) neuron cell death and pruning of synaptic connection after birth, (iii) small world and scale free architecture of neural connectivity, (iv) feature maps and the Kohonen model, (v) self-organized criticality and 1/f frequency scaling.

Keywords: brain, complexity, self-organization
PACS: 87.10.+e,87.19.La,89.75.Fb,87.19.Dd

In physics the reductionist school of thought has long time prevailed, which postulates that the way of understanding laws of nature is by looking at its most elementary building blocks. This point of view has been questioned from the viewpoint of complexity, which says that in nature there are emergent phenomena which cannot be predicted from the laws of elementary constituents - for example the shape of a snow flake cannot be obtained from the laws of hydrogen and oxygen atoms[1]. A prime example of complexity is the human brain. The scientific problem is to understand the formation of such complexity. This leads to the question: Which are the principles that organize the brain to become a complex and also well functioning system? Although the genetic code is known to determine the brain structure to some extent, it is certainly not exclusively responsable. Evidence for this comes, e.g., from the post-natal organization of the visual cortex (see below). It is a biological property that the organization of visual cortex in mammals occurs during a relative short time window shortly after birth during which neural connectivity is established[2, 3]. Most remarkably, this is accompanied by genetically controlled neuron cell death and pruning of synaptic connections[4, 5]. Such apparently paradoxal phenomenon, however, is not an error of nature. This has been studied in a neural network model (adaptive learning variant of Kohonen model) by Pallaver et al.[6], who found that such pruning is actually beneficial as it helps in the organization of inter-neural weights. We may consider such cell death and pruning as a first example of an organizational principle in brain.

In brains of mature adults, functional magnetic resonance imaging (FMRI) has shown evidence for functional networks of correlated brain activity [7]. Eguiluz et al. have measured correlated brain activity with a a local resolution in the order of 1mm. From this they obtained a map of brain activity, having the properties of a Small World Network (SWN)[8], meaning that the network nodes are locally highly connected (large clustering coefficient C) and there are a few direct long range connections (short mean path length L). Also this map displays the property of a scale-free network (SFN)[9] meaning the

CP905, Frontiers of Fundamental Physics (FFP8), Eighth International Symposium
edited by B. G. Sidharth, A. Alfonso-Faus, and M. J. Fullana
© 2007 American Institute of Physics 978-0-7354-0412-0/07/$23.00

the statistical distribution of connection per node follows a power law. Because small-world as well as scale-free networks are widely found in nature (foodwebs of ocean fish, protein networks of yeast, neural connectivity in C.elegans), one may suspect that those network architectures play the role of an organizational principle in nature, and in particular in the brain. This question has been studied in neural network models. SWN neural connectivity has been found in a network of Hodgkin-Huxley neurons to give a fast and coherent system response [10], in an associative memory model to improve restoration of perturbed memory patterns [11] and to reduce learning time and error in a multi layer learning model [12]. Also in an adaptive learning variant of the Kohonen model[6] the neural connectivity architecture was found to be of SWN-type during most of the organizational phase.

Another important principle of neural organization of the brain are the so called feature maps in brain. (i) This can be sets of feature-sensitive cells. An example are cells responding to human faces. (ii) This can also be ordered projections between neuronal layers. Examples are the retinotopic map of the visual field or the somatotopic 'homunculus' in the sensorimotor cortex. (iii) This can also be ordered maps of abstract features. Examples are the color map in the visual area V4, the directional hearing map in the owl midbrain [13] and the target-range map in the bat auditory cortex [14]. More than one response property can be mapped simultaneously within an area. This has been shown in visual, somatosensory and auditory cortex by Hubel and Wiesel [15], Sur et al.[16], Linden and Schreiner[17], and Friedman et al.[18]. No receptive surface exists for abstract features, hence the spatial order of representations must be produced by some self-organizing process in brain. Such process occurs postnatally.

It has been proposed that those maps do smoothly map several response properties onto a low-dimensional (2-dimensional) cortical surface. Because several response properties mathematically correspond to high-dimensional space, this mapping means a dimensional reduction. The smoothness of the map means that the mapping is continuous, i.e. nearby receptors correspond to nearby representation of features in the cortex (a well-known example is the somatotopic 'homunculus' in the sensorimotor cortex). This has led the finish mathematician Kohonen[19] to propose a model of formation of such feature maps (Kohonen model, Self-Organized Map (SOM)). Similar models have been proposed later by Durbin et al.[20], Obermayer et al.[21] and Swindale[22].

The Kohonen model is biologically relevant. This has been established in a number of cases. Obermayer et al.[21, 23], Goodhill[24], Wolf et al.[25], Swindale and Bauer[26] have shown that the SOM gives mappings which resemble the real visual cortex maps, even in some detail. The visual cortex maps in cat have been studied by Swindale et al.[27] They found optimal coverage of the map to be important for cortical map development. This gives support to the concept of dimensional reduction, an essential ingredient of the models of Kohonen[19] and Durbin et al.[20]. Using the Kohonen model Swindale[28] studied how many different feature spaces may be represented in cortical maps. Recently, detailed analysis of brain maps have given further evidence that the Kohonen model is biologically realistic. In a study of attentional impairment and familarity preference in autism Gustafsson[29] employed sucessfully the self-organized map. Sur et al.[30], when investigating the primary visual cortex (V1) of ferret, found a distorsion in the mapping of the visual scene onto the cortex, in agreement with the predictions of the Kohonen model. Very recently, Aflalo et al.[31] investigated origins of complex organization of motor cortex. They found that the SOM map contained many features of actual motor cortex in monkey.

$1/f$ "flicker" noise is another phenomenon widely found in nature. Examples where this occurs are tube amplifiers, resistors, voice and musical broadcasts. There are many others. In the quest for a universal theoretical explanation for $1/f$ frequency behavior, Bak, Tang and Wiesenfeld[32] proposed the sand-pile cellular automaton. A sand-pile (or better a rice-pile) is an example of a system, having meta-stable states, which from time to time declenches avalanches, the frequency and the size of which have self-similar behavior. Bak et al. called it a self-organized critical system (SOC). Self-organized critical states are found for many complex systems in nature, e.g., in earthquakes, avalanches, spreading of forrest fires [33, 34]. Such systems are characterized by some power-law behavior of variables such as event duration or the waiting time between events. $1/f$ noise is usually considered as a footprint of such systems [33]. $1/f$ frequency scaling indicates long-lasting correlations in the system, similar to the behavior of physical systems at critical points.

$1/f$ scaling behavior in the frequencies of electrical signals has been observed in electroencephalograms (EEGs) and magnetoencephalogram (MEGs) from human brain [35, 36]. Also neural avalanches have been seen in experiments from local field potential (LFPs) in *in vitro*[38]. Since the discovery $1/f$ frequency scaling of electrical signals in the brain, people started to wonder if the SOC model might be the underlying organizational principle. There is some evidence in support of the existence of such critical states in brain activity. EEG analysis[37] and avalanche analysis of local field potentials (LFPs)[38] provided evidence for self-organized critical states with power-law distributions. There is also evidence for critical states from the power-law scaling of interspike interval (ISI) distributions computed from retinal, visual thalamus and primary visual cortex neurons[39]. In a neural network model of brain plasticty deArcangelis et al.[40] found that critical states may be associated with frequency scaling consistent with experiments. Moreover, $1/f$ spectra in a dissipative SOC model have been found[41].

However, in order to firmly establish the the SOC model as underlying mechanism of $1/f$ scaling in the brain, it would require to establish the presence of long-living meta-stable states[42]. This has not been experimentally achieved yet. Thus the presence of the SOC mechanism has not been firmly determined yet. In an attempt to answer these questions, Bédard et al.[43] investigated if $1/f$ frequency scaling is present in global variables recorded *in vivo* close to the underlying neuronal current sources (LFPs). They analyzed cortical LFPs which were recorded within cerebral cortex using bipolar extracellular high-impedance microelectrodes [44]. LFP recordings sample localized populations of neurons. This is in contrast to the EEG, sampling much larger populations of neurons [45] and being recorded from the surface of the scalp using millimeter-scale electrodes. LFP signals undergo much less filtering than EEG signals, because the latter scatter through various media, like cerebrospinal fluid, dura matter, cranium, muscle and skin. Thus, finding $1/f$ frequency scaling of bipolar LFPs would be a much stronger evidence that this scaling reflects neuronal activities, as these signals are directly recorded from within the neuronal tissue. In order to distinguish state-dependent scaling properties, data have compared recordings during wakefulness and slow-wave sleep in the same experiments[43].

The following observations were made. LFPs from cat parietal association cortex

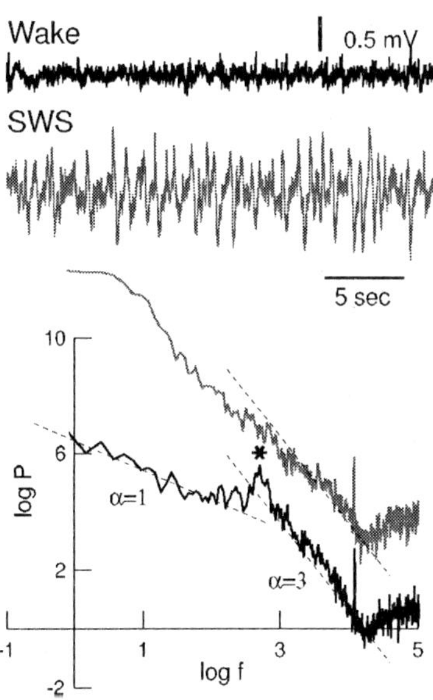

FIGURE 1. Frequency scaling of local field potentials from cat parietal cortex. Top traces: LFPs recorded in cat parietal cortex during wake and slow-wave sleep (SWS) states. Bottom: Power spectral density of LFPs, calculated from 55 sec sampled at 300 Hz (150 Hz 4th-order low-pass filter), and represented in log-log scale (dashed lines represent $1/f^\alpha$ scaling). During waking (black), the frequency band below 20 Hz scales approximately as $1/f$ (*: peak at 20 Hz beta frequency), whereas the frequency band between 20 and 65 Hz scales approximately as $1/f^3$. During slow-wave sleep (gray; displaced upwards), the power in the slow frequency band is increased, and the $1/f$ scaling is no longer visible, but the $1/f^3$ scaling at high frequencies remains unaffected. PSDs were calculated over successive epochs of 32 sec, which were averaged over a total period of 200 sec for Wake and 500 sec for SWS.

show typical landmarks of EEG signals [46]. During waking, LFP signals have low amplitudes and display very irregular (chaotic) temporal behavior (Fig. 1 top). They are dominated by low frequencies around 20 Hz (β frequencies). This pattern of "desynchronized" activity is typically seen during aroused states in human EEG [45]. During slow-wave sleep, LFPs show high-amplitude low-frequency activity (Fig. 1 middle), similar to the "δ frequencies" of human sleep EEG [45]. The power spectral density (PSD) obtained from these LFPs (Fig. 1 bottom) shows a broad-band structure. During wakefulness, one observes in the PSD the existence of two different scaling windows. For low frequencies ($1\,Hz < f < 20\,Hz$) the PSD scales approximately like $1/f$, whereas for higher frequencies ($20\,Hz < f < 65\,Hz$) the PSD scales approximately like $1/f^3$ (Fig. 1, black PSD). During slow-wave sleep, the same $1/f^3$ scaling is observed in the high-frequency band (Fig. 1, gray PSD). These results are consistent with the findings of $1/f$ frequency scaling in the EEG [35], but for cat association cortex LFPs this is observed only during waking and for specific frequency bands.

The finding of a window of $1/f$ frequency scaling in the PSD of cat cortex LFP's reflects self-similarity in the time-evolution of signals. Because the PSD is essentially the same as the auto correlation function, this indicates some long range correlation. It

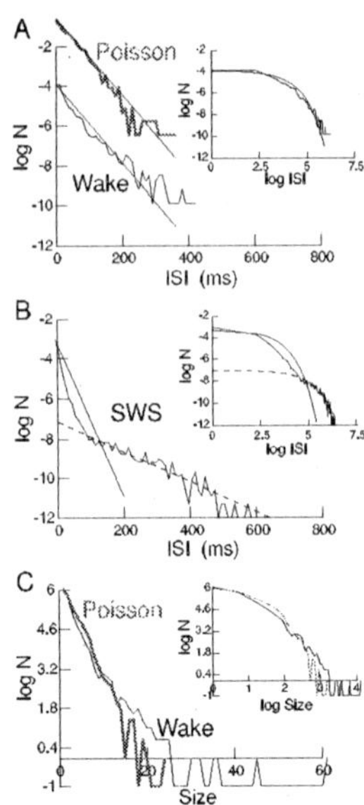

FIGURE 2. Absence of power law distributions in neuronal activity. The logarithm of the distribution of interspike intervals (ISI) during waking (Wake, A, 1951 spikes) and slow-wave sleep (SWS, B, 15997 spikes) is plotted as a function of ISI length, or log ISI length (insets). A poisson process of the same rate and statistics is displayed in A (Poisson; gray curve displaced upwards for clarity). The exponential ISI distribution predicted by Poisson processes of equivalent rates is shown as straight lines (smooth curve in inset). The dotted line in B indicates a Poisson process with lower rate which fits the tail of the ISI distribution in SWS. C. Avalanche analysis realized by taking into account the statistics from all simultaneously-recorded cells in Wake. The distribution of avalanche sizes scales exponentially (black curves), similar to the same analysis performed on a Poisson process with same statistics (gray curves).

may be a signal for the presence of SOC. However, as discussed above, SOC requires the presence of a meta-stable state, which is difficult to detect experimentally. However, in the presence of SOC one would expect self-similarity and a power law behavior not only in the distribution of frequencies but also in the distribution of the inter-spike-intervals (ISI) as well as in the distribution of the size of avalanches. This has been analyzed in Fig. 2. One observes that the distribution of ISIs follows a Poisson distribution and not a power law distribution (Fig. 2 top and middle). The same holds for the distribution of the avalanche sizes (Fig. 2 bottom). This evidence is inconsistent with the SOC model.

What is the explanation of the dynamical behavior of $1/f$ and $1/f^3$ scaling windows observed in LFPs of cat cortex? Bédard et al.[47, 48] have studied models of propagation of electrical signals in brain tissue. In Ref.[48] in a model considering neurons as electrical sources and glial cells standing for a responsive medium, and taking into account resistive and capacitive properties, the model predicted attenuation of high frequencies, consistent with experiments. Moreover, the model was found to be electrically equivalent (in the case of a single neuron source and a single passive glial cell) to an RC-circuit. This suggests that a system of many neurons and many glial cells may be electrically equivalent to a system of coupled RC-circuits. It has been shown by Barnes and Jarvis[49] that a large (infinite) chain of coupled RC-cicuits generates $1/f$ noise. This may give a possible explanation of $1/f$ frequency scaling observed in LFPs of cat cortex, as being a property of the brain tissue. The scaling window showing $1/f^3$ scaling may be due to synaptic noise (showing $1/f^2$ scaling) convoluted with $1/f$ noise of the brain tissue.

OUTLOOK

Recently, experiments have shown that the brain is much more complicated than a network of neurons. Glial cells, which for long time have been thought to play only a passive role (e.g. to restore equilibrium of ion concentrations), have been found to play an active role and, moreover, form their own network interacting via chemical messangers. We are only at the beginning to understand the role of those glial cells. This raises questions like: How are interactions of those cells and its network connections organized? What about interactions between neurons and glial cells? There are many open questions.

ACKNOWLEDGMENTS

H.K. has been supported by NSERC Canada.

REFERENCES

1. R. B. Laughlin, *A Different Universe*, Basic Books, New York, 2005.
2. C. Von der Malsburg, *Kybernetik* **14**, 85 (1973).
3. R. Linsker, *Proc. Natl. Acad. Sci. USA* **83**, 7508 (1986); **83**, 8390 (1986); **83**, 8779 (1986).
4. G.M. Shepherd, *Neurobiology*, Oxford University Press, (1994).
5. E.R. Kandel, J.H. Schwartz, and T.M. Jessel, *Essentials of neuroscience and behavior*, Appleton and Lange (1995).
6. T. Pallaver, H. Kröger, and M. Parizeau, arXiv physics/0609002.
7. V.M. Eguiluz, D.R. Chialvo, G.A. Cecchi, M. Baliki, and A.V. Apakarian, *Phys. Rev. Lett.* **94**, 018102 (2005).
8. D.H. Watts, and S.H. Strogatz, *Nature* **393**, 440 (1998).
9. A. Barbási, and R. Albert, *Science* **286**, 509 (1999).
10. L.F. Lago-Fernandez, R. Huerta, F. Corbacho, and J.A. Sigüenza, *Phys. Rev. Lett.* **84**, 2758 (2000).
11. J.W. Bohland, and A.A. Minai, *Neurocomputing* **38-40**, 489 (2001).
12. D. Simard, L. Nadeau, and H. Kröger, *Phys. Lett. A* **336**, 8 (2005).
13. E.I. Knudsen, and M. Konishi, *Science* **200**, 795 (1978).
14. N. Suga, and W.E. O'Neill, *Science* **206**, 351 (1979).
15. D.H. Hubel, and T.N. Wiesel, *J. Physiol.* **165**, 559 (1963).
16. M. Sur, J.T. Wall, and J.H. Kaas, *Science* **212**, 1059 (1981).
17. J.F. Linden, and C.E. Schreiner, *Cereb. Cortex* **13**, 83 (2003).
18. R.M. Friedman, L.M. Chen, and A.W. Roe, *Proc. Natl. Acad. Sci. USA* **101**, 12724 (2004).
19. T. Kohonen, *Biol. Cybern.* **43**, 59 (1982); **44**, 135 (1982); T. Kohonen, *Proc. of the IEEE*, 1464 (1990); For recent work on Kohonen maps see: U. Seiffert, and L.C. Jain, eds., *Self-Organizing Neural Networks*, Physica Verlag, Heidelberg (2002); E. Oja, and S. Kaski, eds., *Kohonen Maps*, Elsevier, Amsterdam (1999).
20. R.. Durbin, and G. Mitchison, *Nature* **343**, 644 (1990).

21. K. Obermayer, H. Ritter, and K. Schulten, *Proc. Natl. Acad. Sci. USA* **87**, 8345 (1990).
22. N.V. Swindale, *Biol. Cybern.* **65**, 415 (1991).
23. K. Obermayer, G.G. Blasdal, and K. Schulten, *Phys. Rev. A* **45**, 7568 (1992).
24. G.J. Goodhill, *Biol. Cybern.* **69**, 109 (1993).
25. F. Wolf, H.U. Bauer, and T. Geisel, *Biol. Cybern.* **70**, 525 (1994).
26. N.V. Swindale, and H.U. Bauer, *Proc. Roy. Soc. Lond. B* **265**, 827 (1998).
27. N.V. Swindale, D. Shoham, A. Grinvald, and T. Bonhoeffer, *Nature NeuroSci.* **3**, 822 (2000).
28. N.V. Swindale, *Network: Comput. Neur. Syst.* **15**, 217 (2004).
29. L. Gustafsson, and A.P. Papliński, *J. Autism Develop. Disord.* **34**, 189 (2004).
30. H. Yu, B.J. Farley, D.Z. Jin, and M. Sur, *Neuron* **47**, 267 (2005).
31. T.N. Aflalo, and M.S. Graziano, *J. Neurosci.* **26**, 6288 (2006).
32. P. Bak, C. Tang, and K. Wiesenfeld, *Phys. Rev. Lett.* **59**, 381 (1987); *Phys. Rev. A* **38**, 364 (1988).
33. H.J. Jensen, *Self-Organized Criticality. Emergent Complex Behavior in Physical and Biological Systems.* (Cambridge University Press, Cambridge UK, 1998).
34. P. Bak, *How Nature Works* (Springer-Verlag, New York, 1996).
35. W.S. Pritchard, *Int. J. Neurosci.*, **66**, 119 (1992); W.J. Freeman, L.J. Rogers, M.D. Holmes, and D.L. Silbergeld, *J. Neurosci. Methods* **95**, 111 (2000); P.A. Robinson, C.J. Rennie, J.J. Wright, H. Bahramali, E. Gordon, and D.L. Rowe, *Phys. Rev. E* **63**, 021903 (2001).
36. E. Novikov, A. Novikov, D. Shannahoff-Khalsa, B. Schwartz, and J. Wright, *Phys. Rev. E* **56**, R2387 (1997).
37. K. Linkenkaer-Hansen, V.V. Nikouline, J.M. Palva, and R.J. Ilmoniemi, *J. Neurosci.* **21**, 1370 (2001); W.J. Freeman, M.D. Holmes, G.A. West, and S. Vanhatalo, *Clin. Neurophysiol.* **117**, 1228 (2006).
38. J.M. Beggs, and D. Plenz, *J. Neurosci.* **23**, 11167 (2003); **24**, 5216 (2004).
39. M.C. Teich, C. Heneghan, S.B. Lowen, T. Ozaki, and E. Kaplan, *J. Opt. Soc. Am. A* **14**, 529 (1997); A.R.R. Papa, and L. da Silva, *Theory in Biosciences* **116**, 321 (1997).
40. L. de Arcangelis, C. Perrone-Campano, and H.J. Herrmann, *Phys. Rev. Lett.* **96**, 028107 (2006).
41. P. De Los Rios, and Y.C. Zhang, *Phys. Rev. Lett.* **82**, 472 (1999); T. Giesinger, *Biol. Rev.* **76**, 161 (2001).
42. M. Paczuski, P. Bak, and S. Maslov, *Phys. Rev. E* **53**, 414 (1996).
43. C. Bédard, H. Kröger, and A. Destexhe, *Phys. Rev. Lett.* **97**, 118102 (2006).
44. A. Destexhe, D. Contreras, and M. Steriade, *J. Neurosci.* **19**, 4595 (1999).
45. E. Niedermeyer, and F. Lopes da Silva, (editors) *Electroencephalography* (Williams & Wilkins, Baltimore MD 1998).
46. M. Steriade, *Neuronal Substrates of Sleep and Epilepsy* (Cambridge University Press, Cambridge UK, 2003).
47. C. Bédard, H. Kröger, and A. Destexhe, *Biophys. J.* **86**, 1829 (2004).
48. C. Bédard, H. Kröger, and A. Destexhe, *Phys. Rev. E* **73**, 051911 (2006).
49. J.A. Barnes, and S. Jarvis, Jr., *Efficient Numerical and Analog Modeling of Flicker Noise Processes* Nat. Bur. Stand.(U.S.), Tech. Note 604 (June 1971).

A Simulated Study and Thermal Stability of Fullerenes and Derivatives

Consolación Manteca-Diego, Jezabel Doménech,
Gúmer F. Doménech-Manteca

E.I.T. Aeronáutica, Dpto. "Tecnologías Especiales Aplicadas a la Aeronáutica"
Universidad Politécnica de Madrid. Plaza Cardenal Cisneros n° 3, Spain
E-mail:consolacion.manteca@upm.es

Abstract In this communication IRIS 4D, **Silicon Graphics**, working stations have been used. Molecule dynamics and energy calculations have been performed with **Discover** and the graphical representations with **Insight II**. Depending on the Schlegel diagram used, different sets of C-C distances are obtained. 1.51 Å as the average distance between two atoms of carbon defining a hexagon-hexagon edge and 1.43 Å for the hexagon-pentagon ones, in quite close agreement with the experimental. The closest distance between the two different molecules, as determined by the corresponding Van der Walls clouds, can be measured as 3.1Å. One result is the extremely high apparent stability of the C_{60} molecule, 8000 K.

Keywords: Fullerenes, nanotubes, simulated, Schlegel diagram, C_{60}.
PACS: 61.48.+c

INTRODUCTION

The appearance of new allotropic forms of elemental carbon, polyhedrally-shaped cluster, with the general formula C_{20+h}, known in the literature as fullerenes, a great deal of interest has been attracted because of their unusual properties: fivefold local symmetry, new superconductors, transformation into diamond at very high pressures, endohedral or exohedral doping, tubular or onion-shaped nanostructures [1,2] etc. Besides a myriad of organic derivatives, let us say that a new era in carbon chemistry has emerged [3, 4, 5, 6].

Kroto, Smalley, and their co-workers had developed a theory based upon topological considerations to account for the special stability of C_{60} and C_{70}. Chemists have an instinctive delight in symmetry, and the proposed structures, which, in outline form, do look like soccer balls, not only provide tentative rationalization of the stability of the

CP905, *Frontiers of Fundamental Physics (FFP8), Eighth International Symposium*
edited by B. G. Sidharth, A. Alfonso-Faus, and M. J. Fullana
© 2007 American Institute of Physics 978-0-7354-0412-0/07/$23.00

molecules but also have great aesthetic appeal.

Nevertheless, one of the most important limits to perform research in this field is the availability of large quantities of high purity fullerenes: their synthesis is bases on ac discharge between graphite electrodes and the soot produced by these means has to be purified, a mixture of C_{60} and C_{70} being obtained (7). On the other hand, in the fastly growing field of computational chemistry, there is useful software to perform " ab initio" complex calculations, molecular design molecular dynamics and simulation. Moreover, some software of this kind has been developed to be used on PC´s. It is obvious then that simulation may help to understand many features of the compounds under study, thus replacing sample consuming experiments or at least giving some previous ideas and therefore is highly indicated in this field.

METHODS

In this communication IRIX 4D, **Silicon Graphics**, working stations have been used. Molecular dynamics and energy calculations have been performed with **Discover** and the graphical representations with **Insight II**, both computational chemistry programs supplied by Biosym Technologies. At a much more moderate level, the **3.0** version of **Hyperchem** by Addlink Corp. is adequate for use on a 486 PC with these purposes. The calculations are iterative processes which look for local minima of the energy, using, minimization algorithms such as conjugated gradient, Newton-Rampson, etc.

Complementary to this, in molecular dynamics, the conformational space of a given molecule is studied at fixed values of P and T: the procedure is based on the numerical resolution of the Newton equation for motion. All these calculations depend on the field of forces used, which describes the peculiar characteristics of each atom involved. As a previous step, planar connectivity diagrams (Schlegel, Figure 1-A) have to be generated.

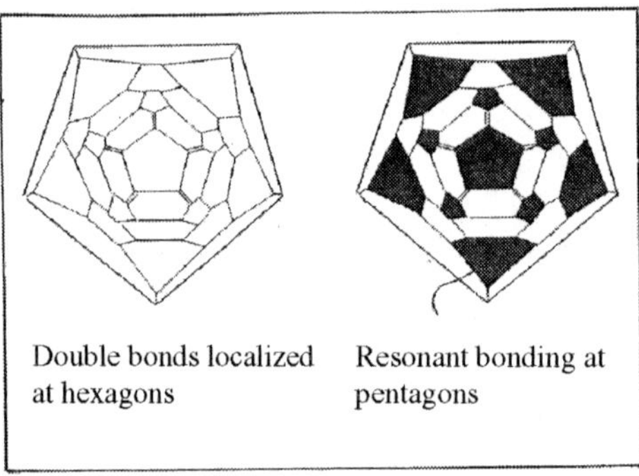

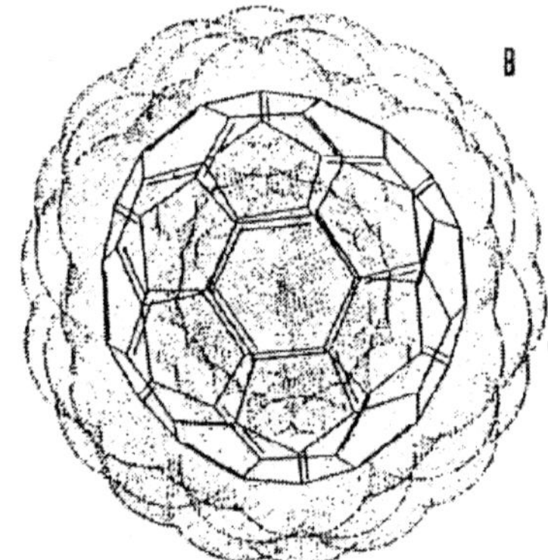

FIGURE 1. (A) Connectivity diagrams used to generate the C_{60} molecule. Observe the localized or delocalized double bonds. (B) Pseudospherical C_{60} represented with its Van der Waals cloud.

RESULTS

Depending on the Schlegel diagram used, different sets of C-C distances are obtained and, although the exact nature of the C-C bonds is not well yet fully understood, model B, which supposes po-delocalization in the pentagons, yields better results: 1.51 Å as the average distance between two atoms of carbon defining a hexagon-hexagon edge and 1.43 Å for the hexagon-pentagon ones, in quite close agreement with the experimental results (see Table I). The pseudosphere diameter is 7.16 Å and the closest distance between the two different molecules, as determined by the corresponding Van der Walls clouds, can be measured as 3.1 Å (Figure 1-B).

This kind of study can be extended to other fullerenes, such as C_{70}, or to nanotubes and to many derivatives as well, in particular to hydrogenated or halogenated species such as $X_n C_{60}$ (X = H, F, Cl, Br; n = 6, 8, 24, 60). The corresponding measured bond distances fit quite well with the calculated values on the basis of covalent single bonding (See Table I).

Table 1. Simulated bond distances (Angstroms)

BOND	COMPOUND	d(calc.)	d(bibl.)
C-C	C_{60}	1.51 (H-H)	1.45
C-C	C_{60}	1.12 (H-P)	1.39
C-C	$X_n C_{60}$	1.55	1.54
C-H	$H_{60}C_{60}$	1.12	......1.09
C-F	$F_{60}C_{60}$	1.39	1.35
C-Cl	$C_{24}C_{60}$	1.90	1.77
C-Br	$Br_{24}C_{60}$	1.94	......2.01

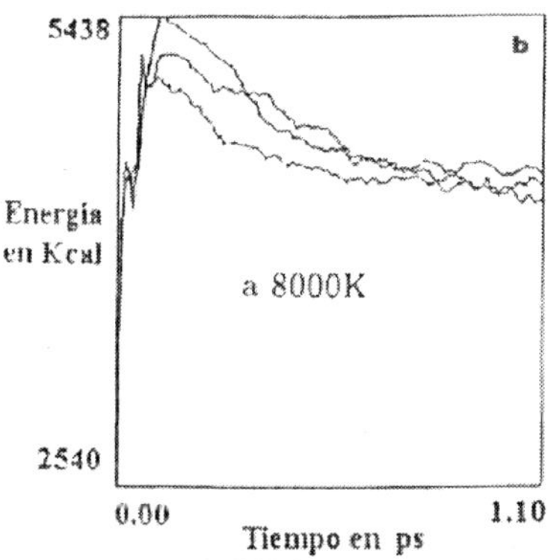

FIGURE 2. Energy vs. time plot of C_{60} at a very high temperature

Regarding thermal behaviour, one very surprising result is the extremely high apparent stability of the C_{60} molecule: even at 8000 K, an asymptotic plot is obtained when plotting energy versus time (see Fig. 2). In spite of the fact that C-C bonding distances increase moderately, the sphere does not blow up as could be expected; this seems to happen at higher temperatures. This result is not at all contradictory to the high reactivity experimentally –i.e. it burns at temperatures much lower than graphite, because this occurs only under vacuum; that is, it refers to intrinsic stability, given by the strength of its bonds and peculiar geometry.

On the contrary, hydrogen or halogen derivatives are much less stable. The molecules are far less rigid than that of pristine fullerene, as a consequence of double bonds breaking. Molecular dynamics let observe how C-C vibrations increase to X atoms; the break limit (3 Å, arbitrarily taken) is reached at fairly low temperatures. This simulated studies seem to indicate that the thermal stability of halogen derivatives follows F> Cl > Br and any at descomposes at temperatures about 400 K, in good agreement with experiments [8] although experimental data vary somehow depending on the authors surely due to chemical instabilities of the compounds studied.

REFERENCES

1.   Diang Lu and Baidurya Bhattacharya. Eng Fract. Mech. **72**, 2037-2071 (2005).
2.   Y. C. Zhang and X. Wang. Int. J. Solids Struct. **42**, 5399-5412 (2005).
3.   H. W. Kroto et al. Nature, **318**, 162 (1985).
4.   Accounts of Chemical Research (special Volume) **98**-175 (1992).
5.   Carbon (special Volume) 30, 1139-1286 (1992).
6   M. S. Dresselhaus et al. Journal of Materials Research. **8**, 2054-2097 (1993).
7   W. Krätschmer et al. Nature 347, 354-358 (1990).
8.   R. Taylor and D. R. W. Walton. Nature, 363, 685-693 (1993).

Spectroscopy Study (IR, ^{1}HNMR, ^{13}C-NMR, MS) of Methyl Cyanide Addition Compound of Boron Trichloride. Donor-Aceptor Bonding and Applications

Consolación Manteca-Diego, Jezabel Doménech, and
Gumer F. Doménech-Manteca

E. I. T. Aeronáutica. Dpto. "Tecnologías Especiales Aplicadas a la Aeronáutica" Universidad Politécnica de Madrid (UPM). Plaza de Cardenal Cisneros nº 3 Madrid (Spain) E-mail:consolación.manteca@upm.es

Abstract. The 1:1 $CH_3 CN: BCl_3$ complex was prepared by reacting equimolecular quantities of $CH_3 CN$ and BCl_3 and was characterized by IR, ^{1}H-NMR, ^{13}C- NMR, mass spectroscopy and elemental analysis. High resolution electron microscopy (HRTEM) and electron energy loss spectroscopy (EELS) suggest that the stoichiometry of the filaments ca..$[BC_2N]$ obtained from pyrolysis of CH_3-$CN:BCl_3$

Keywords: Nanostructures, HRTEM, EELS, CH_3-$CN.DCl_3$
PACS: 61.48.+c

INTRODUCTION

Pyrolysis of $CH_3.CN:BCl_3$ (a) at ca 900-1000°C over C powder generates novel graphitic $B_x C_y N_z$ nanofibres and nanotubes possessing morphologies (E. G curled, branched and bent). In these experiments the metal particles play an important role in the growth since nanotube formation appears to occur at the metal surface. High resolution electron microscopy (HRTEM) and electron energy loss spectroscopy (EELS) studies suggest that the stoichiometry of the filaments ca..$[B C_2N_z]_n$

Their structures were investigated, concentration profiles, along and across the filaments, revealed that B, C and N are not homogeneously distributed within the nanostructures but are separated into pure C and BN domains, this compound provides nanomaterials, which may prove useful as robust nanocomposites and semiconductor nanodevices with enhanced towards oxidation [1, 2, 3].

CP905, *Frontiers of Fundamental Physics (FFP8), Eighth International Symposium*
edited by B. G. Sidharth, A. Alfonso-Faus, and M. J. Fullana
© 2007 American Institute of Physics 978-0-7354-0412-0/07/$23.00

EXPERIMENTAL

The 1:1 CH_3 CN:BCl_3 complex [4] was prepared by reacting equimolecular quantities of CH_3CN and BCl_3 (1 M solution in hexane at $-78°C$ under vacuum). The white solid product was purified by sublimation at 85°C and characterized by UV, IR, ^{1}H-NMR, ^{13}C-NMR, mass spectrometry and elemental analysis (table 1)

Pyrolysis was carried out in a two-stage furnace system fitted with temperature controllers (see figure 1). CH_3–CN:BCl_3 (ca. 100-200 mg) was placed in one end of a quartz tube (6 mm O.D and 60 cm in length) located in the first furnace.

Cobalt powder (Aldrich 99.9 µm particle size) in a quartz boat (6 cm in length was placed in the second furnace.

..Argon was used as a carrier gas and the temperature of the first furnace was slowly increased from room temperature to 500°C. The second furnace was maintained at 1000°C for the duration of the experiment.

Residues scraped from the tube wall in the second furnace were analyzed by HRTEM and EELS and studies suggest that the BC_2N filaments exhibit increasing diameters as a function of distance from the catalytic particles.

RESULTS

TABLE 1. Spectroscopy characterization of CH_3 CN:BCl_3

TECNIQUE	BANDS POSITION	UNITS
IR (KBr)	3261.0 (br), 885.0 (br)	cm^{-1}
^{1}H-NMR (DMSO-d◦	2-0 (s, 3H, Me)	ppm
^{13}C-NMR (DMSO)-d◦	118 (CN), 1. 3 (Me)	ppm
MS (rel. Intensity)	156 (M^+, not detected)	m/e
	115 (BCl^{3+}, 22 80 (BCl^{2+}	m/e
	41 (CH_3-CN)	M/e

CONCLUSIONS

1. Spectroscopy characterization of CH_3- CN: BCl_3
2. Pyrolytically grown BC_2N nanofibres have been studied at the nanoescale level.
3. In spite of the relatively low temperature compared to other synthetic routes employed previously.
4. The resulting BC_2N filaments exhibit increasing diameters as a function of distance from the catalytic particles.

REFERENCES

1. M. Terrones, A. M. Benito, C. Manteca-Diego, W. K. Hsu, O. I. Osman, J.P. Hare, D. G. Reid, H. Terrones, A. K. Cheethan, K. Prassides, H. W. Kroto, D. R. M. Walton, *Chem. Phys. Lett.*, **257**, 576-582 (1996).
2. Ph. Kohler-Redlich, M. Terrones, C. Manteca-Diego, W. K. Hsu, H. Terrones, M. Rühle, H. W. Kroto, D. R. M. Walton, *Chem. Phys. Lett*, **310**, 459-465 (1999).
3. A. W. Laubengayer and D. S. Sears, *J. Am. Chem. Soc.* **67**, 164 (1945).
4. E. Nylas and A. H. Soloway, *J. Am. Chem. Soc*, 2681 (1959).

Are the Truly Constant Constants of Nature?
How is the Real Material Space and its Structure?

José De La Luz Montero García*, Jesús Francisco Novoa Blanco[†]

*Institute for Scientific and Technological Information (IDICT); National Capitol, Havana, Cuba.
E-mail: quantumstructure@yahoo.es
[†] The late.

Abstract. In a concise and simplified way, some matters of authors' theories —Unified Theory of the Physical and Mathematical Universal Constants and Quantum Cellular Structural Geometry—, an only one theoretical main body MN2. This investigation has as objective the search of the last cells that base the existence, unicity and harmony of matter, as well as its structural-formal and dynamic-functional diversity. The quantitative hypothesis is demonstrated that "World is one, is one; but it is one Arithmetic-Geometric-Topological-Dimensional and Structural-Cellular-Dynamic one, simultaneously."
In the Frontiers of Fundamental Physics such last cells are the cells of own Real Material Space of whose whole accretion, interactive and staggered all the existing one at all the hierarchic levels arises, cells these below which make no sense to speak of structure and, therefore, of existence. The cells of the Real Material Space are its "Atoms". Law of Planetary Systems or "4th Kepler's Law".

Keywords: theory; real material space; universal constants; dimension; structure; dimensionality of scale.

1. Dimensional Self-assembly and Deployment of the RMS

Our arithmetic-geometric-topological-physical research[1] of matter brought as result that the own scale of matter has by base a dimensional octet being the primitive dimension characteristic of:

$$length; \qquad \alpha = \left\{ \frac{h \cdot N_A}{c^2 \cdot F} \right\} \left(\frac{G}{\mu_0} \right)^{1/2} ; \qquad \alpha = 3{,}3530 \cdot 10^{-34} \text{ m} \qquad (1)$$

$$timeo; \qquad \beta = \left\{ \frac{h \cdot N_A}{c^3 \cdot F} \right\} \left(\frac{G}{\mu_0} \right)^{1/2} ; \qquad \beta = 1{,}1184 \cdot 10^{-42} \text{ s} \qquad (2)$$

$$mass; \qquad \theta = \left\{ \frac{h \cdot N_A}{F} \right\} \left(\frac{1}{G \cdot \mu_0} \right)^{1/2} ; \qquad \theta = 4{,}5166 \cdot 10^{-7} \text{ kg} \qquad (3)$$

$$electric\ charge; \quad \Delta = \left\{ \frac{h \cdot N_A}{c \cdot F} \right\} \left(\frac{1}{\mu_0} \right) ; \qquad \Delta = 1{,}0978 \cdot 10^{-17} \text{ C} \qquad (4)$$

thermodynamic temperature of energetic hierarchy;

$$\gamma = \left\{ \frac{c^3 \cdot F}{k \cdot N_A} \right\} \left(\frac{\mu_0}{G} \right)^{1/2} ; \qquad \gamma = 4{,}2911 \cdot 10^{31} \text{ K} \qquad (5)$$

$$particle; \qquad \phi = \left\{ \frac{h \cdot N_A^2}{c \cdot F^2} \right\} \left(\frac{1}{\mu_0} \right) ; \qquad \phi = 6{,}8518 \cdot 10^{1} \text{ p} \qquad (6)$$

$$grouping; \qquad \Omega = \left\{ \frac{h \cdot N_A}{c \cdot F^2} \right\} \left(\frac{1}{\mu_0} \right) ; \qquad \Omega = 1{,}1378 \cdot 10^{-25} \text{ kmol} \qquad (7)$$

$$cycle; \qquad \varphi \overset{def}{\equiv} 2\pi ; \qquad \varphi = 2\pi \text{ rad} \qquad (8)$$

CP905, *Frontiers of Fundamental Physics (FFP8), Eighth International Symposium*
edited by B. G. Sidharth, A. Alfonso-Faus, and M. J. Fullana
© 2007 American Institute of Physics 978-0-7354-0412-0/07/$23.00

Everything and any dimension is a property of any and all real material physical system, which is expression of its real material existence and concrete manifestation in the given system of the homologous property of Real Material Space, of which it comprises and to which also this one describes.

If we admitted the orthogonality as future condition indispensable for our theory the description of the real material space inevitably is ten-axial, that is:

3 axes of *length* with unit of measurement α; 1 *temporary* axis with unit of measurement β;

1 *mass* axis with unit of measurement θ; 1 axis of *charge* with unit of measurement Δ;

1 axis of *temperature* with unit of measurement γ; 1 axis of *grouping* with unit of measurement Ω;

1 *particle* axis with unit of measurement ϕ; 1 axis of *cycle* with unit of measurement φ.

2. The Universe has a Measurement or Universal Existential Interval?

That is characterized by four types of dimensional barriers, that is to say:

— Unilateral barriers micro: *length* α and *time* β;

— Bilateral barriers macro: *temperature* γ;

— Barriers of connection type step between the macro and micro:

    mass θ, *electrical charge* Δ, *grouping* Ω and *particle* ϕ;

—Dynamic Systemic barrier: *cycle* φ.

World of the universal constants:

it is infinite by its length and time:

— *length* of $\alpha \rightarrow \infty$ m; *time* of $\beta \rightarrow \infty$ s;

it is limited by its temperature between:

— *temperature* $0 \rightarrow \gamma$ K;

with a separation of two levels (2) whose border is characterized by:

— *mass* $= \theta$ kg;

— *electrical charge* $= \Delta$ C;

with the consequent:

— *grouping* $= \Omega$ kmol;

and it structures of:

— *particle* $= \phi$ p;

in that all local and globally rotate (φ) and gets connected, from the spatial cells to the groups of galaxy, in perfect consonance with the non-existence of the rest (only relatively).

The generalized cause of the gravitation is that in world at local level always rotational processes are happening; to say that the gravity is universal is equivalent to express that the local rotation is generalized and characteristic of the global movement of matter. It is proves through of Law of Planetary Systems or "4th Kepler's Law" shown up by the authors[2].

$$\varphi \cdot \frac{(2)^{\frac{2}{3}}}{10} = 2\pi \cdot \frac{(2)^{\frac{2}{3}}}{10} = 2\pi \cdot \frac{\overline{d}_1}{T_1 \cdot \overline{V}_{0_1}} = 2\pi \cdot \frac{\overline{d}_2}{T_2 \cdot \overline{V}_{0_2}} = \ldots = 2\pi \cdot \frac{\overline{d}_i}{T_i \cdot \overline{V}_{0_i}} \overset{\bullet}{=} 1 \tag{9}$$

The Universe we are shown as well as the unlimited whole, it are whirlwinds among whirlwinds that it transit continuously between pseudohomogeneous and the structured substantially and vice versa, always contextually Euclidean. The hexagonal configuration elongate homolateral B type, allows to modeling a tessellation of the plane, to which we have denominated Quantum Dynamic Hexagonal Network (QDHN) and Radial Arrangement with Spiral Eddy (RASE).

3. Formal Diversity: Dimensionality of Scale

Nature changes of scale, conserving its fundamental dimensional properties arithmetic-geometric-topological.

Nature in dimensional transit of scale from the last structure of the real material space towards the disturbance which we know as Solar System did it (conserving the invariant G) manifestation of the hexagon-icosahedra fundamental structure of world and its basic Euclidean[1]:

$$G = \frac{\alpha^3}{\theta \cdot \beta^2} = c^2 \cdot \frac{\alpha}{\theta} = 4 \cdot \pi^2 \cdot (UA)^3 \cdot M_s^{-1} \cdot Y^{-2} = 6{,}6720 \cdot 10^{-11} \quad [\mathrm{m}^3 \cdot \mathrm{kg}^{-1} \cdot \mathrm{s}^{-2}] \tag{10}$$

$$\left(\frac{UA}{\alpha}\right)^3 \cdot \left(\frac{M_s}{\theta}\right)^{-1} \cdot \left(\frac{Y}{\beta}\right)^{-2} = \frac{1}{4 \cdot \pi^2} \tag{11}$$

4. What is it what Conserve the Laws of Conservation?

The relation length-time, it is immersed in all this theoretical formulation:

$$\alpha^5 \cdot \beta^{-4} = \alpha \cdot c^4 = \frac{h \cdot N_A \cdot c^2}{F} \cdot \left(\frac{G}{\mu_0}\right)^{\frac{1}{2}} = \frac{h \cdot c^2}{2 \cdot \phi \cdot e} = \frac{h \cdot c^2}{2 \cdot \Delta} = \cent = \frac{\sqrt{3}}{2} \cdot \pi \quad [\mathrm{m}^5 \cdot \mathrm{s}^{-4}] \tag{12}$$

What means then that $\alpha^5 \cdot \beta^{-4} = \cent$, of the universal mathematical constant $\cent$, is the topological constant of whole the numerical system and Euclidean property of it? Simply that the Real Material Space is Euclidean.

Likewise, our theoretic conception and its aftermath throw us than the universal constants of Energy (E_{nerg}), of Impulse (I_{mp}), of Moment (M_{om}) and of Charge (Q) also they express the eight-dimensional properties material real space:

$$E_{\mathrm{nerg}} \cdot I_{\mathrm{mp}} \cdot M_{\mathrm{om}} \cdot Q = \frac{h \cdot c^5 \cdot \Delta^3}{2 \cdot \alpha_F^3} = \frac{\theta^3 \cdot \alpha^5 \cdot \Delta}{\beta^4} = \theta^3 \cdot c^4 \cdot \alpha \cdot \Delta = \cent \cdot 10^{-6^2} \quad [\mathrm{kg}^3 \cdot \mathrm{m}^5 \cdot \mathrm{s}^{-4} \cdot \mathrm{C}] \tag{13}$$

Besides, the relation between the constants of magnetic, electrical and gravitational field can be understood as a nexus mass-length:

$$\frac{G}{c^2} = \mu_0 \cdot \varepsilon_0 \cdot \overset{\bullet}{G} \approx \cent^2 \cdot 10^{-28} \quad \mathrm{m} \cdot \mathrm{kg}^{-1} \tag{14}$$

Perhaps, it is an answer, a: why the gravity propagates with the same speed that the one of the light? That is to say, to conserve the topological constant of the real material space.

All real material system is also part indissoluble of the only one and united (multiconnected) real material space with the rest of which it interacts (an uninterrupted chain of microscopic processes type *Big Bang* and *Big Crunch*). These microprocessings have property that its spectrum is looked much like an object in balance to 2.726 K on the zero absolute one, so that, it is possible to be affirmed that it is numerically equal to the Topological Constant of the RMS, they become manifest by means of the Temperature of Background Radiation as fundamentality of material processes, which the space is flat, which its geometry is Euclidean, which universal constants are constants and express the properties of the eight-dimensional RMS.

REFERENCES

[1] Novoa, J., Montero, J. (1999). Realize. Quantum Cellular Structural Geometric Theory; Unified Theory of Physical and Mathematical Universal Constants; Program for the Physical-Arithmetic-Geometric-Topological-Dimensional Unification of Matter with its First Fundamental Equations, Laws, Principles and Postulates. (Introduction). Havana: Library of Congress (U.S. Copyright Office TXu-911-634). Unpublished.

[2] Novoa, J., Montero, J. (2004). Notes for a Geometric Study of Solar System. Quantum Cellular Structural Geometric Solution of Planetary Systems. 4th Kepler´s Law. Hypothesis of Gyroscopic Original Protuberance (HGOP). Havana: Unpublished.

The 4th Thermodynamic Principle?

José De La Luz Montero García*, Jesús Francisco Novoa Blanco†

*Institute for Scientific and Technological Information (IDICT); National Capitol, Havana, Cuba.
E-mail: **quantumstructure@yahoo.es**
† The late.

Abstract. It should be emphasized that the 4th Principle above formulated is a thermodynamic principle and, at the same time, is mechanical-quantum and relativist, as it should inevitably be and its absence has been one of main the theoretical limitations of the physical theory until today.
We show that the theoretical discovery of Dimensional Primitive Octet of Matter, the 4th Thermodynamic Principle, the Quantum Hexet of Matter, the Global Hexagonal Subsystem of Fundamental Constants of Energy and the Measurement or Connected Global Scale or Universal Existential Interval of the Matter is that it is possible to be arrived at a global formulation of the four "forces" or fundamental interactions of nature. The Einstein's golden dream is possible.

Keywords: thermodynamic principle; universal constants; dimension; unification; fundamental interactions.

1. The 4th Thermodynamic Principle

The Real material Space (RMS) is the Dimensional self-assembly and Deployment (eight-dimensional and ten-axial) of the world that has a Measurement or Universal Existential Interval, where the fragmentation of matter is not, then, limitless[1].

Our arithmetic-geometric-topologic physical study of matter brought about the fact that the proper scale of matter has a dimensional octet as a base. Thus the primitive characteristic dimensions are: a self-assembly longitudinal (three-axial) ($\alpha = 3,3530 \cdot 10^{-34}$ m), time ($\beta = 1,1184 \cdot 10^{-42}$ s), mass ($\theta = 4,5166 \cdot 10^{-7}$ kg), of electric charge ($\Delta = 1,0978 \cdot 10^{-17}$ C), of thermodynamic temperature of energetic hierarchy; ($\gamma = 4,2911 \cdot 10^{31}$ K), of particle ($\phi = 6,8518 \cdot 10^{1}$ p), of grouping ($\Omega = 1,1378 \cdot 10^{-25}$ kmol) and of cycle universal ($\varphi = 2\pi$ rad), and precise the Planck's units[1].

The natural scale of temperature implies limiting the fragmentation of matter.

Being the Thermodynamic Temperature of Energetic Hierarchy (γ) calculated by us[1] of:

$$\gamma = \left\{ \frac{c^3 \cdot F}{k \cdot N_A} \right\} \cdot \left(\frac{\mu_0}{G} \right)^{\frac{1}{2}} = 4,2911 \cdot 10^{31} \text{ K} \tag{1}$$

It would need to achieve a concentration of energy which is locally unreachable.

But arrived at that colossal temperature γ, also it is arrived at the discretion of the length (α) and of the time (β), and such barriers are insurmountable, for which our calculations arithmetic-geometric-topological-dimensional and structural-cellular-dynamic-quantum prove[1] that:

$$\alpha = \left\{ \frac{h \cdot N_A}{c^2 \cdot F} \right\} \cdot \left(\frac{G}{\mu_0} \right)^{\frac{1}{2}} = 3,3530 \cdot 10^{-34} \text{ m} \ ; \ \beta = \left\{ \frac{h \cdot N_A}{c^3 \cdot F} \right\} \cdot \left(\frac{G}{\mu_0} \right)^{\frac{1}{2}} = 1,1184 \cdot 10^{-42} \text{ s} \tag{2}$$

$$\alpha \cdot \beta^{-1} = c \ \text{m} \cdot \text{s}^{-1} = 2,99803290 \cdot 10^{8} \ \text{m} \cdot \text{s}^{-1} \tag{3}$$

and such barriers cannot be crossed.. This can be equationally formulated by means of four limit expressions of a tremendous theoretical importance:

$$\lim_{T \to \gamma} \Delta x = \alpha \ ; \ \lim_{T \to \gamma} \Delta t = \beta \ ; \ \lim_{\Delta x \to \alpha} \Delta T = \gamma \ ; \ \lim_{\Delta t \to \beta} \Delta T = \gamma \tag{4}$$

It is the theoretical expressions of the inevitable, physical and topologically coherent-consistent 4th Thermodynamic Principle, which is transformed expressing: It is impossible to reach the top γ

CP905, *Frontiers of Fundamental Physics (FFP8), Eighth International Symposium*
edited by B. G. Sidharth, A. Alfonso-Faus, and M. J. Fullana

of the natural scale of temperature. Equivalently, the ultimate cells of the real material space do not exist independently but are always linked; otherwise neither the universe nor we could exist. Therefore, "The *Big Bang* Never Happened" and the Mach's Principle is right.

The 4th Principle strictly ties to the structure and limits of the universe and with which completes the thermodynamics.

2. Precision of Heisenberg's Uncertainty Principle

The scale of nature and the 4th Thermodynamic Principle[1], specifies this principle as of absolute uncertainty of longitude and time, characteristic of these dimensions of the primitive dimensional octet of matter and therefore, of every magnitude.

$$\alpha \cdot \Delta p_x \geq \frac{h}{4\pi} \Rightarrow \Delta p_x \geq \frac{h}{4\pi \cdot \alpha} = \frac{c^2 \cdot F}{4\pi \cdot N_A \cdot \left(\dfrac{G}{\mu_0}\right)^{1/2}} \tag{5}$$

$$\Delta t \cdot \Delta E \geq \frac{h}{2\pi} \ ; \ \Delta t \geq \beta \Rightarrow \Delta E \geq \frac{h}{2\pi \cdot \beta} = \frac{c^3 \cdot F}{2\pi \cdot N_A \cdot \left(\dfrac{G}{\mu_0}\right)^{1/2}} \tag{6}$$

3. Global Hexagonal Subsystem of Fundamental Constants of Energy

The Primitive Dimensional Octet of Matter, in correspondence with our 4th Principle is possible to deduce rigor yet that the subsystem of energetic universal constants has 6 members $\{P, h, B, k, L, l\}$ and not only two ($h = \dfrac{E_{nerg}}{\phi} \cdot \beta$ and $k = \dfrac{E_{nerg}}{\phi} \cdot \dfrac{1}{\gamma}$), all of them centered in and E_{nerg}, the Great Universal Constant of Energy:

$$E_{nerg} = \theta \cdot \alpha^2 \cdot \beta^{-2} = \theta \cdot c^2 = \frac{h}{\beta} = k \cdot \gamma = \frac{c^2 \cdot h \cdot N_A}{F} \cdot \left(\frac{1}{G \cdot \mu_0}\right)^{1/2} = 4,0593 \cdot 10^{10} \ \text{J} . \tag{7}$$

This great universal constant of energy also is definable by particle:

$$\frac{E_{nerg}}{\phi} = \theta \cdot \alpha^2 \cdot \beta^{-2} \cdot \phi^{-1} = \frac{\theta \cdot c^2}{\phi} = 5,924454 \cdot 10^8 \quad \text{J} \cdot \text{p}^{-1} = 3,6977 \cdot 10^{27} \quad \text{eV} \cdot \text{p}^{-1} \tag{8}$$

From this theoretical development we may conclude that science lacks the very important constant of energy-longitude which we will denote by L: $\ L = E_{nerg} \cdot \alpha .$ $\hspace{2em}$ (9)

which can also be defined by particle (denoted by l): $\ l = \dfrac{L}{\phi} = \dfrac{E_{nerg}}{\phi} \cdot \alpha .$ $\hspace{2em}$ (10)

This constant l is nothing but the key to the unification of the forces of nature.

4. Quantum Hexet of Matter (QHM)

With the Quantum Hexet of Matter (QHM) culminates a cycle of the theoretical development of the physics towards a Universal Generalized Conception Discreet of World[1].

$$LAF = \frac{L \cdot N_A}{F} = l \cdot e \ ; \ BAF = k \cdot \frac{N_A}{F} = \frac{k}{e} \ ; \ PAF = h \cdot \frac{N_A}{F} = \frac{h}{e} \quad [\text{J} \cdot \text{m} \cdot \text{C}^{-1}] \tag{11}$$

Congruent with 4th Principle, the QHM $(l \cdot N_A,\ LAF,\ h \cdot N_A,\ PAF,\ k \cdot N_A,\ BAF)$ is characterized by three lines of development:

a) energetic-temporal line o of Planck and Novoa-Montero:

$$P = F_{nerg} \cdot \beta = (4{,}540\,0 \pm 0{,}008\,6) \cdot 10^{-32}\ \ \text{J} \cdot \text{s}\ ;\ h = \frac{P}{\phi} = (6{,}626 \pm 0{,}013) \cdot 10^{-34}\ \ [\text{J} \cdot \text{s} \cdot \text{p}^{-1}] \qquad (12)$$

(action line, in form absolute and by particle)

b) energetic-thermal-entropic line or of Boltzmann and Novoa-Montero:

$$B = E_{nerg} \cdot \frac{1}{\gamma} = (9{,}460 \pm 0{,}023) \cdot 10^{-22}\ \text{J} \cdot \text{K}^{-1}\ ;\ k = \frac{B}{\phi} = (1{,}380\,6 \pm 0{,}003\,3) \cdot 10^{-23}\ \text{J} \cdot \text{K}^{-1} \cdot \text{p}^{-1} \ (13)$$

(energetic-entropic line, in form absolute and by particle)

c) energetic-longitudinal line or of Novoa-Montero:

$$L = E_{nerg} \cdot \alpha = (1{,}361\,0 \pm 0{,}003\,9) \cdot 10^{-23}\ \text{J} \cdot \text{m}\ ;\ l = \frac{L}{\phi} = (1{,}986\,4 \pm 0{,}005\,7) \cdot 10^{-25}\ \ \text{J} \cdot \text{m} \cdot \text{p}^{-1} \ (14)$$

(line of "volumetric force", in form absolute and by particle)

5. Unified Equation of the Global Strong-Electromagnetic-Weak-Gravitational-Nexus

If at level of the last intimate structure of the real material space the force, more than unified is unique, this means that all is one by its intensity and its dependency with respect to the distance between the interacting entities.

$$F_{gener} = F_{grav.} = F_{mag.} = F_{elect.} = F_{débil} = F_{fuerte} = \frac{L}{F(d)} \qquad (15)$$

where $F(d)$ is a certain function of the distance, presumably $F(d) = d^2$. This way, Generalized the Unique Force then is:

$$\vec{F}_{gener} = \frac{L}{F(d)} \cdot \frac{\vec{d}}{|d|} \qquad (16)$$

then, as a result, we get the Unified Equation of the Global Strong-Electromagnetic-Weak-Gravitational-Nexus[1]:

$$G \cdot \theta^2 = \mu_0 \cdot c^2 \cdot \Delta^2 = \frac{\Delta^2}{\varepsilon_0} = F_0 \cdot f^2 = \frac{g^2}{D_0} = L = \ldots\ [\text{J} \cdot \text{m}] \qquad (17)$$

First of all, it is fundamental to understand that it is not the equivalence of charges (gravitational, electromagnetic, strong and weak) or the equivalence of the corresponding fields, but of the conjunction of both factors. Then, the Einstein's golden dream is possible.

In order to say it summarizing: in the nature only there is an interaction or force that we have come discovering in the four mentioned modalities of field with its respective entities from transmission, approach that is perfectly beyond to the gluonic field of quarks.

REFERENCES

[1] Novoa, J., Montero, J. (1999). Realize. Quantum Cellular Structural Geometric Theory; Unified Theory of Physical and Mathematical Universal Constants; Program for the Physical Arithmetic-Geometric-Topological-Dimensional Unification of Matter with its First Fundamental Equations, Laws, Principles and Postulates. (Introduction). Havana: Library of Congress (U.S. Copyright Office TXu-911-634). Unpublished.

Foundation of quantum mechanics from the principle of relativity

Laurent Nottale

CNRS, LUTH, Observatoire de Paris-Meudon, F-92195 Meudon Cedex, France

Abstract. We briefly recall the main steps by which we suggest to found quantum mechanics and gauge field theories on the principle of relativity, once it is extended to scale transformations of the reference system. The wave functions are constructed as consequences of the nondifferentiability of a continuous space-time, while the Schrödinger and Dirac equations are obtained from its geodesics equations. In this framework, the gauge fields emerge as manifestation of the fractal geometry, and the gauge charges as the conservative quantities which are built from its internal symmetries.

Keywords: Scale relativity, Foundation of quantum mechanics, Gauge field theories
PACS: 03.65.Ta , w1.15.-q

INTRODUCTION

In the theory of scale relativity, one gives up the hypothesis of space-time differentiability. One can prove that a nondifferentiable continuum is fractal [1, 2], i.e., it is explicitly dependent on the scales of resolution. This leads one to extend the principle of relativity to scale transformations. In the present contribution, we summarize the steps by which one recovers, in this framework, the main postulates of quantum mechanics and of gauge field theories. A more detailed account can be found in Refs. [6, 10, 14, 15].

QUANTUM MECHANICS IN SCALE RELATIVITY

From the three main consequences of nondifferentiability [7], namely, (i) infinite number of geodesics, (ii) breaking of the reflexion invariance of the time differential element dt that leads to a two-valuedness of the velocity field, (iii) fractal dimension 2 of the geodesics, one describes the elementary displacements dX on a fractal space [1, 3, 4, 5] as $dX_{\pm} = d_{\pm}x + d\xi_{\pm}$, where $d\xi$ is the "fractal part" and dx the "classical part",

$$d_{\pm}x = v_{\pm}\,dt, \quad d\xi_{\pm} = \eta\,\sqrt{2\mathscr{D}}\,dt^{1/2}. \tag{1}$$

Here η is a stochastic dimensionless variable such that $<\eta> = 0$ and $<\eta^2> = 1$, and $\mathscr{D}$ a parameter that generalizes the Compton scale. The two time derivatives are then combined in a complex derivative [1], $\hat{d}/dt = (d_+ + d_-)/2dt - i(d_+ - d_-)/2dt$. Applying this operator to the position vector yields a complex velocity $\mathscr{V} = \hat{d}x/dt$. Then the geodesics equation, $\hat{d}\mathscr{V}/dt = 0$, can be integrated under the form of a generalized Schrödinger equation [1, 2, 6, 9]. We have recently generalized the proof to the whole velocity field $\tilde{\mathscr{V}} = \mathscr{V} + \mathscr{W}$, including its differentiable and nondifferentiable parts [13,

CP905, *Frontiers of Fundamental Physics (FFP8), Eighth International Symposium*
edited by B. G. Sidharth, A. Alfonso-Faus, and M. J. Fullana

10]. We can build from it a full complex action, $d\mathscr{S} = (1/2)m(\mathscr{V} + \mathscr{W})^2 dt$, then define a full wavefunction as $\tilde{\psi} = \exp(i\mathscr{S}/2m\mathscr{D})$, such that $\tilde{\mathscr{V}} = \mathscr{V} + \mathscr{W} = \nabla\mathscr{S}/m = -2i\mathscr{D}\nabla\ln\tilde{\psi}$. In our framework, this relation keeps a mathematical and physical meaning in terms of fractal functions, explicitly dependent on dt and divergent when $dt \to 0$. Then one builds a generalized full covariant derivative, that reads [10]

$$\frac{\widehat{d}}{dt} = \frac{\partial}{\partial t} + (\mathscr{V} + \mathscr{W}).\nabla - i\mathscr{D}(1 + \tilde{\zeta})\Delta, \tag{2}$$

where $\tilde{\zeta}$ is a stochastic variable of zero mean. Using this covariant derivative, we can finally write a covariant equation which keeps the form of Newton's fundamental equation of dynamics,

$$\frac{\widehat{d}}{dt}\tilde{\mathscr{V}} = -\frac{\nabla\phi}{m}. \tag{3}$$

This equation can be integrated in terms of a generalized Schrödinger equation [13, 10],

$$\mathscr{D}^2\Delta\tilde{\psi} + i\mathscr{D}\frac{\partial\tilde{\psi}}{\partial t} - \frac{\phi}{2m}\tilde{\psi} = 0, \tag{4}$$

which now allows fractal solutions, in agreement with Berry's [11] and Hall's [12] results. But this property is obtained here as a manifestation of space nondifferentiability.

The von Neumann's and Born's postulates are then derived from the identification of "particles" with the various geometric properties of fractal space-time geodesics [6, 10].

Then the account of a new two-valuedness, consequence of nondifferentiability, leads to introduce bispinors in terms of biquaternionic wavefunctions, and to obtain the Dirac [6] and Pauli [14] equations as geodesics equations of a fractal space-time.

GAUGE THEORIES IN SCALE RELATIVITY

The theory of scale relativity can also be applied to the foundation of Abelian [2, 8] and non-Abelian [15] gauge theories. This application is based on a general relativistic description of the internal structures of nondifferentiable space-time geodesics in terms of tensorial scale variables $\eta_{\alpha\beta}(x,y,z,t)$ which may now be function of the coordinates.

We assume for simplicity that the two tensorial indices can be gathered under one common index. The infinitesimal transformation law on the η_α can be written in a linear way as $\eta'_\alpha = (\delta_{\alpha\beta} + \delta\theta_{\alpha\beta})\eta^\beta$. Since the η_α's are now functions of the standard space-time coordinates, this leads us to define a new scale-covariant derivative,

$$d\eta_\alpha = D\eta_\alpha - \eta^\beta \delta\theta_{\alpha\beta} = D\eta_\alpha - \eta^\beta W^\mu_{\alpha\beta}\,dx_\mu. \tag{5}$$

The gauge potentials $W^\mu_{\alpha\beta}$ naturally emerge from this geometrical framework.

The next step amounts to describe how various physical quantities transform under these η_α transformations. To this purpose we generalize to multiplets the relation $\mathscr{V}_\mu = i\lambda\,\psi^{-1}\partial_\mu\psi$ obtained in the Dirac spinor case [6]:

$$\mathscr{V}^\mu_{jk} = i\lambda\,\psi_j^{-1}\partial^\mu\psi_k. \tag{6}$$

The action $dS_{jk} = dS_{jk}(x^\mu, \mathscr{V}^\mu_{jk}, \eta_\alpha)$ also becomes a biquaternionic tensor, so that

$$\partial^\mu S_{jk} = D^\mu S_{jk} - \eta^\beta \frac{\partial S_{jk}}{\partial \eta_\alpha} W^\mu_{\alpha\beta}. \tag{7}$$

This result allows one to define a general non-Abelian group of scale transformations whose generators are $T^{\alpha\beta} = \eta^\beta \partial^\alpha$, yielding the generalized charges,

$$\frac{\tilde{g}}{c} t^{\alpha\beta}_{jk} = \eta^\beta \frac{\partial S_{jk}}{\partial \eta_\alpha}. \tag{8}$$

Knowing that the α, β represent two indices each, this is a large group that contains the standard model $U(1) \times SU(2) \times SU(3)$ as subset [15].

As we have shown in more detail in Ref. [15], the various ingredients of Yang-Mills theories can be recovered in such a framework, but they are now founded on geometric and scale-relativistic first principles.

CONCLUSION

In this contribution, we have recalled the main steps that lead to a new foundation of quantum mechanics and of gauge fields on the principle of relativity itself, once it is generalized to scale transformations of the reference system.

For this purpose, two covariant derivatives have been constructed, which account for the nondifferentiable and fractal geometry of space-time, and which allow to write the equations of motion as geodesics equations. After change of variable, these equations finally take the form of the quantum mechanics and quantum field equations.

ACKNOWLEDGMENTS

I gratefully acknowledge Dr. Sidharth for his kind invitation to the FFP8 symposium.

REFERENCES

1. L. Nottale, *Fractal Space-Time and Microphysics*, World Scientific, Singapore, 1993.
2. L. Nottale, *Chaos, Solitons & Fractals* **7** (1996), 877.
3. G. N. Ord, *J. Phys. A: Math. Gen.* **16** (1983), 1869.
4. L. Nottale and J. Schneider, *J. Math. Phys.* **25** (1984), 1296.
5. L. Nottale, *Int. J. Mod. Phys. A* **4** (1989), 5047.
6. M.N. Célérier & L. Nottale, *J. Phys. A: Math. Gen.* **37** (2004), 931.
7. R. P. Feynman and A. R. Hibbs, *Quantum Mechanics and Path Integrals*, MacGraw-Hill, 1965.
8. L. Nottale, in *Relativity in General*, Diaz Alonso & Lorente Paramo eds., Frontières, 1994, p. 121.
9. J. Cresson, *J. Math. Phys.* **44** (2003), 4907.
10. L. Nottale, in *Proc. 7th Int. Coll. on Clifford Algebra*, Ed. P. Anglès, Birkhauser, (2006) in press
11. M.V. Berry *J. Phys. A: Math. Gen.* **29** (1996), 6617.
12. M.J.W. Hall *J. Phys. A: Math. Gen.* **37** (2004) 9549.
13. L. Nottale, *Chaos, Solitons & Fractals* **10** (1999), 459.
14. M.N. Célérier & L. Nottale, *J. Phys. A: Math. Gen.* **39** (2006),12565.
15. L. Nottale, M.N. Célérier & T. Lehner, *J. Math. Phys.* **47** (2006), 032303.

Stochastic Time

Toru Ohira

Sony Computer Science Laboratories, Inc., Tokyo, Japan 141-0022

Abstract. We present a simple dynamical model to address the question of introducing a stochastic nature in a time variable. This model includes noise in the time variable but not in the "space" variable, which is opposite to the normal description of stochastic dynamics. The notable feature is that these models can induce a "resonance" with varying noise strength in the time variable. Thus, they provide a different mechanism for stochastic resonance, which has been discussed within the normal context of stochastic dynamics.

Keywords: Stochastic Resonance, Time, Delay
PACS: 05.40.-a,02.50.-r,01.55.+b

"Time" is a concept that has gained a lot of attention from thinkers in virtually all disciplines[1]. In particular, our ordinary perception of time is not the same as that of space, and this difference has been appearing in a variety of contemplations about nature. It appears to be the main reason for the theory of relativity, which has conceptually brought space and time closer to receiving equal treatment, continues to fascinate and attract discussion in diverse fields. Also, issues such as "directions" or "arrows" of time are current interests of research[2]. Another manifestation of this difference is the treatment of noise or fluctuations in dealing with dynamical systems. When we consider dynamical systems, whether classical, quantum, or relativistic, time is commonly viewed as not having stochastic characteristics. In stochastic dynamical theories, we associate noise and fluctuations with only "space" variables, such as the position of a particle, but not with the time variables. In quantum mechanics, the concept of time fluctuation is well accepted through the time-energy uncertainty principle. However, time is not treated as a dynamical quantum observable, and a clearer understanding has been explored[3].

Against this background, our main theme of this paper is to consider fluctuations of time in classical dynamics through the presentation of a simple model. There are variety of ways to bringing stochasticity to some temporal aspects of dynamical systems. The model which we present is one way, it is an extension of delayed dynamical models[4, 5, 6, 7, 8]. With stochastic time, we have found that the model exhibits behaviors similar to those investigated in the topic of stochastic resonance[9, 10, 11], which are studied in a variety of fields[12, 13, 14, 15, 16]. The difference is that the phenomena are induced by noise in time rather than by noise in space.

The general differential equation of the class of delayed dynamics with stochastic time is

$$\frac{dx(\bar{t})}{d\bar{t}} = f(x(\bar{t}), x(\bar{t} - \tau)). \tag{1}$$

Here, x is the dynamical variable, and f is the "dynamical function" governing the dynamics. τ is the delay. The difference from the normal delayed dynamical equation appears in "time" $\bar{t}$, which contains stochastic characteristics. We can define $\bar{t}$ in a variety

CP905, *Frontiers of Fundamental Physics (FFP8), Eighth International Symposium*
edited by B. G. Sidharth, A. Alfonso-Faus, and M. J. Fullana

of ways as well as the function f. To avoid ambiguity and for simplicity, we focus on the following dynamical map system incorporating the basic ideas of the above general definition.

$$
\begin{aligned}
x_{n_{k+1}} &= f(x_{n_k}, x_{n_k-\tau}), \\
n_{k+1} &= n_k + \xi_k
\end{aligned}
\tag{2}
$$

Here, ξ_k is the stochastic variable which can take either $+1$ or -1 with certain probabilities. We associate "time" with an integral variable n. The dynamics progress by incrementing integer k, and n occasionally "goes back a unit" with the occurrence of $\xi = -1$. Let the probability of $\xi_k = -1$ be p for all k, and we set $n_0 = 0$. Then naturally, with $p = 0$, this map reduces to a normal delayed map with $n_k = k$. We update the variable x_n with the larger k. Hence, x_n in the "past" could be "re-written" as n decreases with the probability p.

Qualitatively, we can make an analogy of this model with a tele–typewriter or a tape–recorder, which occasionally moves back on a tape. A schematic view is shown in Figure 1A. The recording head writes on the tape the values of x at a step, and "time" is associated with positions on the tape. When there is no fluctuation ($p = 0$), the head moves only in one direction on the tape and it records values of x for a normal delayed dynamics. With probability $0 < p$, it moves back a unit of "time" to overwrite the value of x. The question is how the recorded patterns of x on the tape are affected as we change p.

The dynamical function is chosen to be a negative feedback function (Figure 1B) and the concrete map model that we will study is given as follows.

$$
x_{n_{k+1}} = x_{n_k} + d\delta \left(-\alpha x_{n_k} - \frac{2}{1 + e^{-\beta x_{n_k-\tau}}} + 1 \right),
\tag{3}
$$

with α, β and $d\delta$ as parameters. With both α and β positive and no stochasticity in time, this map has the origin as a stable fixed point with no delay. We consider the case of $\alpha = 0.5$, $\beta = 6$, and $d\delta = 0.1$. By a linear stability analysis, the critical delay τ_c, at which the stability of the fixed point is lost, is given as $\tau_c \sim 6$. The larger delay gives an oscillatory dynamical path. We have found, through computer simulations, that an interesting behavior arises when delay is smaller than this critical delay. The tuned noise in the time flow gives the system a tendency for oscillatory behavior. In other words, adjusting the value of p controlling ξ induces an oscillatory dynamical path. Some examples are shown in Figure 1C. With increasing probability for the time flow to reverse, i.e., with p increasing, we observe oscillatory behavior both in the sample dynamical path as well as in the corresponding power spectrum. However, when p reaches beyond an optimal value, the oscillatory behavior begins to deteriorate. The change in the peak heights is shown in Figure 1D. This phenomenon resembles stochastic resonance. A resonance with delay and noise, called "delayed stochastic resonance"[20], has been proposed for an additive noise in "space". Analytical understanding of the mechanism is yet to be explored for our model. However, this mechanism of stochastic time flow is clearly of a different type and new.

We would like to now discuss a couple of points with respect to our model. First, we can view this model as a dynamical model with non-locality and fluctuation on time

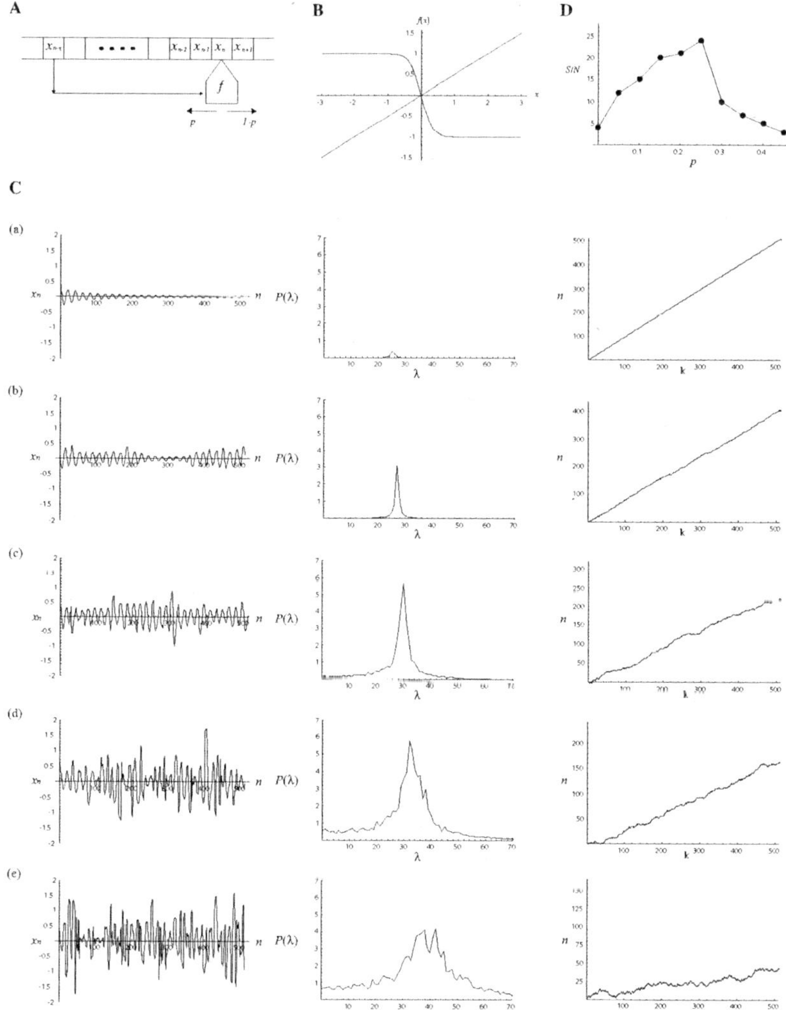

FIGURE 1. **A**: A schematic view of the model. **B**: Dynamical functions $f(x)$ with parameters as examples of simulations presented in this paper. Negative feedback function with parameters $\beta = 6$. Straight line has slope of $\alpha = 0.5$. **C**: Dynamics (left) and power spectrum (middle) of delayed dynamical model with stochastic time flow (right). This is an example of the dynamics and associated power spectrum through the simulation of the model given in Eq. (3) with the probability p of stochastic time flow varied. The parameters are set as $\alpha = 0.5$, $\beta = 6$, $d\delta = 0.1$, $\tau = 5$ and the stochastic time flow parameter p are set to (a) $p = 0$, (b) $p = 0.1$, (c) $p = 0.25$, (d) $p = 0.35$, and (e) $p = 0.45$. We used the initial condition that $x_n = 0.5(n \leq 0)$, and $n_0 = 0$. The simulation is performed up to $k = 5000$ steps. At that point the values of x_n for $0 \leq n \leq L$ with $L = 512$ are recorded. The 50 averages are taken for the power spectrum on this recorded x_n. The unit of frequency λ is set as $\frac{1}{L}$, and the power $P(\lambda)$ is in arbitrary units. **D**: The signal to noise ratio $\frac{S}{N}$ at the peak height as a function of the probability p of stochastic time flow. The parameter settings are the same as in C.

axis. Both factors are familiar in "space", but not on time. We may extend this model to include non-locality and fluctuations in the space variable x. Proceeding in this way, we have a picture of dynamical systems with non-locality and fluctuations on both the time and space axes. The analytical framework and tools for such a description need to be developed, along with a search for appropriate applications.

Another way might be to extend the path integral formalism. The question of whether this extension bridges to quantum mechanics and/or leads to an alternative understanding of such properties as time-energy uncertainty relations also requires further investigation. Also, there is a theory of elementary particles with a fluctuation of space–time, where the noise term is added to the metric[21]. If we can connect our ideas here to such a theory remains to be seen.

Finally, if these models do capture some aspects of reality, particularly with respect to stochasticity in the time flow, this resonance may be used as an experimental indication for probing fluctuations or stochasticity in time. We have previously proposed "delayed stochastic resonance"[20], a resonance that results from the interplay of noise and delay. It was theoretically extended[22], and recently, the effect was experimentally observed in a solid-sate laser system with a feedback loop[23]. It is left for the future to see if an analogous experimental test could be developed with respect to stochastic time.

REFERENCES

1. P. Davies, *About Time*, (Simon and Schuster, New York, 1995).
2. S. F. Savitt, *Time's Arrows Today*, (Cambridge Univ. Press, Cambridge, 1995).
3. P. Busch, "The Time-Energy Uncertainty Relation," in *Time in Quantum Mechanics* (J. G. Muga, R. Sala Mayato and I. L. Egusquiza, eds.) 69-98 (Springer-Verlag, Berlin, 2002).
4. M. C. Mackey and L. Glass, Science **197**, 287–289 (1977).
5. K. L. Cooke and Z. Grossman, J. Math. Anal. and Appl. **86**, 592–627 (1982).
6. J. G. Milton, et al., J. Theo. Biol. **138**, 129–147 (1989).
7. T. Ohira and T. Yamane, Phys. Rev. E **61**, 1247–1257 (2000).
8. T. D. Frank and P. J. Beek, Phys. Rev. E **64**, 021917 (2001).
9. K. Wiesenfeld, and F. Moss, Nature **373**, 33–36 (1995).
10. A. R. Bulsara and L. Gammaitoni, Physics Today **49**, 39–45 (1996).
11. L. Gammaitoni, P. Hänggi, P. Jung, and F. Marchesoni, Rev. Mod. Phys. **70**, 223–287 (1998).
12. B. McNamara, K. Wiesenfeld and R. Roy, Phys. Rev. Lett. **60**, 2626–2629 (1988).
13. A. Longtin, A. Bulsara and F. Moss, Phys. Rev. Lett. **67**, 656–659 (1991).
14. J. J. Collins, C. C. Chow and T. T. Imhoff, Nature **376**, 236–238 (1995).
15. F. Chapeau-Blondeau, Sign. Process **83**, 665–670 (2003).
16. I. Y. Lee, X. Liu, B. Kosko and C. Zhou, Nano Letters **3**, 1683–1686 (2003).
17. L. Glass and M. Mackey, *The Rhythms of Life*, (Princeton Univ. Press, Princeton, 1988).
18. J. L. Cabrera and J. G. Milton, Phys. Rev. Lett. **89** 158702 (2002).
19. J. L. Cabrera and J. G. Milton, Chaos **14**, 691-698 (2004).
20. T. Ohira and Y. Sato, Phys. Rev. Lett. **82**, 2811–2815 (1999).
21. Y. Takano, Prog. Theor. Phys. **26**, 304–314 (1961); Y. Takano, *ibid*, **38** (1967) 1185–1186.
22. L.S. Tsimring and A. Pikovsky, Phys. Rev. Lett. **87** 250602 (2001).
23. C. Masoller, Phys. Rev. Lett. **88** 034102 (2002).

The Common Origin of Gravitation, General Relativity, Electromagnetism and Quantum Theory

B.G. Sidharth

International Institute for Applicable Mathematics and Information Sciences Hyderabad (India) &
Udine (Italy) B.M. Birla Science Centre, Adarsh Nagar, Hyderabad - 500 063 (India)

Abstract. We argue that a non commutative geometry at the Compton scale is at the root of mass, Quantum Mechanical spin and QCD and electromagnetic interactions as well as gravitation, in a formulation that mimics linearized General Relativity. It thus leads to a reconciliation of linearized General Relativity and Quantum Theory.

Keywords: Linearized General Relativity, Spin, Interaction, Mass
PACS: 04.00.-m, 12.10

INTRODUCTION

It is well known that General Relativity or Gravitation has resisted a unified description with Quantum Theory for over seven decades now. This has lead to a consideration of non differentiable space time. Modern Fuzzy Space time and Quantum Gravity approaches deal with a non differentiable space time manifold. There is a minimum space time cut off, which, as shown by Snyder leads to what is nowadays called a non commutative geometry [1, 2, 3, 4, 5, 6]. The new geometry is given by

$$[dx^\mu, dx^\nu] \approx \beta^{\mu\nu} l^2 \neq 0 \tag{1}$$

We would like to make (1) the starting point and stress that this is true whenever there is a minimum space time interval. While equation (1) is true for any minimum cut off l, it is most interesting and leads to physically meaningful relations including a rationale for the Dirac equation and the underlying Clifford algebra, when l is at the Compton scale (Cf.ref.[3]). From this point of view, (1) leads to Quantum Theory. Conversely, as shown in detail, (Cf.ref.[3]), Relativistic Quantum Theory leads to (1). In any case given (1), the usual invariant line element,

$$ds^2 = g_{\mu\nu} dx^\mu dx^\nu \tag{2}$$

has to be written in terms of the symmetric and nonsymmetric combinations for the product of the coordinate differentials. That is the right side of Equation (2) would become

$$\frac{1}{2} g_{\mu\nu} \left[(dx^\mu dx^\nu + dx^\nu dx^\mu) + (dx^\mu dx^\nu - dx^\nu dx^\mu) \right],$$

In effect we would have

$$g_{\mu\nu} = \eta_{\mu\nu} + k h_{\mu\nu} \tag{3}$$

CP905, *Frontiers of Fundamental Physics (FFP8), Eighth International Symposium*
edited by B. G. Sidharth, A. Alfonso-Faus, and M. J. Fullana
© 2007 American Institute of Physics 978-0-7354-0412-0/07/$23.00

So the noncommutative geometry introduces an extra term, that is the second term on the right side of (3). It has been shown in detail by the author that (1) or (2) lead to a reconciliation of electromagnetism and gravitation and lead to what may be called an extended gauge formulation [7, 8, 9, 10]. We will return to this later.

The extra term in (3) leads to an energy momentum like tensor but it must be stressed that its origin is in the non commutative geometry (1). All this of course is being considered at the Compton scale of an elementary particle.

COMPTON SCALE CONSIDERATIONS

As in the case of General Relativity [11, 12], but this time remembering that neither the coordinates nor the derivatives commute we have

$$\partial_\lambda \partial^\lambda h^{\mu\nu} - (\partial_\lambda \partial^\nu h^{\mu\lambda} + \partial_\lambda \partial^\mu h^{\nu\lambda})$$

$$-\eta^{\mu\nu}\partial_\lambda \partial^\lambda h + \eta^{\mu\nu}\partial_\lambda \partial_\sigma h^{\lambda\sigma} = -kT^{\mu\nu} \tag{4}$$

It must be reiterated that the non commutativity of the space coordinates has thrown up the analogue of the energy momentum tensor of General Relativity, viz., $T^{\mu\nu}$. *We identify this with the energy momentum tensor.*

Remembering that $h_{\mu\nu}$ is a small effect, we can use the methods of linearized General Relativity [11, 12], to get from (4),

$$g_{\mu\nu} = \eta_{\mu\nu} + h_{\mu\nu}, h_{\mu\nu} = \int \frac{4T_{\mu\nu}(t - |\vec{x} - \vec{x}'|, \vec{x}')}{|\vec{x} - \vec{x}'|} d^3 x' \tag{5}$$

It was shown several years ago in the context of linearized General Relativity, that for distances $|\vec{x} - \vec{x}'|$ much greater than the distance $\vec{x}'$, that is well outside the Compton wavelength, we can recover from (5) the electromagnetic potential (Cf.ref.[13] and references therein). We will briefly return to this point.

Let us now see what happens when $|\vec{x}| \sim |\vec{x}'|$. In this case, we have from (5), expanding in a Taylor series about t,

$$h_{\mu\nu} = 4 \int \frac{T_{\mu\nu}(t, \vec{x}')}{|\vec{x} - \vec{x}'|} d^3 x' + (\text{terms independent of} \vec{x}) + 2$$

$$\int \frac{d^2}{dt^2} T_{\mu\nu}(t, \vec{x}').|\vec{x} - \vec{x}'| d^3 x' + 0(|\vec{x} - \vec{x}'|^2) \tag{6}$$

The first term gives a Coulombic $\frac{\alpha}{r}$ type interaction except that the coefficient α is of much greater magnitude as compared to the gravitational or electromagnetic case, because in an expansion of $(1/|\vec{x} - \vec{x}'|)$, all terms are of comparable order. The second term on the right side of (6) is of no dynamical value as it is independent of $\vec{x}$. The third term however is of the form constant $\times r$. That is the potential (6) is exactly of the form of the QCD potential [14]

$$-\frac{\alpha}{r} + \beta r \tag{7}$$

In (7) α is of the order of the mass of the particle as follows from (6) and the fact that $T^{\mu\nu}$ is the energy momentum tensor given by

$$T^{\mu\nu} = \rho u^{\mu} u'' \qquad (8)$$

where u^{μ} represented the four velocity. Remembering that from (1), we are within a sphere of radius given by the Compton length where the velocities equal that of light, we have equations

$$\left| \frac{du_v}{dt} \right| = |u_v| \omega \qquad (9)$$

$$\omega = \frac{|u_v|}{R} = \frac{2mc^2}{\hbar} \qquad (10)$$

Alternatively we can get (9) from the theory of the Dirac equation itself [15], viz.,

$$i\hbar \frac{d}{dt}(u_i) = -2mc^2(u_i),$$

Using (8), (9) and (10) we get

$$\frac{d^2}{dt^2} T^{\mu\nu} = 4\rho u^{\mu} u^{\nu} \omega^2 = 4\omega^2 T^{\mu\nu} \qquad (11)$$

Equation (11) too is obtained in the Dirac theory (loc.cit). Whence, as can be easily verified, α and β in (7) have the correct values required for the QCD potential (Cf. also [13]). (Alternatively βr can be obtained, as in the usual theory by a comparison with the Regge angular momentum mass relation: It is in fact the constant string tension like potential which gives quark confinement and its value is as in the usual theory [16]). Let us return to the considerations which lead via a non commutative geometry to an energy momentum tensor in (4). We can obtain from here the origin of mass and spin itself, for we have as is well known (Cf.ref.[12])

$$m = \int T^{00} d^3 x$$

and via

$$S_k = \int \varepsilon_{klm} x^l T^{m0} d^3 x$$

the equation

$$S_k = c < x^l > \int \rho d^3 x.$$

While m above can be immediately and consistently identified with the mass, the last equation gives the Quantum Mechanical spin if we remember that we are working at the Compton scale so that

$$\langle x^l \rangle = \frac{\hbar}{2mc}.$$

Similarly (5) above exactly as in the theory of General Relativity, and as is well known leads to the gravitational potential,

$$\phi(r) = -Gm/r$$

Once again we reiterate that all these above relations, including the mass, spin and gravitational potential have their origin in the (noncommutative) geometry (1). Returning to the considerations in (1) to (4) it follows that (Cf.ref.[7])

$$\frac{\partial}{\partial x^\lambda}\frac{\partial}{\partial x^\mu} - \frac{\partial}{\partial x^\mu}\frac{\partial}{\partial x^\lambda} \text{ goes over to } \frac{\partial}{\partial x^\lambda}\Gamma^\nu_{\mu\nu} - \frac{\partial}{\partial x^\mu}\Gamma^\nu_{\lambda\nu} \tag{12}$$

Normally in conventional theory the right side of (12) would vanish. Let us designate this non vanishing part on the right by

$$\frac{e}{c\hbar}F^{\mu\lambda} \tag{13}$$

We have shown here that the non commutativity in momentum components leads to an effect that can be identified with electromagnetism and in fact from expression (13) we have

$$A^\mu = \hbar\Gamma^{\mu\nu}_\nu \tag{14}$$

where A_μ as noted can be identified with the electromagnetic four potential and the Coulomb law can be deduced for $|\vec{x} - \vec{x}'|$ in (5) much greater than $|\vec{x}'|$ that is well outside the Compton scale (Cf.ref.[3] and also ref. [13]). To see this in the light of the usual gauge invariant minimum coupling (Cf.ref.[13]), we start with the effect of an infinitesimal parallel displacement of a vector in this non commutative geometry,

$$\delta a^\sigma = -\Gamma^\sigma_{\mu\nu}a^\mu dx^\nu \tag{15}$$

As is well known, (15) represents the effect due to the curvature and non integrable nature of space - in a flat space, the right side would vanish. Considering the partial derivatives with respect to the μth coordinate, this would mean that, due to (15)

$$\frac{\partial a^\sigma}{\partial x^\mu} \rightarrow \frac{\partial a^\sigma}{\partial x^\mu} - \Gamma^\sigma_{\mu\nu}a^\nu \tag{16}$$

Letting $a^\mu = \partial^\mu\phi$,, we have, from (16)

$$D_{\mu\nu} \equiv \partial_\nu\partial^\mu \rightarrow D'_{\mu\nu} \equiv \partial_\nu\partial^\mu - \Gamma^\mu_{\lambda\nu}\partial^\lambda$$

$$= D_\mu - \Gamma^\mu_{\lambda\nu}\partial^\lambda \tag{17}$$

Now we can also write

$$D_{\mu\nu} = (\partial^\mu - \Gamma^\mu_{\lambda\lambda})(\partial_\nu - \Gamma^\lambda_{\lambda\nu}) + \partial^\mu\Gamma^\lambda_{\lambda\nu} + \Gamma^\mu_{\lambda\lambda}\partial_\nu$$

So we get

$$D_{\mu\mu} - \Gamma^\mu_{\lambda\lambda}\partial_\mu = (p^\mu)(p_\mu)$$

Further we have

$$D'_{\mu\mu} = D_{\mu\mu} - \Gamma^{\mu}_{\lambda\mu}\partial^{\lambda}$$

Thus, (17) gives, finally,

$$D'_{\mu\nu} = (p_{\mu})(p_{\nu})\text{or}, \frac{\partial}{\partial x^{\mu}} \rightarrow \frac{\partial}{\partial x^{\mu}} - \Gamma^{\nu}_{\mu\nu}$$

Comparison with (14) establishes the required identification.
It is quite remarkable that equation (14) is mathematically identical to Weyl's unified formulation, though this was not originally acceptable because of the adhoc insertion of the electromagnetic potential. Here in our case it is a consequence of the geometry - the noncommutative geometry (Cf.refs.[13] and [17] for a detailed discussion).
It was also described in detail how in the usual commutative space time the Dirac spinorial wave functions conceal the noncommutative character (1) [3].
Indeed we can verify all these considerations in a simple way as follows:
First let us consider the usual space time, in which the Dirac wave function is given by

$$\psi = \begin{pmatrix} \chi \\ \Theta \end{pmatrix},$$

where χ and Θ are two component spinors. It is well known that under reflection while the so called positive energy spinor Θ behaves normally, on the contrary $\chi \rightarrow -\chi, \chi$ being the so called negative energy spinor which comes into play at the Compton scale [18]. That is, space is doubly connected. Because of this property as shown in detail [8], there is now a covariant derivative given by, in units, $\hbar = c = 1$,

$$\frac{\partial \chi}{\partial x^{\mu}} \rightarrow [\frac{\partial}{\partial x^{\mu}} - nA^{\mu}]\chi \tag{18}$$

where

$$A^{\mu} = \Gamma^{\mu\sigma}_{\sigma} = \frac{\partial}{\partial x^{\mu}} log(\sqrt{|g|}) \tag{19}$$

Γ denoting the Christofell symbols.
A^{μ} in (19)is now identified with the electromagnetic potential, exactly as in Weyl's theory except that now, A^{μ} arises from the bi spinorial character of the Dirac wave function or the double connectivity of space time. In other words, we return to (14) via an alternative route.
What all this means is that the so called ad hoc feature in Weyl's unification theory is really symptomatic of the underlying noncommutative space time geometry (1) Given (1) (or (3)) we get both gravitation and electromagnetism in a unified picture, because both are now the consequence of space time geometry. We could think that gravitation arises from the symmetric part of the metric tensor (which indeed is the only term if $0(l^2)$ is neglected) and electromagnetism from the antisymmetric part (which manifests itself as an $0(l^2)$ effect). It is also to be stressed that in this formulation, we are working with noncommutative effects at the Compton scale, this being true for the Weyl like formulation also.

DISCUSSION

We have noted that nearly ninety years of effort has gone in to get a unified description of electromagnetism and gravitation starting with Hermann Weyl's original Gauge Theory. It is only in the recent years that approaches in Quantum Gravity and Quantum Super Strings, amongst a few other theories are pointing the way to a reconciliation of these two forces. These latest theories discard the differentiable space time of earlier approaches and rely on a lattice like approach to space time, wherein there is a minimum fundamental interval which replaces the point space time of earlier theories. Indeed as Hooft has remarked, "It is some what puzzling to the present author why the lattice structure of space and time had escaped attention from other investigations up till now...." [13, 19, 20, 14]. In fact we had recently shown that within this approach, it is possible to get a rationale for the de Broglie wavelength and Bohr-Sommerfeld quantization relations as well [17]. Nevertheless, the link with the gauge theories of other interactions, based as they are, on spin 1 particles, is not clear, because the graviton is a spin 2 particle (or alternatively, the gravitational metric is a tensor).

In this latter context, we will now argue that it is possible for both electromagnetism and gravitation to emerge from a gauge like formulation. In Gauge Theory, which is a Quantum Mechanical generalization of Weyl's original geometry, we generalize, as is well known, the original phase transformations, which are global with the phase λ being a constant, to local phase transformations with λ being a function of the coordinates [21]. As is well known this leads to a covariant gauge derivative. For example, the transformation arising from $(x^\mu) \rightarrow (x^\mu + dx^\mu)$,

$$\psi \rightarrow \psi e^{-ie\lambda} \tag{20}$$

leads to the familiar electromagnetic potential

$$A_\mu \rightarrow A_\mu - \partial_\mu \lambda \tag{21}$$

The above transformation, of course, is a symmetry transformation. In the transition from (20) to (21), we expand the exponential, retaining terms only to the first order in coordinate differentials.

Let us now consider the case where there is a minimum cut off in the space time intervals. As is well known this leads to a noncommutative geometry (Cf.ref.[19])

$$[dx_\mu, dx_\nu] = O(l^2) \tag{22}$$

where l is the minimum scale. From (22) it can be seen that if $O(l^2)$ is neglected, we are back with the familiar commutative space time. The new effects of fuzzy space time arise when the right side of (22) is not neglected. Based on this the author had argued that it is possible to reconcile electromagnetism and gravitation [7, 8, 22, 23]. If in the transition from (20) to (21) we retain, in view of (22), squares of differentials, in the expansion of the function λ we will get terms like

$$\{\partial_\mu \lambda\} dx^\mu + (\partial_\mu \partial_\nu + \partial_\nu \partial_\mu)\lambda \cdot dx^\mu dx^\nu \tag{23}$$

where we should remember that in view of (22), the derivatives (or the product of coordinate differentials) do not commute. As in the usual theory the coefficient of dx^μ in the first term of (23) represents now, not the gauge term but the electromagnetic potential itself: In fact, in this noncommutative geometry, it can be shown that this electromagnetic potential reduces to the potential in Weyl's original gauge theory [21, 7].

Without the noncommutativity, the potential $\partial_\mu \lambda$ would lead to a vanishing electromagnetic field. However Dirac pointed out in his famous monopole paper in 1930 that a non integrable phase $\lambda(x,y,z)$ leads as above directly to the electromagnetic potential, and moreover this was an alternative formulation of the original Weyl theory [24, 25].

Returning to (23) we identify the next coefficient with the metric tensor giving the gravitational field:

$$ds^2 = g_{\mu\nu}dx^\mu dx^\nu = (\partial_\mu\partial_\nu + \partial_\nu\partial_\mu)\lambda dx^\mu dx^\nu \tag{24}$$

In fact one can easily verify that ds^2 of (24) is an invariant. We now specialize to the case of the linear theory in which squares and higher powers of $h^{\alpha\beta}$ can be neglected. In this case it can easily be shown that

$$2\Gamma^\beta_{\mu\nu} = h_{\beta\mu,\nu} + h_{\nu\beta,\mu} - h_{\mu\nu,\beta} \tag{25}$$

where in (25), the Γs denote Christofell symbols. From (25) by a contraction we have

$$2\Gamma^\mu_{\mu\nu} = h_{\mu\nu,\mu} = h_{\mu\mu,\nu} \tag{26}$$

If we use the well known gauge condition [11]

$$\partial_\mu\left(h^{\mu\nu} - \frac{1}{2}\eta^{\mu\nu}h_{\mu\nu}\right) = 0, \text{ where } h = h^\mu_\mu$$

then we get

$$\partial_\mu h_{\mu\nu} = \partial_\nu h^\mu_\mu = \partial_\nu h \tag{27}$$

(27) shows that we can take the λ in (23) as $\lambda = h$, both for the electromagnetic potential A_μ and the metric tensor $h_{\mu\nu}$ (26) further shows that the A_μ so defined becomes identical to Weyl's gauge invariant potential [26].

However it is worth reiterating that in the present formulation, we have a noncommutative geometry, that is the derivatives do not commute and moreover we are working to the order where l^2 cannot be neglected. Given this condition both the electromagnetic potential and the gravitational potential are seen to follow from the gauge like theory. By retaining coordinate differential squares, we are even able to accommodate apart from the usual spin 1 gauge particles, also the spin 2 graviton which otherwise cannot be accommodated in the usual gauge theory. If however $O(l^2) = 0$, then we are back with commutative space time, that is a usual point space time and the usual gauge theory describing spin 1 particles.

We had reached this conclusion in ref.[19], though from a different, non gauge point of view. The advantage of the present formulation is that it provides a transparent link with conventional theory on the one hand, and shows how the other interactions described by non Abelian gauge theories smoothly fit into the picture.

It may also be pointed out that the author had argued that a fuzzy space time input explains why the purely classical Kerr-Newman metric gives the purely Quantum Mechanical anomalous gyromagnetic ratio of the electron [27, 28], thus providing a link between General Relativity and electromagnetism. This provides further support to the above considerations.

Finally, we would like to mention that it is possible to argue that within the context of Black Hole Thermodynamics, there is a transition from the transient Planck scale to the stable elementary particle Compton scale (Cf.ref.[29] for details).

REFERENCES

1. Snyder, H.S., *Physical Review*, Vol.72, No.1, July 1 1947, p.68-71.
2. Amati, D., in *Sakharov Memorial Lectures*, Eds. L.V. Kaddysh and N.Y. Feinberg, Nova Science, New York, 1992, pp.455ff.
3. Sidharth, B.G., *The Universe of Fluctuations*, Springer, Dordrecht, 2005.
4. Kempf, A., in *From the Planck length to the Hubble radius*, Ed. A. Zichichi, World Scientific, Singapore, 2000, pp.613ff.
5. Madore, J., *An Introduction to Non-Commutative Differential Geometry*, Cambridge University Press, Cambridge, 1995.
6. Madore, J., *Class.Quantum Grav.* 9, 1992, p.69-87.
7. Sidharth, B.G., *Nuovo Cimento B* 116B (6) 2001, p.4ff. 1
8. Sidharth, B.G., *Nuovo Cimento B* 117B (6) 2002, pp.703ff.
9. Sidharth, B.G., *Int.J.Mod.Phys.E.*, 14 (2) 2005, 215ff.
10. Sidharth, B.G., *Gravitation From a Gauge Like Formulation*, to appear in Annales Fondation L. De Broglie.
11. Ohanian, C.H., and Ruffini, R., *Gravitation and Spacetime*, New York, 1994, pp.64ff.
12. Misner, C.W., Thorne, K.S., and Wheeler, J.A., *Gravitation*, W.H. Freeman, San Francisco, 1973, pp.819ff.
13. Sidharth, B.G., *Chaotic Universe: From the Planck to the Hubble Scale*, Nova Science, New York, 2001.
14. Lee, T.D., *Particle Physics and Introduction to Field Theory*, Harwood Academic, 1981, pp.391ff.
15. Dirac, P.A.M., *The Principles of Quantum Mechanics*, Clarendon Press, Oxford, 1958, pp.4ff, pp.253ff.
16. Sidharth, B.G., Physics/0608222.
17. Sidharth, B.G., Annales Fondation L. De Broglie, 29 (3), 2004, 1.
18. Bjorken, J.D., and Drell, S.D., *Relativistic Quantum Mechanics*, Mc-Graw Hill, New York, 1964, p.39.
19. Sidharth, B.G., *Annales de la Fondation Louis de Broglie*, 27 (2), 2002, pp.333ff.
20. Hooft, G.'t., arXiv/gr-qc/9601014.
21. Jacob, M., (Ed.), *Gauge Theory and Neutrino Physics*, North Holland, Amsterdam, 1978.
22. Sidharth, B.G., *Found.Phys.Lett.*, August 2002.
23. Sidharth, B.G., *Found.Phys.Lett.*, 15 (6), 2002, pp.577-583.
24. Dirac, P.A.M., in *Monopoles in Quantum Field Theory*, Eds. N.S. Craigie et al., World Scientific, Singapore, 1982.
25. Sidharth, B.G., *Il Nuovo Cimento*, 118B (1), 2003, pp.35-40.
26. Bergmann, P.G., *Introduction to the Theory of Relativity*, Prentice-Hall, New Delhi, 1969, 245ff.
27. Sidharth, B.G., *Gravitation and Cosmology* 4 (2) (14), 1998, p.158ff.
28. Sidharth, B.G., *Found.Phys.Lett.*, 16 (1), 2003, p.91-97.
29. Sidharth, B.G., *From the Planck Scale to the Compton Scale*, to appear in *Int.J.Mod.Phys.E.*, arXiv/physics/0608222.

IV. SCIENCE AND SOCIETY

Professional Development and Personal Evolution: Creativity as a Tool

Miguel Alfonso García

SUMMA 112 Training Department
(Servicio Urgencia y Emergencias Medicas De Madrid)
Emergencies and Medical Urgencies Service of Madrid
C/Antracita 2 Bis. CP 2804   Madrid Spain.

Abstract. Creativity is a quality we all have; we continually use it without often realising it. We only need to be conscious of it and try to enhance its development and growth. Here I offer some ideas that I hope will be useful to whoever is interested.

ARE WE CREATIVE?

I would like to begin by making a clarification and by posing a question to those present. I am not here as an economist which is my academic training or as an emergency technician which is my present profession. I am not well recognised or a creative example of a company famous for its advertising success, I only collaborate as a volunteer in risk prevention drug dependency programs for adolescents.

I would like to take this opportunity to point out the curious irony of this situation. The ideas that I contribute for the prevention of drug dependency in adolescents are not by any means economically rewarded, but, if I had a great idea that would urge adolescents to smoke more or drink certain marketed products, I would surely earn a large sum for it. A strange fact...

And this is the most validity I can offer to those in attendance—my absence of fame and recognition on topics of creativity—since my goal today is not to further stimulate the social fever of our times, or the search for models of fame and success, above all economic success, to follow and imitate. My goal is to pass along my deep conviction that all of us are creative from the moment we are born. The problem lies in realising this reality, and above all in beginning to be conscious of it so that we may effectively use it in our lives, since social or economic recognition is not necessary to bring about this intrinsically human quality.

My question to those present is: How many of you consider yourselves to be creative? Only two people raise their hands...fine, then the necessity of my speech is justified and, I hope at the end a larger number of the audience is able to raise their hand, hopefully all those present.

First I would like to remind everyone of the meaning of the word "create", from the Latin word "creare", it is to produce something from nothing, in a figurative sense.

On the other hand, creativity could be the ability to create, the capacity for creation.

CP905, *Frontiers of Fundamental Physics (FFP8), Eighth International Symposium*
edited by B. G. Sidharth, A. Alfonso-Faus, and M. J. Fullana
© 2007 American Institute of Physics 978-0-7354-0412-0/07/$23.00

Creativity is usually related with art, science or its discoveries, and with advertising and its economic interests in large advertising campaigns.

But creativity exists in all fields of human life. Let us look at some examples. Our own perception of ourselves, our image, is a creative act in itself. I believe that my image, the image I want to offer to others and the distinct roles we play in our life are continuous acts of creativity. Today I am not a father but when my child is born a new role will be created in my life, and with it I will create and configure my personality and image. Every time there is a loss in my life, every time a role disappears, I have to recreate myself and redefine my personality and who I am in an act of continual creation.

Let's look at some simple guidelines that can help us discover and benefit from the creativity that exists in all of us:

-View problems as opportunities: Opportunities for what? -Opportunities for personal growth, for developing abilities and qualities that we already have in our latent state but that we still don't know about ourselves. Problems help us to develop this "something more", that in a different situation we would never have had to bring out from our conscience. We must come to terms with the word "problem".

-Select a focused perspective: The angle from which we look at problems or situations is important and, above all, it is important to know how to vary this angle. That change in perspective is in itself a creative act, to be able to generate new points of view and focus points about the same problem, instead of stagnating in only one. There is always a different way of contemplating the same situation or problem. It provides us with different information and points encounters that create new scenarios so that creative solutions can appear.

- Develop techniques: this is essential regardless of how skilled or able we are in an innate function or activity; it is fundamental not only to trust in the qualities or talents we posses, but we should also train and develop our techniques for maximum optimization and expression of our original potential in whatever activity we are carrying out. We must polish the tools we posses so that our talents can be manifested to their fullness.

- Lose the fear of making mistakes and breaking the mould: errors are always a guarantee of learning. Recognized and incorporated mistakes provide us with an excellent experience for learning and advancing our path. The fear of mistakes, being interpreted as failures, block us and hold us back when we have to make and act on new and creative decisions. Breaking moulds, creating new paths, acting bravely and alone in our first steps on unknown ground…it is much easier to walk on paths already taken than to go off the path having to support loneliness, doubt and critique from the rest of the group, as well as confront our own doubts about whether or not that road will bring us good fortune. Creativity requires moments of walking alone on our own path in life.

Thomas Edison conducted 2,000 experiments before inventing the light bulb. A young reporter asked him about why there were so many failures. Edison responded: "I didn't fail once. I invented the light bulb. It just happened to be a process with 2,000 steps."

-Search for more than one answer, choose yours and pay the price; in whatever situation more than one answer can always be found. We should not be content with only one, once we discover the options we choose ours and we pay the price of our choice. There is always a price to pay: the opportunity cost of the one option to leave the others behind.

-Convert the winning or losing scenarios into winning or learning; our leisure education is centred in competition. Albert Einstein said on a particular occasion: "The greatest pain caused by the capitalist system is the undermining of the being human. This dishonour is detrimental to the entire educational system because it teaches the student to compete excessively and instructs him to worship economic success as training for a future professional life".

Loss and failure are two paths directed towards a sure education and evolution; it is not about searching for them but taking advantage of them when they appear. Whatever loss produced in our life is an opportunity for creativity since, when we lose something or someone important in our life our identity is changed, our relationship with the subject of loss disappears, our expectation vanishes creating a vacancy. It obligates us to once again redefine our image, the meaning and direction of our life, and to create a new "I" without that person or situation, therefore, filling that vacancy is also an act of creativity -To create a new meaning for my existence in a different context.

To a degree, I accept losses and the rediscovery of who I am after these losses appear. I submerge myself in a more profound, self-conscious process, each time creating or discovering a concept of myself more intimate and authentic, independent of external factors, roles and etiquettes, that before were considered an integrate and essential part of my identity.

-Search for the most potential place; every human should do it, as there is always a place with the most potential for developing our creativity to the maximum. This place appears in many forms or can even be a difficult or adverse situation. It is good to be conscious when we are in it, but what really counts is simply being there. On occasions it is the place that finds us. A well-known Nobel award winner fractured his leg skiing and during his hospital stay discovered a good part of the work for which much later he would win the award. He could have used the time to complain and curse his bad luck; instead, he converted it into a highly potential situation. Sometimes a catastrophe is a highly potential place to bring many hidden qualities of being human to the surface. When what is already known no longer serves us, when what is already established collapses, when a relationship no longer functions, a crisis arises and the crisis is the precursor of creativity or something new arising.

- If you do not like a situation you have three options; change it, leave, or accept it. We are only fanatics of trying to change things and people: on occasions change can be the right path. Leaving, sometimes, can be a wise decision. But when the first two options have not worked or simply are not possible, acceptance, not resignation or passivity, but acceptance of the situations or people as they are creates an environment that in the absence of fighting and resistance, favours the flow of creativity. Even seeing reality for what it is can be an act of creativity in itself.

Educating a child is one of the greatest lessons in creativity. In the first instance, the merit of creation is natural. But the truly creative human rises to the act of educating and training a child. The criteria and method useful whit a child doesn't work with another, nor could they continue with the same child throughout different stages of its growth as a person. This process needs lots of creativity.

If by misfortune there is a loss of a child, the most painful test that one can emotionally experience appears. It is the greatest void we can feel, the greatest loss of our meaning of existence, but at the same time, although it seems incredible, a deeply and authentic creative opportunity is born with the acceptance and integration of this experience. A last gift from our child remains -the possibility to discover the most profound and hidden aspect of human existence, the acceptance of life in all its dimensions, with beginnings and ends, and in this new scenario of deep acceptance there will be no searching for creativity…simply it will flow

"A new world can be created by entrusting our wisdom to the youth."

ACKNOWLEDGMENTS

Dedicated, with all my respect, to the human beings that search for answers and meaning to the pain of their losses. In other words: dedicated to all human beings that share this world and its profound mystery.

About the Undecidable Thing of Onthologic Truth of the Reality of the Fundamental Physics

J.M. Amaya, M.V. Carbonell, E. Martínez & M. Flórez

*Departamento de Física y Mecánica Fundamentales y
Aplicadas a la Ingeniería Agroforestal
Universidad Politécnica de Madrid, España*

Abstract. In the philosophy of contemporary science, different currents from thought have been arising, each one of which, it has developed methodologies that include a set of normative rules for the evaluation that must decide the acceptance or rejection of the scientific theories already elaborated. These methodologies, rivals to each other, are: "Inductivism", "Conventionalism", "Falsacionism" and the "Theories like structures" (Kühn and Lakatos [1]). These evaluative methodologies of the quality and validity of the scientific knowledge (or metatheories of science) associate a epistemological validation as soon as until limit or border of science forms, it presents/displays and it describes the observable world as well as the one that there is behind the appearances. Consequence of it has been the appearance of different valorative interpretations from the representation that the scientific theories give us of which reality is called, such as "Scientific realism", "Antirealism", "Conjectural realism", "Structural realism" etc that has based their theses on the problem on language, truth and reality.

Keywords: onthology, reality, scientific theories, fundamental physics
PACS: 01.70.+w, 43.10.Mq

ABOUT THE PROBLEM OF THE TRUTH OF THE SCIENCE

In the XVIIIth century the German philosopher Kant that so many influence practised in the European thought, raised the Aristotelian logic and the analytical geometry to the categorical range of universal and necessary truths what, on the other hand, affected negatively in the development of the non Euclidean geometries delaying its appearance in the scene of the ideas. At the end of the XIXth century and at the beginning of the XXth, a crisis arises in both mathematics and physics science. In mathematics gives place to the development of the mathematical axiomatic formalized systems, and in physics, to the advent of the quantum mechanics and the theory of the relativity. Consequently, these universal necessary truths that the philosopher Kant had postulated were remaining buried. The philosopher Kant distinguishes between the phenomenon (what demonstrates) that he calls "the thing in itself" or noumeno (what does not demonstrate) and he says: "it is what will be the thing in itself" cannot be an object of experience with knowledge of what they are the things in itself, these, with independence of the science, it is not possible [2]. A dualism of meaning of the term truth of the science appears, epistemological truth or empirical knowledge truth, that corresponds to the objective phenomenon (of the appearance) and ontological truth (or of the last reality) does not demonstrate that it corresponds to the noumenum (or "the ting in itself") Kantian. The noumenum term is stayed as "well-considered thing" or intelligibly. Is it

CP905, *Frontiers of Fundamental Physics (FFP8), Eighth International Symposium*
edited by B. G. Sidharth, A. Alfonso-Faus, and M. J. Fullana
© 2007 American Institute of Physics 978-0-7354-0412-0/07/$23.00

so possible to raise at least the ontological truth of the reality in the nature, chance as an "entity" limit? The response remains opened to be raised ultimately.

THE PROBLEM OF THE TRUTH INDEED IN THE FORMAL SCIENCES

In 1931 a logical Austrian mathematician called Kart Gödel demonstrated a theorem that takes his name, for which any mathematical system formalized (axioms and rules of inference) if it was consistent, necessarily have that incomplete; this is, some enunciated exist whose truth or falsehood are not theorems, in consequence, such terms which truth or falsehood is indemonstrable with the methods that the system allows are undecidibles. Even it managed to demonstrate that the consistency of the own system is indemonstrable and therefore undecidible. Gödel demonstrated that arithmetical propositions can construct them that being real they are not demonstrable inside the system. These formulae, which truth is unquestionable but which are formally indemonstrable, are called themselves "G Gödel formulae". The axioms can be extended by these formulae G that will include more theorems but that, in turn, more undecidibles will generate and, as consequence, in all axiomatic formalized system axioms have to coexist, theorems and undecidibles and this logical limitation of these mathematical systems it is for principle, for what Gödel managed to demonstrate logically the blemish of the logic. In philosophy of the mathematics there coexist three currents of thought: formalism, idealism (Platonism) and intuitionism. The mathematical truth comes out the demonstratibly and establishes itself in an absolute level that is independent from any symbolic construction. Gödel's principle was placed in this type of Platonic idealism of the mathematical entities. In the strict intuitionism a mathematical proposition or it is neither real nor false less that it admits a mental construction (intuitionist) and eliminates one of the two [3].

THE PROBLEM OF THE TRUTH INDEED IN THE SCIENCES OF NATURE

As science of the nature distinguishes the physics that takes part the natural sciences. In the philosophy of the contemporary science, the problem of truth depends on evolutionary methodologies of the quality and validity of the scientific knowledge such as "inductivism", falsacionism", conventionalism", "structuralism", relativism, and so on. In some on which, this problem or is defined by principle or is not questionable by inoperative or anti-pragmatic. As consequent of this methodologies has there have been arising different valorative interpretations of the reality of the science such as "science realism", "antirealism", "critical realism" structural realism" and so on. From all of these interpretations of the reality, which one raises the problem of the scientific truth with conceptual and philosophical rigour is the "science realism" in its critical version exposed by the philosopher of science I. Niiniluoto and R. Tuomela. They based their researches about the Karl Popper ideas who postulated that normative character of the science is the incessantly critical search of the truth understanding this way the scientific progress [4]. Niiniluoto defends the increase of the realistic scientific content to ontological level and

Tuomela places, ideally, this ontological realism in the limit of a succession of theories that converge on the truth and says: "a succession of theories T1, T2,... Tn (as set of legal terms of reference) should have a conjuntivism limit that is "the truth" in a scientific particular field. If the science changes this form (or of other one) towards theories increasingly truly, it is an opened question. We neither can know her to priori, and even we nor have very good methods (and assurances) to estimate it to posteriori. The thesis of the "scientific realism" critical sound: 1) "Things in itself" exist, objects which existence does not depend on any mind. 2) This things in itself are knowledgebly partially for successive approximations. 3) This knowledge is reached by means of the experiment and the theory. 4) The knowledge in fact is opened (not final). 5) Any knowledge of a thing by self is deformed and symbolically [4]. From the scientific critical realism and from his progressive associate dynamics, there is deduced that in the supposition of a scientific theory was the unique real it does not arrange of any methodology, not from inside the science not from out of it, which demonstrates his truth, since always there would exist the possibility of being falsed. Therefore, in any scientific theory his falsehood is demonstrable so as to find a falser, but his truth, in the ideal case to it be (no falser will exist), it is an undecidable in the sense of Gödel's theorem, is indemonstrable. This argumentation is valid if there is applied Taring's theorem of the computation that is founded on Gödel's theorem, therefore the falsation of a theory is computable, while his absolute truth (in the ideal case it be) is not computable.

REFERENCES

1. I. Lakatos, *Historia de la Ciencia y sus reconstrucciones racionales*. Madrid, Editorial Tecnos, 3ł Edición 1993, pp. 15-23.
2. I. Kant, *Crítica de la razón pura*. Méjico, 1991 p. XXXVII.
3. Penrose, 2002 R. T. Wang, "La mente nueva del emperador" in *En torno a la cibernética, la mente y las leyes de la física*, 2ł Edición, Méjico: FCE, 2002, pp.134-142.
4. A. Rivadulla, *Filosofía actual de la ciencia*. Madrid, Editorial Tecnos, 1986, pp.280-282.
5. A. F. Chalmers, *¿Qué es esa cosa llamada ciencia?* Madrid, Siglo XXI editores, 2000, pp. 213-230.
6. J.M. Amaya, *En torno al problema de la verdad científica sobre el mundo*. XI Conference of History, Philosophy of the Engineer, Science and Technology. 2005.

The Path of the Entrepreneur

Javier de Alfonso

Vocesenlared.com / Voicesonthenet.com, San Bernardo 114, 4º-I, Madrid 28015, Spain

Abstract. A definition of entrepreneur and a decalogue of personal reflections taken from the experience of an entrepreneur specialized in new communication technologies.

Keywords: Definition, Decalogue, Reflections, Path, Entrepreneur.

PRESENTATION

First of all I would like to thank you for the opportunity to take part in this conference and to be able to share some remarks and conclusions taken from my path as an entrepreneur specialized in new communication technologies.

I am currently promoting two business projects:

VOCESENLARED.COM   (www.vocesenlared.com/www.voicesonthenet.com) which was awarded the NETI award from the Madrid Business Institute for the best innovative technology-based project. It is the first interactive study of recording voice-overs on the internet. It is based on an online store that allows people to select voices as well as receive estimates and due dates in real-time; and, in a web-management platform which coordinates a growing network of professional voice-over talents equipped with home-studios and a sub-network of managers which controls the texts sent by the clients and the recordings returned by the voice-over talents.

PORVOZ (www.porvoz.com), which develops the products "barofvoice", "barofvoices", "barofvideo" and "barofvideos" for instantly publishing audio and video on the Internet from a telephone and "voiceshuttle", an automatic call service that uses telephone databases to send personalized and interactive voice messages.

PERSONAL DEFINITION OF ENTREPRENEUR

Some of you will begin the path of an entrepreneur, some for vocational reasons, and others as a necessity of self-employment. The type of company where our parents worked, in many cases during their entire life, is disappearing. Today's market demands complete flexibility: globalization, dislocation, fragmentation, and small-scale operations. In Spain these concepts translate into the existence of 2,000,000 companies with approximately 5 employees and 3,500 companies with more than 250 employees.

CP905, *Frontiers of Fundamental Physics (FFP8), Eighth International Symposium*
edited by B. G. Sidharth, A. Alfonso-Faus, and M. J. Fullana
© 2007 American Institute of Physics 978-0-7354-0412-0/07/$23.00

We often ask ourselves what makes us bear the fear, the insecurity and sometimes the pressure of family and friends reminding us of the difficulty in succeeding with our project. Sometimes there is not any money to pay for the invested effort. In fact, if they paid us a good salary for half of what we do many of us would feel like we are being exploited. Pride, a yearning for freedom, not tolerating the boss, doing things in our own way, the desire to improve things...I believe what characterizes entrepreneurs is that "We relinquish safety in exchange for possibility". In other words, "Entrepreneurs don't need to have a floor; they need to not have a roof". In a certain way, we have a necessity to feel that we put all our abilities to the test without having someone else take credit for our wise decisions or for our errors.

The points emphasized in business school detail a series of abilities that the entrepreneur needs for starting a business, almost as a means to achieving an end. I would state it like this: If one attempts to be a businessperson during a sufficient amount of time, he or she will end up developing all of these abilities. Enterprising initiative is an evolutionary path that when used well will make you steadfast; it will develop your emotional and intellectual abilities and in short will bring you personal maturity and balance. Therefore, once we achieve an economic abundance we will most likely make good use of it.

PERSONAL DECALOGUE OF CONCLUSIONS

The Entrepreneur

As we have stated, big companies tend to organise themselves on a small-scale, in small human teams that are easier to organise and coordinate. At the head of every team there is usually someone in-charge with an entrepreneurial profile, who simultaneously has a wide margin of freedom as well as a large quantity of company resources. This has sky-rocketed the demand for professionals with an entrepreneurial spirit. The entrepreneur was born years ago. However, today business schools have perfected their training methods to satisfy this growing demand so that now the entrepreneur is self-made.

Training

It is an obsession among those of us who share a technical profile, very valuable for selecting objectives and developing plans. But objectives are only reached with steadfastness and willpower. The frantic rhythm at which everything evolves makes it impossible to stay up to date, so it becomes necessary to subcontract whatever is not core-business.

The Idea

Original or copy? Both are equally valid. It is not necessary to be constantly innovating; many times it is better to copy while incorporating small improvements in the product or service. We must keep in mind that the more innovative our product or service is, the less amount of initial competition there will be; but, the market penetration period will be longer because the market is not accustomed to its consumption. More and more it is the market that is setting the rules, the client is the main character and clients today classify companies based on the quality of their products and services. Luxurious facilities and spectacular presentations are taking a flat second and sometimes give the client the feeling that they are being over-charged to pay for them.

Financing

There are three principle sources for financing –the 3Fs: friends, family and fools-, banks and capital risk companies. In Spain there is no betting on ideas, therefore the banks only loan money in exchange for guarantees and the capital risk companies impose such strict conditions that we could omit the word "risk" just by referring to them. If our project does not require a large investment it is always preferable to resort to the 3Fs and make maximum use of our personal resources.

The Business Plan

This orientation document helps clarify many aspects of our business, and is essential for the capital risk companies. Nevertheless, except with very large sized projects, the target's deviations from reality are usually enormous, and the speed at which everything changes makes it very difficult to make predictions beyond one year. Frequently, an innovating new technologies project reaches the breakeven point after 4 or more years.

The Objective

Money should not be the objective; instead it should be seen as a consequence and a means. When seen as the goal, money is usually poorly utilised because there is a lack of maturity gained from positive projects that have required us to develop our abilities. "Easy come easy go". When choosing objectives one must be ambitious while at the same time being as realistic as possible to avoid accumulating unfulfilled objectives. Although it seems contradictory, it is possible to reach a target with a great deal of interest and effort, and simultaneously detach ourselves from the result, which is often determined by multiple factors that we do not control.

The Energy Source

We must enjoy the process and forget "the goal". Moreover, there isn't just one goal, but many goals. I remember years ago when I would go swimming and would cover 25 meters at full speed but I would finish really tired. I wondered how it was possible that the person in the lane next to me was able to cover 2,500 meters without stopping and at a speed not much slower than mine. With time I understood that the difference was quite simple: I was swimming with the goal of getting to the end of the swimming pool and the process turned into a big effort, whereas the other swimmer was enjoying the water temperature and his physical sensations with every stroke; in fact it would bother him to reach the end of the swimming pool because it meant he had to turn around. If you feel like you get tired, like you are putting forth too much effort, the fact is you are not doing it well. When we do any task there is always a proportion between two basic energies: love and fear. Do I do this because I like to do it or because I am afraid of the consequences if I do not do it? We could define love as that invisible ingredient that our mothers use when cooking that makes the same dish better because they cooked it. When we make it so that love is the greater of the proportion, things usually turn out better. We are often unable to do what we like, but we can always like what we do a little more. By living in the present as consciously as possible and taking note of the things we do to add life to our tasks, we make better decisions, and with sufficient perspective we realise that the result is always positive. We have to put forth effort, but it is more important to develop the mentality "What can I give" instead of "What can I get". A linear ratio between effort and results does not exist.

The Competition

This world is a disaster because we compete with each other. The world will work much better when it is based on collaboration, with everyone contributing what they best know how to do. Meanwhile, we can focus the competition inward, for facing self-denial everyday and doing what has to be done instead of what we would like to do. Normally there is more than enough of a market for everyone, therefore we can concentrate on doing things the best we can and whatever everyone else does will be secondary. Small companies have a rapid decision making ability which makes it easier to create short-term strategic alliances. Behind everything there are always people, and people are important. Advances are made much more quickly by developing emotional intelligence, choosing our alliances well and positively collaborating with them, instead of designing competitive strategies.

Communication

We are saturated. We receive hundreds of advertising impacts everyday, consciously and unconsciously. We avoid taking in the message because the effective advertising resources draw too much attention and make us forget the informative part. We must communicate in a simple and direct way, centred on solutions which support our product or service.

Luck

"Luck" depends on chance, "Good Luck" we create ourselves.

REFERENCES

1. Alex Rovira y Fernando Trias, *La Buena Suerte*, Ediciones Urano, 2005.

Science or Technoscience

Màrius Josep Fullana i Alfonso[1]

Institut de Matemàtica Multidisciplinar, Universitat Politècnica de València,
Camí de Vera s/n, 46022, València, Països Catalans

Abstract. This paper tries to give an alternative point of view of the use of Science nowadays. It also presents some reflexions to the scientists on the meaning of our work. It connects Science with other activities of the Economy and Society. Some general ideas are presented on the basis of previous work. The paper defends that workers of Science could make their job in a different way in the benefit of the whole humankind and of the preservation of Earth. So as to act this way a new socioeconomic system should be built.

Keywords: Philosophy of science; History of science; Science and society; Social systems
PACS: 01.70.+w, 43.10.Mq; 01.65.+g; 01.75.+m; 89.65.−s

A BRIEF HISTORY ON THE ORIGIN OF SCIENCE

Science is an invention of the XVI and XVII centuries. Before this epoch, in the occidental culture we could speak about pre-science and proto-science. In other cultures, like in China, India, Ancient Egypt, Middle East, Arab-Muslim Countries or America, there are also similar phases, not necessarily in the same time or in the same way, but there is not enough space here to talk about them. Nevertheless, we consider them as important as the occidental ones. So as to better understand the present state of Science it is necessary to study the evolution of human knowledge. It is worthwhile to make a short overview of its evolution and we recommend the reader to make a deep critical study of this point by him or herself. Let us first consider the long evolution in the so-called pre-history when the human being learned to find the optimal equilibrium between tool and productivity, weight and distance, and effectiveness, matter and time. This was a fight for survival, a global and systemic process of self creation and evolution. Recent studies suggest that this process was made in a friendly and sociability way of the human groups existing against the image given of the aggressive hunter [1]. The relation between hand and mind is necessary to understand the anthropogenesis and the social production of intelligence, thought and progressive knowledge [2]. This cooperative way of living may be clearly seen for example in the historic sequence of the learning of the use of energies and in its corresponding techniques. This leads us to talk about pre-science. In different places and cultures, about four or five thousand years ago, with the introduction of the work with metals, humankind developed the knowledge to use tools and to make every time much more complicated technical works. Let us remark some features that appeared in this period. This was the time when the patriarchal society appeared: the exploitation of men over women united with the growth of pre-science [3]. The benefits that men

---

[1] E-mail: mfullana@mat.upv.es

CP905, *Frontiers of Fundamental Physics (FFP8), Eighth International Symposium*
edited by B. G. Sidharth, A. Alfonso-Faus, and M. J. Fullana
© 2007 American Institute of Physics 978-0-7354-0412-0/07/$23.00

obtain with this domination are very important. On this basis, the patriarchal society had a plus of time, energy and work strength that men could use to accelerate the process of appropriation of work strength of other communities and, at the end, of their own people. A few men became the owners of the majority of the products of work, the owners of knowledge, and the slavery system began. To keep this oppression the art of war was developed. The importance of all this point in our discussion is in the fact that the essential features have been kept in the next development of all society and Science as a part of it till present.

Let us tell some few words about the Ancient Greece. They made a magnificent advance in Natural History, Mathematics, Logic, Dialectics, Sociology, Social History, Physics, Medicine and Ethics. We do not consider it Science in the modern sense, but some authors have called it proto-science. The internal and external limitations of their economical system did not allow making the jump to Science. The idealism captured knowledge which representative figures were Plato and Aristotle. There is not space here to explain the posterior period of Roman Empire, Feudalism in Europe, the brightness period of Arab-Muslim Countries, the evolution in China, India or America. Let us just remark that the features of exploitation of women, people and classes remains.

SCIENCE OR TECHNOSCIENCE

As it is well known, the birth of Science as we understand it, began in Italy at the end of XV century until the middle of XVI with figures like Leonardo da Vinci or Copernicus, then it continued between the middle of XVI century and the middle of XVII, with figures like Kepler or Galileo, and was definitely established at the end of XVII century with figures like Newton or Leibniz. However this growth was intimately related with the birth of a new socioeconomic system in Europe: capitalism. The technical expansion caused by the foundation of this system and the consequent evolution made a close relation between science and technology. This change connected even more economy, technology, war and knowledge. The European States were every time more dependent of the new class leading this process, the bourgeoisie. This is the time of development of chemistry and the establishment of the principles of thermodynamics. This is the time of the birth of new capitalist States as the USA.

Although at the beginning it was a revolutionary thought, the deterministic mechanism was founded with the figures of Descartes or Kant as the main philosophers. And this thought established the cause-effect principle and the no-change thought. This system needs the submission and exploitation of women, people and, in general, a new class also born in it, the working class. This system incorporated the former exploitation mentioned above in an even stronger way. The connection between machinery and great industry is to make maximum the productivity of the working time and therefore to intensity the exploitation and the disciplinary control on the basis of the technological innovation [4]. And this socioeconomic system produces hunger, destruction, death and disease, as we can see around the world, but not only this, the own life is in danger, and the climate change is an example of it (for instance, the destruction of all the West Mediterranean Coast with all the intensive building and the disappearance of vegetation).

As we are concerned, we define technoscience as the complex formed by great corporations and states so as to guarantee the control of research and the technological advances in the benefit of a few people. This does not mean that there is no possibility to do Science out of this control but it is difficult. Although it is not possible to explain in detail all the history of this phenomenon, this appears as a consequence of the evolution of the present socioeconomic system which dominates a great part of the planet [5], [6]. Just two examples show how this works: the control on medicines of the great pharmaceutical corporations to the extreme of making a lot of death on not allowing VIH medicines for countries of the so-called 'third world. The second example is the research in weapons which is the main research in the world, and which allows the domination of the great industrialised countries on the rest of the world. Connected with this evolution, the majority of scientists have become workers of science, and every time more we have less control on our work. Moreover, due to the patriarchal system the majority of scientists are men, thinking as men; although this situation is changing a little. Besides, they are imbedded with the occidental way of thinking defending the occidental domination of the world.

IS IT POSSIBLE A DIFFERENT WAY OF MAKING SCIENCE?

We have three main ways of thought in Occident: The ancient contemplative slavery one sensitized by Aristotle which arose when the democratic-slavery system was destroyed and repressed the pre-Socratic method, with its praxis conception of technique and knowledge; the Cartesian rationality and the Newtonian Science which defeated the medieval Weltanschauung and the alternative rationalities of the revolutionary movements emerging. The third one is to be developed in this section. It is the practical-historical and dialectical rationality synthesized by Marx, which stays for an emerging paradigm related with the new conceptions of Science as quantum physics, relativity, theory of chaos and complexity, the new advances in biology and so on, which overcomes the static, closed and non-historical Cartesian rationality [7]. It is closed connected, and it is continuously evolving, with the resistance of working class, oppressed people and women to the exploitation suffered by capitalism. It is also connected with the historical way of thinking contradicting oppression in society since its foundation, which has had different expressions in history, as the collision between materialism and idealism in philosophy, dialectics and mechanism in epistemology, or collectivism and property in Science nowadays. This way of thinking contains the physics of complexity and the dialectic of order/disorder, chance/necessity, quantity/quality, and essence/phenomenon and so on. It explains the conditions that affect Society, and Science as a part of it, through the dialectic between productive forces and social relationships of production, and that, in capitalism, all Science is converted in Applied Science as an exchange product, as a product for consumption and not for use, not for searching the needs of human beings. It is important to notice the difference between exchange and use value as the expansion of the first one is in the basis of the present exploitation of humankind [8].

This third critical-scientific method explains changes and the possibility of behavior of humankind as a whole in the future evolution of earth, of course in the limits imposed by matter. But what for? Let us develop this a little more. In this sense, Science would have another meaning which be constructed as it goes. Science would really be the historical-practical capacity of knowing and changing the problems to which we face, so as to decrease pain and hunger, the time of hard and painful work and in order to increase free time, dedicated to the emancipating pleasure and laugh, to the own culture, the artistic creation and the affective and sexual relationships: in short, in order to appropriate of our self future (for instance, in my case, the possibility of living in a free Catalan Countries, without the present patriarchal and exploitation system). One can see some examples of acting in this way: the decision of States like Cuba, India or South Africa of making free VIH medicines against private medicine. Besides, Anthropology has shown that many human communities have lived and live without inner exploitation, and that their collective thinking system acts very well not only to easy the securing of food, and other needs, but also to decide among all their members how to avoid the origin of exploitation, how to socialize the products of work, how to keep a basic social equality in a very rich difference of personalities and individual characters.

What I have tried to briefly explain in this paper is the necessity of changing the way of acting. This need arises as the only possibility to preserve most life in the Earth, including human life. My intention is to remember the possibility to scientists and to the readers in general, in working in this direction. Of course, this is difficult to see and more difficult to make, but it is time to change the course. This must be done with the participation of every one of us. One necessary step is to finish exploitation. In the way of democratisation of Science there should be some steps to make: among all scientific community and in its connection with society. But, of course, this should only be possible if the way of living changes, if we are able to build a new society, a new socioeconomic system, with the participation of every one of us. It is worthwhile to return Knowledge and Science to people. Science will be enriched with the participation of all women, people and workers and will produce quality of life for everyone, with clean technologies, with nature as a *friend* and not an enemy, with reunification of mind and hands. Then work will be a way of creative pleasure and not a punishment of gods. This way is not still imaginable as is to be constructed but has the potential of making a different and better world, a world like all of us wish to build, always in the limits of matter.

REFERENCES

1. I. Eibl-Eibesfeldt, *Amor y odio. Historia natural del comportamiento humano*, Salvat, Barcelona, 1994, pp. 238-239.
2. F. Engels, "El papel del trabajo en la transición del mono al hombre", in *Dialéctica de la naturaleza. OME-36*, CRITICA Grupo Editorial Grijalbo, Barcelona, 1979, pp. 164-177.
3. G. Lerner, *La creación del patriarcado*, Crítica, Barcelona, 1990.
4. K. Marx, *El Capital, II*, Edicions 62, Barcelona, 1983.
5. Boltxe, *Algunas relaciones entre capitalismo, globalización y tecnociencia*, (2001), http://www.boltxe.info/berria/?p=460
6. I. Gil de San Vicente, *Algunas consideraciones sobre ciencia, tecnología y emancipación*, (2000), http://www.lahaine.org/paisvasco/consideraciones_inaki.pdf
7. J. Zelený, "Transformaciones en la fundamentación gnoseológica de la ciencia actual" in *Dialéctica y conocimiento*, Cátedra, Colección Teorema, Madrid, 1982.
8. K. Marx, *El Capital, I*, Edicions 62, Barcelona, 1983.

Music and Computers

Miguel G. Palomo

Amusic, Music Technology Books
Plaza del Conde del Valle de Suchil 7, 28015 Madrid, Spain

Abstract. This paper introduces the main concepts supporting the use of computers to record, generate, edit, process, mix and play sound, and music in particular.

Keywords: computer, sound, music, digital audio, sound synthesis, sequencers, MIDI.
PACS: 43.75.Wx

COMPUTERS

Electronic, digital computers are devices designed to represent objects. They are made up of a *Central Processing Unit* (CPU), central and external *memory* and some elements to interact with the final user: *I/O peripherals* like the keyboard, the screen or the printer. At a basic functional level, the memory is made up of a finite number of elements that can be in one of two possible states, and they are said to hold a *bit* of information. Duly organized in packets of a certain size and with an associated address, different combinations of the status of these bits can be mapped very flexibly to different sets of objects (numbers in digit form for example). When these representations are the elements to be transformed, or the result proper of the transformations, they are called *data*. When the objects represented are the operations to be performed on the data, they are called *instructions* and are usually grouped in *programs*. Some of these programs manage the whole of the system resources, and they are said to constitute the *Operating System* of the computer [1].

Instructions and data are fetched from memory into the CPU, where operations are performed in synchronization with an electronic clock. The external memory is usually larger than the central one, and does not need in general to be powered to keep the data.

Representable numbers in digit form

Owing to the limitation in the number of bits of *any computer*, when the objects to be represented are numbers in digit form (i.e.: pi as 3.14159... and not as "pi", for example), only a *finite quantity* of numbers, each with a *finite quantity* of decimals are representable, i.e., only a finite quantity of numbers taken from of a *finite* subset of $\mathbf{Q}$ is representable. We will refer to this set as $\mathbf{Q}^*$, and due to the different memory capacities of different computers on one side, and to the different existing ways to represent numbers using bits on the other, the elements of $\mathbf{Q}^*$ and the quantity of

CP905, *Frontiers of Fundamental Physics (FFP8), Eighth International Symposium*
edited by B. G. Sidharth, A. Alfonso-Faus, and M. J. Fullana
© 2007 American Institute of Physics 978-0-7354-0412-0/07/$23.00

numbers that are representable will depend in general on the particular computer system under consideration.

For the purpose of this paper, we are interested in the possibilities of computers to record, create, edit, process, mix and play music, and as music is sound with a particular structure, we will first turn our attention to it.

SOUND

Sound is the name for pressure waves in elastic media like air, water or certain solids. For these sound waves to be audible, their frequency must be in the range 20Hz to 20,000Hz. In air, for example, the variations happen around the atmospheric pressure, whose standard value is 100KPa. An additional condition for sound to be audible regards the amplitude of the pressure wave, which must be above 20µPa rms. At the other end, amplitudes of around 65Pa can damage the listener's ears. By means of microphones, sound waves can be turned into electrical waves, and these ones back into sound with the help of amplifiers and loudspeakers.

Theoretically, sound pressure (and its associated voltage signal) can be represented as a continuous function of time, typically assigning 0 to the value around which the pressure oscillates, with positive portions corresponding to overpressures and negative ones to depressures. With amplitude and time taking values on $\mathbf{R}$, this continuous representation rely upon the Real Numbers Theory, that states that some real numbers (the so called *irrational* ones) have to be represented in digit form as a non-repeating, arbitrarily large sequence of digits belonging to a particular base. Furthermore, continuous functions like the one described have input and output sets that contain also arbitrarily large quantities of both irrational and rational numbers. This is the limit case of the more practical ones in which, due in part to limitations in the equipment used to convert pressure into voltage, both the time and the voltage take indeed values in just another *finite* subset of $\mathbf{Q}$, which we will call $\mathbf{Q^{**}}$. Though $\mathbf{Q^{**}}$ is obviously smaller than $\mathbf{R}$, it is in general much larger than $\mathbf{Q^*}$. In practice, this means that these so-called *analog* signals can't be represented on computers.

Digital vs. Analog Sound Signals

A possible solution is to give up some precision by transforming the analog signal into another one, a *digital* signal, one in which both the time and the amplitude take values only in $\mathbf{Q^*}$. Signal Theory shows that this does work if certain conditions easily met by current technology are satisfied. In practice, to record sound on a computer, the transformation implies the use of an analog-to-digital converter (ADC), a system usually contained in a single integrated electronic circuit connected to the computer. The ADC takes *samples* of the analog signal (produced by a microphone, an electric guitar or an electronic keyboard, for example) at a rate that is at least twice the highest frequency component in the signal, and rounds off the values obtained so that all of them become members of $\mathbf{Q^*}$. Note that the number of samples taken, which depends both on the *sample rate* and the duration of the signal, must also fit in the computer's memory. A related process is the construction of an analog signal from the original samples by means of a digital-to-analog converter (DAC), which is carried out by holding each sample for the duration of the interval of time that separates samples, producing in this way a staircase-like signal. This signal is later smoothed out by means of a filter. If the conditions mentioned above are met, the human ear will be unable to tell the original signal from the reconstructed one.

OPERATIONS ON DIGITAL AUDIO

Once in memory, a suite of programs allows us to edit the signal (erase, split and splice portions of it) and to process it with Digital Signal Processing (DSP) techniques, adding effects like filtering, reverberation, delay, etc., all in order to achieve a particular musical result. Other interesting operations are described below.

Sampling and sequencing

We can also *sample* the different notes of a particular instrument, tweak them if necessary and organize the whole into *a sound library*. Once in the computer's memory, samples can be triggered to produce music, thus *playing the sounds* of that instrument from the computer. This can be achieved either by sending the triggering messages live from a keyboard (see MIDI below), or by having a computer program known as *sequencer* to automate the recording and dispatching of the triggering messages. When we have samples from several instruments held in memory in this way, a sequencer can trigger them in a timely manner, thus opening the door to compositions for several instruments. A sequencer organizes visually the parts of the intervening instruments as a stack of tracks, and provides many tools for the composer to record, arrange and mix a piece of music. Sequencers, samplers and sound libraries are common nowadays among musicians that use computers [2].

Sound Synthesis

Another interesting possibility is the programmatic generation of sound signals, i.e., the synthesis of sound via the direct calculation of the value of each sample at a time, followed by the corresponding DAC conversion. The calculations may be based on a particular physical model, for example the differential equation that governs the vibration of a string, or may stray from physical constrains to produce a rich variety of sounds. One could apply, for instance, *substractive synthesis* on a signal having a rich spectral content; by removing certain frequencies with the help of filters, one can end up obtaining very interesting sounds. *Additive synthesis,* on the other side, proceeds by adding several simple signals to get a final sound.

MIDI

Good standards are always a boon to users and manufacturers. Back on the seventies, electronic musical equipment was becoming increasingly affordable [3]. The need to make keyboards, synthesizers, sequencers and computers -all from different manufacturers- talk to each other in a standard way gave birth to the MIDI specification in 1983. MIDI is an acronym for *Musical Instruments Digital Interface*, and refers to a suite of specifications covering the way devices connect to each other and the kind of messages they exchange. For example, a keyboard connected to a computer sends the MIDI message *note-on* each time a key is depressed. The message carries several numbers that identify the note assigned to the key, the intensity with which the note was played, and a channel of communication from among 16 possible ones [4]. If there is a synthesizer or a sampler player program running on the computer, tuned to the channel specified by the message, it will trigger the corresponding note with the corresponding intensity. After conversion in the DAC, its sound can be heard through loudspeakers or headphones connected to the computer.

If there is instead a sequencer program running, the message can be time-stamped according to the sequencer's clock and recorded on the computer's memory. As each *note-on* message is followed by the corresponding *note-off* message upon the releasement of the key, the sequencer can easily figure out the duration of the sound from the time-stamping information of both messages, and optionally display that duration in a score as a particular note.

The sequencer is thus the heart of the computer-based recording studio. Besides providing a useful visual interface to the user and the possibility to record and play MIDI messages from and to external devices, high-end modern sequencers come with integrated samplers and programmable synthesizers like the ones described in this paper. Many are also capable of recording and reproducing digital audio, thus integrating in a single environment all elements conducive to modern music production. On the downside, if the computer is not fast enough, the overhead to handle many tracks of MIDI and audio data (which includes their retrieval from memory, processing, mixing, etc.) may render the overall process so slow that the computer delivers the music not in real time, but with some latency. For a fixed workload, the delays involved get shorter with each improvement in technology.

ACKNOWLEDGMENTS

The author wishes to thank professor Antonio Alfonso Faus for his kind support on the occasion of the 8th International Symposium: "Frontiers of Fundamental Physics" held in Madrid in October 2006.

REFERENCES

1. Gregorio Fernández Fernández and Fernando Saéz Vacas, *Fundamentos de los ordenadores Vol. I.*, Madrid: E.T.S.I.T. Ciudad Universitaria, 1978.
2. Palomo, Miguel, *El Estudio de Grabación Personal*, Madrid: Amusic, 1995
3. http://en.wikipedia.org/wiki/Musical_Instrument_Digital_Interface
4. Giulio Clementi, *Non solo MIDI*, Ancona (Italy): Berben Edizioni Musicali, 1989.

Cosmic Origin and Theology of the Revelation

José Francisco Guijarro

STD, Madrid, Spain

Abstract. All along cultural history man has asked himself about the origin of man, the origin of life ante the origin of the cosmos. Regarding the question about the origin of the cosmos, any theological research must settle before any other goal the question of its language: what we understand as scientific, mythical or theological language. The biblical texts on Creation are analyzed in their historical, cultural and theological context. It is concluded that the fundamental religious meanings of the biblical texts are not in opposition to scientific interpretation of cosmic origin.

Keywords: Philosophy of Science; History of Science; Science and Society; Cosmogony
PACS: 01.70.+w, 43.10.Mq; 01.65.+g; 01.75.+m; 96.10.+i

INTRODUCTION

Among the questions raised by men thorough out cultural history an important place must be given to the questions about the origins:

Man's origin, framed within a much wider problem, such as

Life's origin, itself included in the broader subject of

Cosmic origin.

Judeo-christian revelation gives a global religious perspective on these questions. A question which cannot get an exhaustive answer —even science does not give such answer–, but it provides the subject matter for a theological answer which we may summarize in three words: *anything without God*. The light in the darkness supplied by a vision of faith based upon revelation implies the intervention of a personal God, even if we do not know in every its particulars the way how this personal intervention be performed.

QUESTION OF LANGUAGE

This makes unavoidable a previous question, the question of the various available languages to treat such problems and relative solutions, which may be reduce to three ones: scientific, mythic and theological languages: only taking this distinction into account can we establish an interdisciplinary dialogue capable of allowing us to advance through the complementary contributions of human learning which often uses similar words for different realities, leading to contradictory conclusions about a goal which, at least as hypothesis, cannot be excluded to be the same.

a) The **scientific** language

Positive science, due to the limitations self-imposed from its own nature and methodology, is reduced to analysis and attaining possible conclusions based exclusively on empirical observation. Because of the fact that it is impossible to have any experimental

CP905, *Frontiers of Fundamental Physics (FFP8), Eighth International Symposium*
edited by B. G. Sidharth, A. Alfonso-Faus, and M. J. Fullana
© 2007 American Institute of Physics 978-0-7354-0412-0/07/$23.00

knowledge about the world before man, and more before life, and even more before the existence of the universe, it must be excluded any possibility of direct scientific knowledge of these events; science will be forced to apply a regressive extrapolation in time to the process which we know are taking place at present. This requires taking as a hypothesis, impossible of verification, that those processes have taken place in a homogeneously from the chronological moment we wish to describe to the point to which we extend our empirical observation.

Then, all our speculations about we may calculate from the data obtained are grounded on an unreliable base.

The empirical observation of the universe does not have any other physical means to know what is not immediate to us than the various radiations, wave radiations for the time being, and the mechanistic observation of the effects of the force of gravitation, whose nature is unknown to us. Evidently, nobody can discard future discoveries of new radiations able of providing new information about events further in time and space to us; in every time men must proceed with the means at their disposal.

It was *Galileo Galilei* (1563-1642) who, using a simple telescope, observed that around the planet Jupiter there were moving bodies (he was able to see only four ones) which moved the same distance laterally. In one oscillation they disappeared behind the central body, while in the next one, in the opposite direction, they projected a "shadow" on the planet. Galileo did interpreted his observations as due to "satellites", or Jupiter's companions, which were describing closed orbits around the principal astral body, which once moved in front, and next behind the planet. He also observed that Saturn had rings around, observable with a simple telescope when they reflected sun light in the direction of the Earth. And he discovered that Venus, as in the case with the Moon, presented different phases depending on its position with respect to the Sun. He concluded that they were opaque bodies, with luminosity supplied by the sun light reflected towards the Earth.

This conclusion, drawn from an empirical observation came to confirm the hypothesis of the polish astronomer and mathematician *Niklas Koppernik* (Copernicus) (1473-1543) set forward from calculations but without any possible experimental confirmation with the means at his disposal, as given in *De revolutionibus orbium coelestium*, concluded in 1507, and published only after his dead. According to him, the Sun was the unmovable center of the universe, around which all other celestial bodies moved. This broke definitively Claudius Ptolomeus former Hellenistic cosmological understanding.

It implied for the catholic Copernicus a theological confrontation with Luther: the hypothesis that it was the Earth which moved around the Sun. The Reformer, and then Calvinus after him, based their claim on the literal understanding of the book of Joshua, in which he commanded the Sun to stay still, and no to the Earth, if this was moving.

Later, on some independent ways, Newton, in England, and Leibnitz, in Germany, concluded the need of a force, labeled "gravitation", which has not as yet been determined how is constituted and how does propagate. A force which affects to all bodies in space, interacting among themselves and moving constantly, in a seemingly perfect equilibrium, describing closed orbits without colliding with one another.

It was concluded later from the "Doppler effect" that celestial bodies (galaxies) are moving away from each other and that the longer the distance, the faster the speed going away. From the fact that the Universe is in a phase of expansion, it has been possible to calculate the "age of the Universe", i.e. the time needed to arrive at its present dimensions, starting from an origin assuming to be zero (the initial great explotion, or "big bang") and also assuming that the expansion has been always, as it is now, uniformly accelerated, something impossible to check.

There are two conclusions:

- That the universe is composed of countless celestial bodies interacting among themselves through the generational force, a force whose nature and transmission is experimentally unknown; and
- That the same universe finds itself *at present* in a phase of expansion.

b) The **mythic** language

All along cultural history it is possible to observe an almost universal trend to give anthropomorphic explanations of natural events, easily observable, but no so easily explainable, giving them very often in poetic language. These stories made up each, and formulated in their time and respective culture, are called **myths**.

For a better description of what it is the doctrine of the catholic faith on creation, it is necessary and foremost to have into account the cultural framework of the cosmogonic myths of the two great cultures neighboring the Jewish people, Mesopotamia and Egypt. It is evident from the analysis of the texts which have survived up to our time that the sections of the judeo-christian revelation relative to the origin of the world life and man basically accept the cosmogonic expressions of the neighboring cultures, entering into polemics with them only about one fundamental point: the substitution of the many different gods intervening in those mythic âĂĺJstoriesâĂĺ by a God who presents itself as the only true God.

c) The **theological** language

What constitutes the exclusive and unique nucleus of the Catholic Theology of Creation is the rotund affirmation that at the origin of all reality as we know it now is, exclusively, the intervention of a unique personal God. There is no room within the limits of the theological language to formulate how it toke place historically, if we can speak in those terms, the event with which God intervened - unfathomable to our knowledge. If they do not exclude the intervention of a personal God, any hypothesis proposed in scientific or mythical language can be taken in consideration.

The development of the revelation on creation, which took place at a time in history at which mythic language was predominant, accepts with no difficulty the language without arriving to a doctrinal or theological compromise with the contents.

It must be noted that theology, as the science of revelation, starts with the experience of e personal God intervening in the history of the humanity, and as any human experience —even that founded on an empirical knowledge developing in a positive science— is an individual act, which can be communicated to others only as *exposing* something, never as *imposing* anything; accepting it is a free act by he who is receiving it. ·

BIBLICAL TEXTS ON CREATION

a) Narrations of the book of Genesis

The Old Testament presents *two* different ample narrations on the creation of the world, which make up the three first chapters of the book of Genesis, besides many minor allusions to the fact of creation distributed throughout the different books, allusions also found in different passages of the New Testament. But, because it is not possible to disguise the true meaning of all creation which is man, in order that he make take advantage of creation to glorify God the Creator, in biblical narrations the creation of universe, life and man are inseparable.

Let us begin with the oldest text (*Genesis 2*).

"This is the story of the creation of heaven and earth.

In the day the Lord God Created earth and heaven, there were no tickets on earth, neither grass on the fields...

[...]

The Lord God took man and placed him in the Garden of Eden to guard it and cultivate him".

As we can see, basically, it is a narration of the origin of man, aside from any reference of the cosmological framework in which the origin of humanity takes place, a question, then, not even proposed.

In this oldest text, it is evident, first of all, that every action is attributed directly to the Lord God (Yahweh Elohîm): five times this name is repeated, as the grammatical subject of seven verbal actions, in such a brief text:

Make earth and heaven;

Not yet **had sent** rain to earth;

Modelled man and **inspired** in him the breath of life;

Planted a garden and **placed** man in it;

Made grow from the soil all kinds of trees.

It can be seen I the text a profound dependence on the mesopotamic culture: the word used in the hebrew text to express "garden" (*gan*) is ethimologically related to the verb *ganan*, which means to *protect*, to put fences; and the place in which it is located, "Eden", is related ethimologically with the accadic *edinu*, fertile and irrigated field, located towards "orient", i.e. indicating clearly while generically the region, or the *geographic context* in which the creation of man by God takes place, *located* (*yosem*, from the root *swn*) in it. The Greek translation, a few centuries after, introduces the word that, through the Latin one, goes over to different modern languages to design this "garden" as **paradise**, a term of evident Persian ethimology: this makes clear the permanence of the mesopotamic provenance of this narration.

All of this underlies the cultural dependence of the "mode" of the biblical narration, while its content - which is the only thing recognized as revealed by the Church - is an insistent monotheism.

But the archaic narration of the origin of man lacks something: the extension of the creation of the rest by the same unique God.

In contrast with the spontaneous freshness of the more archaic text, the more recent one (considered of a "priestly" level) presents a rigid structure with repeated formulae: six times we find the sentence "and God said...", after a command and its fulfillment;

immediately after the "checking" on the part of God of what he has done, giving it a name and assigning to it a purpose, concluding with the "o.k." of the Creator: "and God saw that it was good", and finally the transition from a journey to the next one, to keep track of the priestly and "cultic" purpose intended, which is saying that the seventh day God rested: "and there was afternoon and there was morning, day ..."

This text begins with an introduction:

"In beginning, God created heaven and earth.

The earth was nothing and void,

And darkness on the face of the abyss,

And spirit of God fled over the waters".

His text is intentionally schematic: lacking article on every undetermined thing (beginning, darkness, spirit); and being determined, on the other hand, with respect to "heaven" and "earth", because it does not contemplate neither other heaven nor another earth; with this duality it is implied everything the author of the text considers a whole: he uses a verb which is repeated throughout the Hebraic Bible to denote an action which is exclusive of God (*bara*, to create), repeating two words which are practical synonymous to tell what in our modern philosophical language we would call *the nothing*, using an active participle to describe a continuous action.

Then starts the set of schematic statements:

"And God said: let be light. And light was.

And God saw the light, which [was] good.

And God ser apart light from darkness,

And God called day to the light,

And to the darkness called night.

And there was afternoon and there was morning:

Day One".

The word "light" appears first without article; but once light is an existing thing, is treated grammatically as a determinates reality, as was the darkness. It is curious to note also that in this case the cardinal number is used, instead of the ordinal: "One" instead of "first"; while in all remaining schematic statements, the ordinal is used.

Thereafter the priestly justification of the schematic character of the account comes forward, introducing no less than God's example as exemplar of the sabbatical rest.

b) Other Ancient Testament's texts

Among the many other allusions in the bible to the act of creation deserve to be briefly mentioned the references in the Psalms, in the Book of Joshua ben Sirac (written in deep Hellenistic times) and in the second Book of the Maccabeus.

It is paradigmatic the text of Psalm 19, 2, which has decorated the dome of so many observatories of astronomic institutions linked to the Catholic Church:

"The heavens proclaim God's glory,

The firmament is His handy work".

No mythic element here. Only the conviction that every object of human observation is God's handy work.

From a philosophical and cultural perspective, the latest book of the Old Testament arrives to us in a greek translation written by the grand son of its author, Joshua ben Sirac, who wrote its original hebrew about 180 B.C. Having into account the literary genre of its original text, no mythic expressions are used in it, such as those given in the

Genesis, centering itself exclusively in the religious content and message. So we find in it the monotheistic reivindication of the origin of the world, of life and of man. Some examples may suffice:

"The Lord created man in the beginning and delivered him to his free will; if you wont, you will keep the commandments, because it is prudent to fulfill his will" (chap. 15).

"When God in the beginning created all his works and made them to exist, assigned their own role to each one; determined their activity and their dominion for ever; they do not faint nor become weary of their obligation. No one disturbs its companions; they never disobey God's orders" (chap. 16).

"The Lord made up man from earth and made him to come back to Him; He gave him his days counting them and gave man dominion on the earth; He infused in him a power like His own power and made him in His own image; impressed terror in all living animals [...]. He formed (in man) mouth and tongue, eyes and ears, and mind to understand; He filled him with intelligence and wisdom and taught him what is good and evil" (chap. 17).

"He who lives for ever created the Universe: the Lord is the only one without blame, there is no other besides Him. He conduces the universe with the palm of His hand, and everybody obeys His will: He is all powerful and universal King to separate the sacred from the profane" (chap. 18).

In some even later biblical text, the second book of the Maccabei, we find a language unthinkable in the two narrations of Genesis: when the mother of seven brothers damned to death if they did not desert their faith in God and their religious traditions, she exhorts the youngest one to follow the example of his older brothers, saying: "look to heavens and earth, deem everything they contain and see that God created all from nothing, and man has the same origin" (II Maccabei 7, 28). Abstract concepts such as the all and the nothing were not understanding in Semitic thoughts before hellenization, but they fit in a book which is accepted as part of the revelation.

CONCLUSION

The texts of the Bible regarding Creation have been analyzed in their historical, cultural and theological context and it has been concluded that their fundamental meaning, which is basically religious, is not incompatible to any historical or today's scientific interpretation of cosmic creation.

Have Ethic Issues Changed in Professions?

Erik Luepke-Estéfan

Centro de Especialidades Médicas Clínica San Juan
28500 Arganda - Madrid - España

Abstract. Professions are important today due to the growing number and their development. Furthermore there is a technological development unimaginable in the previous centuries. At the beginning it was recognized that there were three professions: Priest, Ruler and Doctor, representing the classical conception of Universe divided into "Macrocosmos", "Mesocosmos" and "Microcosmos" respectively. Modern age means the beginning of a change in this classical conception; that has been arguable, until the actual view that it is difficult to define what an ethical behaviour is in the professionals. This presentation tries to show some of the difficulties and conflicts presented by the technological and professional development.

Keywords: Professions

INTRODUCTION

WHAT IS A PROFESSION? To practise a profession is to exercise with voluntary inclination the art to exert a discipline in which we have specialized. Therefore to profess has the force of the vocation, the force of which is exerted because it is felt inside: it is kept inside with certain passion and possibility of development and improvement. [1] In the development and exercise of the profession, as we can see from the definition, there is a delivery spirit, besides a searching for the truth (which would explain at least partially the development of each profession and the perfectionism of each) Trying to answer this question, I am noticing three aspects interconnected: 1) The principalist ethic conception that takes into consideration the fact that human behaviour is based in a limited set of principles. [2] As a consequence we would have a very rigid ethic. 2) On the other hand we would have the casuistic ethic that comes from each of the cases. [2] Here we would find the lax ethic. 3) Besides, we cannot disregard the recognition of autonomy [2] and the freedom of conscience [3]. With these elements we will try to outline a reasoning that is useful to define what we will denominate an ethics of minimums.

BRIEF HISTORICAL REVIEW. At the beginning three professions were recognized in Occident, as symbols for the conception of the Universe they had at that time. The conception was divided in macro cosmos (exerted by the priest), the microcosms (exerted by the king-governor) and the microcosms (exerted by the doctor). The classic conception of the universe was based on understanding the laws of the universe and trying to govern itself by them. The professions in addition had an internal regulation.

It is with the development of the concept of autonomy in the modern age and its crystallization in the political systems of the contemporary age [4] when this system falls, and another one is developed in which the man can influence in his surroundings. In the land of the politician this crystallizes in the development of the present democratic systems; nevertheless in some professions, in particular in the medical one it takes more

CP905, *Frontiers of Fundamental Physics (FFP8), Eighth International Symposium*
edited by B. G. Sidharth, A. Alfonso-Faus, and M. J. Fullana
© 2007 American Institute of Physics 978-0-7354-0412-0/07/$23.00

in crystallizing this system (this takes place fundamentally in second half of century XX with the development of the informed consent).

Having this, the present situation is that the professions have an external regulation unlike what happened previously where the regulation was mainly internal (in fact this was one of the assignments of the professional schools). Even so, at the moment we find collectives that are regulated fundamentally in an internal form [5].

On the other hand and to complicate still more this panorama, we are ourselves immersed in a great technological development that creates new ethical problems in the use of the technology (in the field of the medicine the increase of conflicts is quite significant on the matter) in many cases unsolved legally or with legislating differences in different countries on identical questions.

FUNDAMENTATION

I have wanted to take as part from the present fundamentation the principles of personal autonomy, charity, non-malfeasance and justice that below are defined in a quite schematic form: Personal autonomy [6] "Is the free government of oneself " Beneficence: In essence it consists in trying to provide wellness for the other counting with which the individual understands by wellness [6] [7] Non-malfeasance: [6] Defined in negative terms: not to make damage nor to inflict evil. [8] Justice [9] we could define it as giving each one what it corresponds to one another [8].

Therefore we can make an empirist fundamentation (Anglo-Saxon philosophy) more emotiviste [10] and consecuencialist, or one principialist fundamentation (philosophies of continental Europe) more rationalist and deontological [11] [12] that it grants more importance to justice like absolute principle. In this way, the ethical Anglo-Saxons turn out to be utilitarists, (giving importance to principle of autonomy) which differs of the Central Europeans who usually are not it.

At this point I consider it interesting to mention the Kantian imperative: It is based in the treatment to all human beings as goals in themselves and not like means to all Humanity as kingdom of the goals [8]. He speaks about previous obligations to the empiric autonomy of the persons and these obligations can be synthesized in two principles: the one of non-malfeasance and the one of justice which would oblige previously to the empiric persons autonomy [8]. In Spain this is the base of the argumentation which is known as ethics of minima [8].

Using the principles already defined at the beginning of this paragraph, we shall talk about principles of Level I (Non-malfeasance and justice) and principles of Level II (autonomy and beneficence) [6]. Following this, as principles of level II and level II meet against, the ones of level I must have priority (and inside this level, we shall intend to give priority to the principle of non-malfeasance). In this pattern of argumentation, the principles of level I which are known as ethics of minima, based on the "application of the general law stating that we are all equal and deserve an equal consideration and respect to the rules of social life(...)" [13]. Minima ones are the principles gathered in the penalty code (in the case of the principle of non-malfeasance) and in the civil code (in the case of principle of justice) to which all persons are obliged even against their will. Likewise, the principles of Level II are the correspondents to the ethics of maxima, which is related to happiness and self development principles. Mainly they are defined Şby a set of persons or communities as per their free arbitrio and value system, they follow then to a particularity perspectiveŤ. In the same line run most of the religious precepts (for example Ideal of holiness in the catholic religion) [6]. Finally as an important element in the exposed fundamentation is a principle or basic premise taken from the principalism of Beauchamp and Childres, that is to say that Şrelevant moral judgements, being the starting initial point, are certain by themselves, auto evident, being accepted by everyone without major justification, because they belong to our moral tradition and have been defended by morally prudent and judicious personsŤ [6]. Having all these elements, we could face the ethic conflicts when they arise (what means that even in the searching of a solution to them prudentially made, not always we are certain with the result, in the same way that a doctor thoroughly intends to diagnose his patient correctly, neither is always certain). Using these principles (beneficence, non-malfeasance, justice and autonomy with the articulation above mentioned) and recognizing moral principles already known (for example: Şyou will not killŤ) we can face a conflict obliging us to make an exception to this recognized moral principle (for example Self Defence Legitimacy) Here I will not mention Beauchamp and Childres which includes the deliberation [14] these authors show that to follow a moral principle two premises must be accomplished. a) That the infraction is justified due to the necessity of the circumstances (that is to say, no alternative actions could be taken morally preferable) and b) that the form of the selected infraction is the least infraction.

ETHIC REGULATION

Each profession should have a deontological behaviour code. During ages this has been internally regulated. The main critic received to these codes is the fatherhood. In politics field, it meant to disconsider the persons able of knowing what they want for themselves. Also both happened in social and health area which individuals have been considered "moral deficients" having a "disorder" or illness. This has made that an external regulation is established for the said professions.

Nevertheless, this does not mean that the very same professions have nothing to do regarding their own ethics. Each profession has to face certain difficulties of their own and generally unknown to the general justice (if this difficulty/conflict is not cleared up by a convenient perusal). In biomedicine field it is clearly shown. As it has been already commented at the beginning of the exposure, there are a series of difficulties with a "legal vacuum" or there are different legislations in different countries about identical questions (as examples, we can mention legislation about abortion, euthanasia, mother cells, experiments with embryo, etc. in different countries).

In my opinion the general law should guarantee a series of minimum, applying to person defence and human life in all its extension. To be able to distinguish about difficult decisions, it is necessary to have the collaboration of the corresponding professionals in the moment of judging these decisions, which hardly can be understood outside the problem. Likewise in several Faculties of Sciences and Arts, they miss the program to study the subject of deontology in the professions.

CONCLUSIONS

Professions are at the service of man. This information can be supported taking into account the Kant imperative of treating each man as a goal in himself and not as a mean. It is important for me to have in mind the existence of Şmoral relevant juicesŤ which are likewise reasonable, as a starter to define what we consider ethic in the professions. Using principles of beneficence, autonomy, non-malfeasance and justice and its articulation in two levels (level I non-malfeasance and justice; level II autonomy and beneficence) can help to solve ethic difficulties generated at present in the professions. Likewise, the present conflicts obliges (besides the using of the principles quoted previously) prudence and deliberation (defined in this work as Aristotle) and last, it seems important to me to point out the importance of the teaching of deontology in the corresponding universities. With these conclusions, I dare to say that the ethic criteria in the professions have not changed, what has changed is the way of understanding and basing ethics.

ACKNOWLEDGMENTS

My thanks to Prof. Antonio Alfonso-Faus and the FFP8 Symposium organization for letting me participate as a speaker, presenting this subject of utmost importance in the construction of a more humanitarian and worthy society for human beings.

REFERENCES

1. Alfredo Gorrochotegui, Ser profesor universitario: Reflexiones acerca de la docencia universitaria. CONCIENCIACTIVA21, número 11, enero 2006 (http://www.concienciactiva.org/ConcienciActiva21/conciencia11/06.pdf).
2. J.J. Ferrer, y col. Bioética: Un diálogo plural. Homenaje a Javier Gafo Fernández, S.J. Universidad Pontificia ICAI - ICADE Comillas Madrid 2002. (Sección IV Cáp. "Sobre la fundamentación de la Bioética")
3. I.Kant Texto 3 "Respuesta a la pregunta: Qué es la ilustración?" http://www.terra.es/personal/ofernandezg/ilustrac.htm http://www.terra.es/personal/ofernandezg/fundamen.htm
4. Enciclopedia Wikipedia/La Revolución Francesa http://es.wikipedia.org/wiki/Revoluci%C3%B3n_Francesa
5. (conflictos deportivos en el ámbito de los tribunales deportivos vs. tribunales ordinarios). Jusport el web jurídico del deporte: http://www.iusport.es/opinion/javier_latorre_charleroi.htm. "20 minutos" Cft: http: //www.20minutos.es/columna/86824/0/gonzalo/martinez-fresneda,
6. P. Simón "Teoría del consentimiento informado" Ed Triacastela Madrid 2000 Cáp. segundo
7. Informe Belmont, de 30 de Septiembre de 1978 (The National for the protection of humans subjects of biomedical and behavioral research). http://iier.isciii.es/er/pdf/er_belmo.pdf)
8. A. Conceiro. Bioética para clínicos Ed Tracastela Madrid 1999 Cáp. primero
9. Platón, La República, lib. 4 http://www.monografias.com/trabajos11/platonn/platonn.shtml#cuatro
10. Hume, Investigación sobre los principios de la moral (fragmentos de obras de Hume), 1751 http://www.webdianoia.com/moderna/hume/textos/hume_moral.htm
11. Real Academia de la lengua española: http://www.rae.es
12. M. Arnal, http://www.elalmanaque.com/febrero/22-2-eti.htm
13. Gracia, 1999: 29-30; 1991I: 128. (Cft P. Simón "Teoría del consentimiento informado" Editorial Triacastela Madrid 2000 Cáp. 2)
14. Aristóteles: Ética a Nicomano De la deliberación. http://filosofia.org/cla/ari/azc01063.htm

Ethics and Law

Andrés Vilacoba Ramos[1]

*Avenida de San Luis, 95 B–H,
28033, Madrid, Spain*

Abstract. Ethics are the set of moral rules that govern human conduct. Hegel, for his part, asserted that ethicity implied the full realization of freedom, as well as the suppression of it as arbitrariness. In this paper, we point out that, through the relation between Law and Ethics, we can discover how high are the Ethics of a society, as well as the adherence of its members to it.

Keywords: Ethics. Laws

INTRODUCTION

Ethics are the set of moral rules that govern human conduct. Hegel, for his part, asserted that ethicity implied the full realization of freedom, as well as the suppression of it as arbitrariness. If we take a look inside the legal field, we find that the Honourable Francisco José Hernando, President of the General Council of the Judiciary (Consejo General del Poder Judicial - CGPJ), has stated that, without ethics, Law is a "ship adrift" that can become an instrument in the service of power.

The legal system itself reveals that the legal profession has gradually purged values safeguarded by deontological rules that are necessary not only for the right to defence, but also for the protection of the best interests of the State, which today is proclaimed as a social, democratic one under the rule of law (Code of Conduct of the Spanish Legal Profession (Código Deontológico de la Abogacía Española), approved by the Plenary on 27th September 2002.).

Ethical rules exist to enable the best possible solution for the problems of coexistence in society (RALLO JULIÁN, Juan Ramón: Todo un hombre de Estado Bitácora). Individuals act in society, and ethical rules must be symmetrical and functional, conform to human nature and enable development and progress. This is the meeting point between Ethics and Law. The liberation of the individual in the context of every sensitive, immediate and natural bond starts from a reality objectified by Law, the starting point of which is the person and the property.

ETHICAL AND LEGAL RULES

We must point out a basic distinction between ethical and legal rules. The former ones contain a requirement of a universal nature, since they show us the way to conduct

---

[1] E-mail: vilacoba@icam.es

CP905, *Frontiers of Fundamental Physics (FFP8), Eighth International Symposium*
edited by B. G. Sidharth, A. Alfonso-Faus, and M. J. Fullana
© 2007 American Institute of Physics 978-0-7354-0412-0/07/$23.00

ourselves in order to behave humanly. Every one must assume them as his/hers freely and consciously; they are binding on our conscience, even in those cases in which there is no risk of punishment for breaking them. When we don't comply with them, we think our conduct is not ethical, and we feel a certain sorrow or regret. The latter ones, on the other hand, do not need to be accepted in consciousness, since they rely on the support of the public force to enforce their observance. This characteristic is due to their origin, which lie in the authorities and the community, and their scope comprises all the community members. Their infringement involves a response by the courts, and the offender must serve the pertinent punishment.

We can start from the premise that ethics are not invented nor chosen democratically, and that ethical rules pursue the avoidance of evil and the achievement of good. Then, the question arises about the concept of good and evil. Logically, for each one of us, the answer can be terribly different; therefore, we can't obviate that the objectivity of the ethical rule must always have into account the subjectivity of the human being.

Other fundamental issue would be the attainment of the purpose of the ethical rules, this is, the good. Since we have legal rules, we can take into consideration if, through them, this worthy objective can be attained. To do so, we must start from the fact that rules, legal or ethical, are expressed as permissions, obligations or prohibitions in respect of conducts. For that reason, it can be gathered that our actions, from the perspective of the rules, are sometimes voluntary (permissive rule), sometimes obligated (prohibitive rule and obligatory rule). In our legal system, we have clear examples of what has been said above: - Permissive rule: I can join together with whom I want (positive right of association) or prevent someone who is unwanted from joining up (negative right of association) (Organic Law 1/2002, of 22nd march, regulating the right of association (L.O.1/2002, 22 de marzo, reguladora del Derecho de Asociación). - Prohibitive rule: It is forbidden to enter into the national territory without the appropriate visa (Organic Law 4/2000, of 11th January, on the rights and freedoms of foreigners in Spain and their social integration (Ley Orgánica 4/2000, de 11 de enero, sobre derechos y libertades de los extranjeros en España y su integración social), and Royal Decree 2393/2004, of 30 th December, approving the Regulations of the Organic Law 4/2000, of 11th January, on the rights and freedoms of foreigners in Spain and their social integration (Real Decreto 2393/2004, de 30 de diciembre, por el que se aprueba el Reglamento de la Ley Orgánica 4/2000, de 11 de enero, sobre derechos y libertades de los extranjeros en España y su integración social). - Obligatory rule: It is obligatory to pay the taxes (Taxation Laws).

Nevertheless, there is still a main issue pending: £can legal rules force people to do good? The answer must be necessarily negative. Indeed, legal rules can only prevent people from doing evil, but they can't force them to do good. This premise can be refuted by referring to the offence of failure to render assistance, although the purpose of the rule creating this offence is not forcing to do good, but alleviating the consequences of an already existing evil in order to avoid a worse one.

CONCLUSIONS

Ethics and Law converge on several points that can be summarized as follows:

- Both of them regulate the relations between human beings through a set of rules.
- Both kinds of rules, ethical and legal, are of an imperative nature.
- Both of them respond to the same social need: the regulation of human relations in order to ensuring a certain level of social cohesion.
- Both of them change if the content of their social function also changes (Chapter II, section 3.2. of the Civil Code (Código Civil) provides that rules must be interpreted in accordance with the genuine sense of the words therein and in relation with the context, the historical and legislative background and the social reality of the time when they are to be applied, fundamentally considering their spirit and purpose).

Nevertheless, both ways -ethical and legal- diverge with respect to the ideas listed below:

- Ethics demand an internal conviction on the part of the individual that Law does not require.
- The imposition of ethical rules comes from the inside of the individual, whereas legal rules are imposed externally. Ethical coercion is fundamentally internal; in the legal field, the coercion is external.
- Law has an official nature, since it is based on a set of codified rules, whereas Ethics lack this official nature.
- Ethics affect all the relations of coexistence between human beings, and Law is the vehicle through which human relations are regulated in order to ensure the State's good governance.
- Ethics are intrinsic to the person and, therefore, do not need the existence of a social order; Law's raison d'être, on the contrary, is the organization of a society. The individual, as an independent entity, does not need Law (This is clearly illustrated in literature, DEFOE, DANIEL: Robinson Crusoe).
- The different relation of Ethics and Law to the State explains, in turn, the different situation of both ways of human conduct in a same society. Since morality is not linked necessarily to the State, in a same society there can be a morality that corresponds to the current governmental authority and a morality that enters into contradiction with it. This is not the case with Law: since Law is necessarily linked to the State, there is only one law or legal system for the whole society, even though such a system lacks the moral support of all of its members.

Despite the differences between Ethics and Law, their connection and their inherent historical relation are permanent. Ethics expand, at the expense of Law, as persons observe the basic rules of coexistence voluntarily, not through coercion. It would be possible to state that the ethical sphere reduces gradually the legal sphere, since, as a society progresses, legal rules are increasingly replaced by moral rules. In conclusion, may we point out that, through the relation between Law and Ethics, we can discover how high are the Ethics of a society, as well as the adherence of its members to it.

REFERENCES

1. A. Viñas, *Ética y derecho*.
2. G. Jakobs amd L.M. Reyna Alfaro, *Derecho penal, ética y fidelidad al Derecho: estudio sobre las relaciones entre Derecho y Moral en el funcionalismo sistémico*.
3. G. Peces-Barba Marínez, *La dignidad de la persona desde la filosofía del derecho*, 2002.
4. M. Utande Igualada, *Raíces de lo ilícito y razones de licitud. El Iusnaturalismo de John Finnis*, 2006.
5. J.A. Pinto Montanillo, *Raíces de lo ilícito y razones de licitud. Interrogantes de la filosofía jurídica de Bobbio, a la luz del derecho natural*, 2006.
6. B. Castro Cid, *Manual de teoría del derecho. Derecho, moral y usos sociales*, 2004.
7. R. Asís Roig, *Justicia, migración y derecho. Derechos humanos, inmigración y solidaridad razonable*, 2006.

V. POSTERS

MASS BOOM VERSUS BIG BANG: WAS EINSTEIN RIGHT IN HIS STATIC MODEL FOR THE UNIVERSE?

THE BIRTH OF THE SCIENTIFIC COSMOLOGY

- EINSTEIN'S RELATIVITY THEORY
- FIELD EQUATIONS OF GENERAL RELATIVITY
- EINSTEIN'S COSMOLOGICAL EQUATIONS

THE COSMOLOGICAL PRINCIPLE

HOMOGENEOUS AND ISOTROPIC UNIVERSE

STATIC, FINITE IN SIZE, CURVED AND CLOSED.

UNLIMITED AND WITHOUT A CENTER

SIZE (constant unit) $ct = 10^{28}$ cm

THE HUBBLE LAW

WE CONSIDER THE EXPANSION OF THE UNIVERSE, TO EXPLAIN THE REDSHIFT, AS AN APPARENT EFFECT INTERPRETED FROM THE POINT OF VIEW OF OUR STATIC LOCAL SYSTEMS (LABS) SEEN AS "ABSOLUTE OBSERVERS".

WE REVERSE THE INTERPRETATION BY CONSIDERING A CONTRACTING QUANTUM WORLD (CONTRACTING LABS) INMERSED IN A STATIC UNIVERSE, (THE INITIAL EINSTEIN MODEL).

MACH'S PRINCIPLE

Energy of a mass m equal to the gravitational potential energy of m with respect to the rest of the masses of the Universe M. Then $GM/c^2 = ct$ (gravitational radius of the Universe is the same as its size)

Our result (Mass Boom) gives (from the Action Principle)

$$G = c \quad \text{and} \quad M = t$$

(see references internet: search Mass Boom or Alfonso-Faus)

The theory of gravity quanta (published in Physics Essays) predicts its linear emission, with negative energy, and therefore a linear increase of all gravitational masses. The Mass Boom effect.

TIME VARIATION OF THE SPEED OF LIGHT

GIVEN THE MASS BOOM (M=t), CONSERVATION OF MOMENTUM IMPLIES TIME VARIATION OF THE SPEED OF LIGHT r (first anounced by the author in 1987 at the ERE in Tenerife, I.A.C.)

From $Mc = $ constant we get

$$c = 1/t$$

Australian astronomers confirm this law. The speed of light decreases linearly with the cosmological time t.

QUANTIZED TIME

Initial "tics" imply a very high speed of light then

THERE IS NO NEED OF INFLATION

THE HORIZON PROBLEM IS SOLVED

HUBBLE'S REDSHIFT INTERPRETED AS A SHRINKAGE OF THE QUANTUM WORLD

Planck's "constant" becomes equal to the speed of light c.

Quantum sizes proportional to

$$h = c = 1/t$$

THE MASS OF THE QUANTUM OF GRAVITY (Physics Essays, Alfonso-Faus 1999)

$$m_g = h/c^2 t = 1/ct = 1 \text{ (constant)}$$

Unit of mass 10^{-66} gr

THE QUANTUM OF TIME
$$t_g = h/Mc^2 = 1 \text{ (constant)}$$

Unit of time 10^{-104} sec

AGE OF THE UNIVERSE 10^{122}

HARMONIZATION OF GENERAL RELATIVITY AND QUANTUM MECHANICS

A first step:

the quantum of gravity, unlocalized, size ct = 1, the size of the Universe, constant, corresponding to the initial model proposed by EINSTEIN

$M = t = S$ (entropy)

Our definition of S:
number of emitted gravity quanta = M/m_g

Hawking-Bekenstein-Mass Boom
SAME RESULT
$$k/hc \, GM^2 = ER/hc = kM/m_g$$

ALL EQUAL TO COSMOLOGICAL TIME t

OKHAM'S RAZOR DEMANDS A CHANGE OF PARADIGMS

We eliminate the Big-Bang problems and start to work with a new frame for cosmology:

THE MASS BOOM

(INITIATING A MARRIAGE BETWEEN GENERAL RELATIVITY AND QUANTUM MECHANICS)

ANOUNCEMENT OF NEXT 8th SYMPOSIUM "FRONTIERS IN FUNDAMENTAL PHYSICS"

11-14 September 2006 MADRID, SPAIN

SEE THE CIBELES, THE PRADO MUSEUM, AND THE HISTORICAL NEARBY CITIES

ALFONSO-FAUS
UNIVERSIDAD POLITÉCNICA DE MADRID, SPAIN

Quantum and Classical Resonances

F. J. Arranz*, R. M. Benito† and F. Borondo**

*Grupo de Sistemas Complejos. Departamento de Ingeniería Rural. ETSI Agrónomos.
Universidad Politécnica de Madrid. 28040 Madrid (Spain).
†Grupo de Sistemas Complejos. Departamento de Física y Mecánica. ETSI Agrónomos.
Universidad Politécnica de Madrid. 28040 Madrid (Spain).
**Departamento de Química, C–IX.
Universidad Autónoma de Madrid. Cantoblanco, 28049 Madrid (Spain).

Keywords: Quantum chaos; Quantum–classical correspondence; Avoided crossings; Periodic orbit; Molecular Vibrations
PACS: 03.65.-w, 05.45.Mt

The Poincaré–Birkhoff and Kolmogorov–Arnold–Moser theorems describe the general structure and evolution, as energy increases, of phase space of nonlinear Hamiltonian systems [1]. Particularly, this structure involves the existence of chains of islands associated to periodic orbits or *classical resonances*. In 2D systems, each resonance is characterized by the frequency ratio $\omega_x : \omega_y$ (order of the resonance), where ω_i is the frequency for the i–th coordinate. Then, as energy increases, the appearance of a sequence of resonances: $\omega_{x,1} : \omega_{y,1}$, $\omega_{x,2} : \omega_{y,2}$, ..., is observed.

On the other hand, in quantum mechanics an eigenenergies correlation diagram versus a system parameter can be computed, showing different series of avoided crossings, due to the noncrossing rule between eigenstates of same symmetry, associated to *quantum resonances*. The order of the resonance at each avoided crossings is given by $\omega_x : \omega_y = |\Delta n_y| : |\Delta n_x|$, where Δn_i is the difference between quantum numbers, corresponding to the i–th coordinate, of both eigenstates involved in the avoided crossing.

In this context, we have found for a realistic 2D model for the vibrations of the LiCN molecule, series of quantum resonances (each avoided crossing of a series having the same order of resonance) in the correlation diagram of eigenenergies versus Planck's constant, $\hbar$. As this parameter increases, we observe the appearance of the series of resonances: 1:6, 2:14, 1:8, 2:18, 1:10, 1:10, 1:8. Moreover, for the same system, we observe a *similar* sequence of classical resonances (only "regular" periodic orbits are considered): 1:6, 1:7, 1:8, 1:9, 1:10, 1:10, 1:8 ...This is a very interesting result on the correspondence between classical and quantum mechanics, since it shows the importance of periodic orbits for this problem. This result also shows the usefulness of correlation diagram E $\hbar$ for understanding quantum–classical correspondence [2].

This work was supported by MEC-Spain (MTM2006–15533 and CSD2006–32).

REFERENCES

1. M. A. Lieberman and A. J. Lichtenberg, *Regular and Chaotic Dynamics*, Springer, New York, 1992.
2. F. J. Arranz, F. Borondo, and R. M. Benito, *Phys. Rev. Lett.* **80**, 944 (1998).

CP905, *Frontiers of Fundamental Physics (FFP8), Eighth International Symposium*
edited by B. G. Sidharth, A. Alfonso-Faus, and M. J. Fullana
© 2007 American Institute of Physics 978-0-7354-0412-0/07/$23.00

WATERHAMMER EFFECT IN MICROGRAVITY ENVIRONMENT

R. Elvira, A. Monzón, J. Fernández-Cabrera and I. Riesgo

E.U.I.T. Aeronáutica. Univ. Politécnica de Madrid.
Plaza Cardenal Cisneros N° 3. E-mail: ruben.elvira@gmail.com

ESA PARABOLIC FLIGHT CAMPAIGN

* Every year 120 students are given the opportunity to take part in ESA's student parabolic flight campaign

* Each of the 30 experiments carried out on board the flight is designed by teams of 4 students studying in one of the 17 ESA Member States

* Flights take place on an Airbus-300 specially equipped to perform parabolic flights

* Novespace helps and supports the work of the students in the Flight Campaign

ZERO G WATER HAMMER EFFECT

WATER HAMMER EFFECT
* Over-pressure caused in a pipe across which one liquid fluids and suddenly it is stopped by a quick-closing valve
This over-pressure induces stresses that may cause failures in the pipe

IN ZERO G
* How will be the Water Hammer Effect in microgravity?
* It will be smaller that the Effect in gravity conditions?

We suppose it but we need experiment it

OBJETIVE

GOAL
* To make a comparison between the strength of the water hammer effect on ground conditions and microgravity conditions

INDUSTRIAL APPLICATION
* Improvement design, increase safety, decrease costs and weight

Minor Overpressure
Minor Thickness
Minor Weight
Minor Cost   30.000 € per Kg in orbit

EXPERIMENT

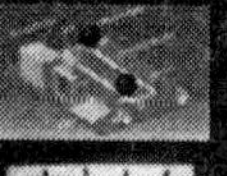

PROCEDURE

1. Mass flow across one pipe
2. Closing the flow by a quick-closing valve
3. Measure the over-pressure
4. Registration
5. Comparison between microgravity and on ground data

THE FLIGHT

30 parabolas
20 second of microgravity per parabola
5 hours of flight

RESULTS

CONCLUSIONS
* The Water Hammer Effect in microgravity is less strong than on gravity conditions
* But we need a deeper investigation for assure it

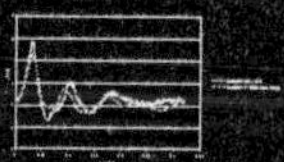

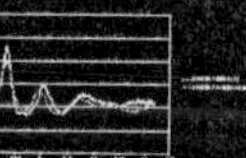

FUTURE RESEARCH LINES

ADITIONAL GUIDELINES
* Cuantitative relation Water Hammer Effect & Gravity
* Thickness reduction
* Influence of temperature (heating of the pipe)
* Test with other liquids

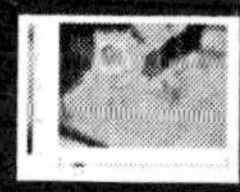

TEAM   SpaceHammers
EXPERIMENT   Zero G Water Hammer Effect
COUNTRY   Spain

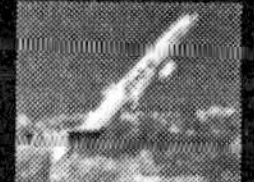

∩OVE∫PAC∈

Field Quantization in *E-B* Phase Space

Peter Enders

Fischerinsel 2, D-10179 Berlin, Germany; Enders@dekasges.de

Abstract. A novel scheme of field quantization is proposed that attempts to avoid the drawbacks of normal-mode quantization and of common photon wave mechanics.

Keywords: Field quantization, normal-mode quantization, photon wave mechanics
PACS: 03.70.+k

INTRODUCTION

Drawbacks of normal-mode field quantization (*cf* Schleich, *Quantization in Phase Space*, Wiley 2001, Ch.10): Infinite zero-point energy; different treatment of space and time variables; difficulties in interpreting the results as particle properties.

Drawbacks of standard quantum mechanics of electromagnetic field [recent review: O. Keller, *Phys. Rep.* 411 (2005) 1-232]: Maxwell-Lorentz equations for $\mathbf{E}$, $\mathbf{B}$ are manipulated towards formal analogy to Schrödinger equation, <u>without</u> giving physical needs to do so; $\mathbf{E}$, $\mathbf{B}$ are normalized for reinterpretation as wave function, <u>without</u> physical motivations, again.

Here, analogously to our axiomatic derivation of the Schrödinger equation (Enders & Suisky, *Int. J. Theor. Phys.* <u>44</u> (2005) 161-194; Enders, *Von der klassischen Physik zur Quantenphysik*, Springer 2006), Newton's, Euler's and Helmholtz's representations of Classical Mechanics are exploited to yield Schrödinger-like equations for wave functions $\Psi(\mathbf{E};\mathbf{x},t)$, $\Phi(\mathbf{B};\mathbf{x},t)$.

ALTERNATIVE WAVE MECHANICS FOR FIELDS

Analogously to the quantization of the harmonic oscillator, the elm. field energy density is represented as $\mathcal{E}=\int(\varepsilon_0/2)\mathbf{E}^2|\Psi(\mathbf{E};\mathbf{x},t)|^2 d^3\mathbf{E}+\int(1/2\mu_0)\mathbf{B}^2|\Phi(\mathbf{B};\mathbf{x},t)|^2 d^3\mathbf{B}$. This results in a quantized momentum density: $\Pi=(n+\frac{1}{2})\Pi_0$. Is there such an elementary Π_0 (and corresponding elementary energy density, $\mathcal{E}_0=c\Pi_0$)?

Advantages: No need for modes without real boundary conditions (frequency and wavelength are external parameters of a wave!); spatial and temporal dependencies are, (i), treated on equal footing, (ii) independent of field quantization, (iii), thus, possibly, able to realize Einstein's idea (1905) of a "granular" field structure; no infinite ground state energy; no renormalization necessary(?).

PROSPECT

Not only artificial bare particles, but also artificial bare Hamiltonians become superfluous (*cf* Stefanovich, *Relativistic Quantum Dynamics*, arXiv:physics/0504062).

CP905, *Frontiers of Fundamental Physics (FFP8), Eighth International Symposium*
edited by B. G. Sidharth, A. Alfonso-Faus, and M. J. Fullana
© 2007 American Institute of Physics 978-0-7354-0412-0/07/$23.00

Global dynamics of nonrigid triatomic molecular systems of three degrees of freedom

C. G. Giralda*, R. M. Benito*, J. C. Losada† and F. Borondo**

*Grupo de Sistemas Complejos. Dpto. Física y Mecánica. ETSI Agrónomos. Universidad
Politécnica de Madrid. Avda. Complutense s/n. 28040 Madrid.
†Grupo de Sistemas Complejos. Dpto. Tecnología de la Edificación. EU Arquitectura Técnica.
Universidad Politécnica de Madrid. Avda. Juan de Herrera, 6. 28040 Madrid.
**Dep. de Química, C–IX. Univ. Autónoma de Madrid. Cantoblanco, 28049–Madrid (Spain).

Keywords: Classical mechanics; Nonlinear dynamics; Chaos theory; Frequency analysis
PACS: 05.45.Ac, 05.45.Tp

Composite Poincaré surface of section is an excellent tool to unveil the phase space structure of dynamical systems. Unfortunately, this method is only practical for 2D Hamiltonian systems, and as soon as a third mode is considered new ways of visualization are required. The frequency analysis (FA) method, described here, is ideally suited for this task.

In this respect, triatomic molecules can be considered as Hamiltonian systems formed by a collection of three coupled, anharmonic oscillators, whose dynamics obey the usual equations of motion, in terms of the corresponding Hamiltonian function, $H(q_1, q_2, q_3, P_1, P_2, P_3)$, expressed in suitable (internal, normal or Jacobi) coordinates and their conjugate momenta. From numerically computed trajectories, a complex dynamical function, $f_k(t) = q_k(t) + iP_k(t)$ $(k = 1, 2, 3)$, for each dimension is constructed and Fourier analyzed. Now by taking the three fundamental frequencies, $\omega_1, \omega_2, \omega_3$ corresponding to the biggest coefficients, a frequency map can be constructed:

$$\mathbf{FM} : B^2 \in \mathbb{R}^2 \longrightarrow \Omega \in \mathbb{R}^2$$
$$\left(P_1^0, P_2^0\right) \longrightarrow \left(\frac{\omega_2}{\omega_1}, \frac{\omega_3}{\omega_1}\right), \tag{1}$$

where B^2 is the domain of initial momenta (for fixed values of the rest of coordinates and momenta) and Ω is a set of frequency values obtained by FA. The results of this map are now suitable for standard $x - y$ visualization.

At low values of the excitation energy the frequency map is regular, and frequency lines are easily identified. On the contrary, at higher energies, invariant tori are destroyed, and spread points due to the presence of chaotic motions appears in the corresponding regions of phase space. Moreover, the underlying structure of this region shows up, and many resonance lines are detected. The time evolution of different trajectories, starting at different regions of phase space, can then be followed. For chaotic trajectories the Arnold diffusion is observed.

Finally, it should be stressed that the global picture provided by FA makes research in intramolecular dynamics very efficient, allowing to target regions of dynamical interest.

This work was supported by MEC–Spain (MTM2006–15533 and C3D2006 32).

CP905, *Frontiers of Fundamental Physics (FFP8), Eighth International Symposium*
edited by B. G. Sidharth, A. Alfonso-Faus, and M. J. Fullana

FULLERENES AND DERIVATIVES: A SIMULATED STUDY

__C. Manteca-Diego__, J. Doménech, G. Doménech-Manteca and R. Elvira

Escuela Ingeniería Técnica Aeronáutica, Universidad Politécnica de Madrid
Dpto. Tecnologías Aplicadas a la Aeronáutica

Plaza Cardenal Cisneros Nº 3, España

INTRODUCTION

Since the recent discovery of the new, polyhedrally shaped, pristine forms of carbon with the general formula C_{2n}, known in the literature as fullerenes, a great deal of interest has been attracted because of their unusual properties: fivefold local symmetry, new superconductors, transformation into diamond at very high pressures, endohedral or exohedral doping, tubular or onion-shaped nanostructures, etc. Besides a myriad of organic derivatives, let us say that a new era in carbon chemistry has emerged (1,2,3,4).

METHODS

In this communication IRIX 4D, Silicon Graphics, working stations have been used. Molecular dynamics and energy calculations have been performed with Discover and the graphical representations with Insight II, both computational chemistry programs supplied by Biosym Technologies. At a much more moderate level, the 3.0 version of Hyperchem by Addlink Corp. is adequate for use on a 486 PC with these purposes. The calculations are iterative processes which look for local minima of the energy using minimization algorithms such as conjugated gradient, Newton-Rampson, etc.

Complementary to this, in molecular dynamics, the conformational space of a given molecule is studied at fixed values of P and T; the procedure is based on the numerical resolution of the Newton equation for motion. All these calculations depend on the field of forces used, which describes the peculiar characteristics of each atom involved. As a previous step, planar connectivity diagrams (Schlegel, Figure 1-A) have to be generated.

RESULTS

Depending on the Schlegel diagram used, different sets of C-C distances are obtained and, although the exact nature of the C-C bonds is not yet fully understood, model B, which supposes po-delocalization in the pentagons, yields better results: 1.51 A as the average distance between two atoms of carbon defining a hexagon-hexagon edge and 1.43 A for the hexagon-pentagon ones, in quite close agreement with the experimental results (see Table I). The pseudosphere diameter is 7.16 A and the closest distance between the two different molecules, as determined by the corresponding Van der Waals clouds, can be measured as 3.1 A (Figure 1-B).

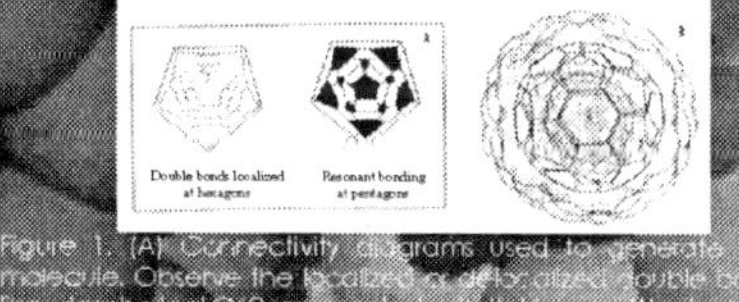

Figure 1. (A) Connectivity diagrams used to generate the C_{60} molecule. Observe the localized or delocalized double bonds. (B) Pseudospherical C60 represented with its Van der Waals clouds.

This kind of study can be extended to other fullerenes, such as C_{70}, or to nanotubes and to many derivatives as well, in particular to hydrogenated or halogenated species such as X_nC60 (X = H, F, Cl, Br; n = 6, 8, 24, 60). The corresponding measured bond distances fit quite well with the calculated values on the basis of covalent single bonding (See Table I).

TABLE I SIMULATED BOND DISTANCES (ANGSTROMS)

BOND	COMPOUND	d (calc.)	d (bibl)
C-C	C_{60}	1.51 (H-H)	1.45
C-C	C_{60}	1.42 (H-T)	1.38
C-C	X_nC_{60}	1.55	1.54
C-H	H_nC_{60}	1.12	1.09
C-F	$F_{48}C_{60}$	1.39	1.35
C-Cl	Cl_xC_{60}	1.90	1.77
C-Br	$Br_{24}C_{60}$	1.94	2.01

Regarding thermal behaviour, one very surprising result is the extremely high apparent stability of the C_{60} molecule: even at 8000 K, an asymptotic plot is obtained when plotting energy versus time (see Fig. 2). In spite of the fact that C-C bonding distances increase moderately, the sphere does not blow up as could be expected: this seems to happen at higher temperatures. This result is not at all contradictory to the high reactivity experimentally i.e. it burns at temperatures much lower than graphite (6)-, because this occurs only under vacuum; that is, it refers to intrinsic stability, given by the strength of its bonds and peculiar geometry.

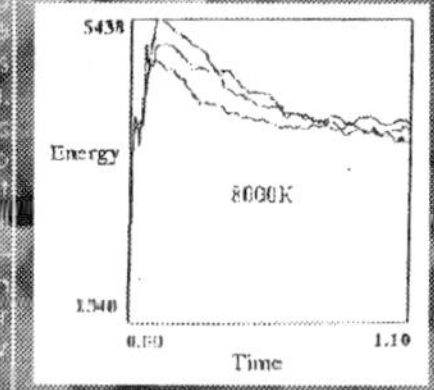

On the contrary, hydrogen or halogen derivatives are much less stable. The molecules are far less rigid than that of pristine fullerene, as a consequence of the double bonds breaking. Molecular dynamics let observe how C-C vibrations increase in amplitude just for those atoms of carbon linked to X atoms; the break limit (3 A, arbitrarily taken) is reached at fairly low temperatures. This simulated studies seem to indicate that the stability of halogen derivatives follow F > Cl > Br and any of them decomposes at temperatures about 400 K, in good agreement with experimental data, although experimental data vary somehow depending on the author, surely due to chemical instabilities of the compounds studied.

REFERENCES

1- Qiang Lu and Baidurya Bhattacharya, Eng Fract. Mech. 72, 807-823 (2005)
2- Y. C. Zhang and X. Wang. Int. J. Solids Struct. 42,5399-5412 (2005)
3- H. W. Kroto et al. Nature, 318, 162 (1985)
4- Accounts of Chemical Research (special Volume) 99, 175 (1992)
5- Carbon (special Volume), 30, 1139-1286 (1992)
6- M. S. Dresselhaus et al. Journal of Materials Research, 8, 2054-2097 (1993)
7- W. Krätschmer et al. Nature 347, 354-358 (1990)
8- R. Taylor and D. R. W. Walton. Nature, 360, 685-693 (1993)

Madrid (IAA), Septiembre de 2006

SYNTHESIS AND APPLICATIONS OF METHYL CYANIDE ADDITION COMPOUND OF BORON TRICHLORIDE, Donor-Aceptor Bonding.

C. Manteca-Diego[1], J. Domenech[1], G. Domenech-Manteca[1]

1. E.I.T. Aeronáutica, Dpto. Tecnologías Especiales Aplicadas a la Aeronáutica . Univ. Politécnica de Madrid. Plaza Cardenal Cisneros Nº 3. E-mail: consolacion.manteca@upm.es

INTRODUCTION

PYROLYSIS OF $CH_3CN \cdot BCl_3$ (4) AT ca. 900-1000°C OVER Co POWDER GENERATES NOVEL GRAPHITIC $B_xC_yN_z$ NANOFIBRES AND NANOTUBES POSSESSING MORPHOLOGIES (E.G. CURLED, BRANCHED AND BENT). IN THESE EXPERIMENTS THE METAL PARTICLES PLAY AN IMPORTANT ROLE IN THE GROWTH SINCE NANOTUBE FORMATION APPEARS TO OCCUR AT THE METAL SURFACE. HIGH RESOLUTION ELECTRON MICROSCOPY (HRTEM) AND ELECTRON ENERGY LOSS SPECTROSCOPY (EELS) STUDIES SUGGEST THAT THE STOICHIOMETRY OF THE FILAMENTS IS ca. $[BC_3N]$.

THEIR STRUCTURES WERE INVESTIGATED, CONCENTRATION PROFILES, ALONG AND ACROSS THE FILAMENTS, REVEALED THAT B, C AND N ARE NOT HOMOGENEOUSLY DISTRIBUTED WITHIN THE NANOSTRUCTURES BUT ARE SEPARATED INTO PURE C AND BN DOMAINS. THIS COMPOUND PROVIDES NANOMATERIALS WHICH MAY PROVE USEFUL AS

EXPERIMENTAL

THE 1:1 $CH_3CN \cdot BCl_3$ COMPLEX (4) WAS PREPARED BY REACTING EQUIMOLECULAR QUANTITIES OF CH_3CN AND BCl_3 (1 M SOLUTION IN HEXANE) AT −78°C UNDER VACUUM. THE WHITE SOLID PRODUCT WAS PURIFIED BY SUBLIMATION AT 85°C AND CHARACTERISED BY UV, IR, ^{1}H-NMR, ^{11}C-NMR, MASS ESPECTROMETRY AND ELEMENTAL ANALYSIS (TABLA I)

PYROLYSIS WAS CARRIED OUT IN A TWO-STAGE FURNACE SYSTEM FITTED WITH TEMPERATURE CONTROLLERS (SEE FIG. I). $CH_3CN \cdot BCl_3$ (ca. 100-200 mg) WAS PLACED IN ONE END OF A QUARTZ TUBE (6 mm O.D. AND 60 cm IN LENGTH) LOCATED IN THE FIRST FURNACE.

COBALT POWDER (ALDRICH 99.9% < 2 μm PARTICLE SIZE) IN A QUARTZ BOAT (6 cm IN LENGTH) WAS PLACED IN THE SECOND FURNACE.

ARGON (40 ml/min) WAS USED AS A CARRIER GAS AND THE TEMPERATURE OF THE FIRST FURNACE WAS SLOWLY INCREASED FROM ROOM TEMPERATURE TO 500°C. THE SECOND FURNACE WAS MAINTAINED AT 1000°C FOR THE DURATION OF THE EXPERIMENT.

RESIDUES SCRAPED FROM THE TUBE WALL IN THE SECOND FURNACE WERE ANALYSED BY HRTEM AND EELS.

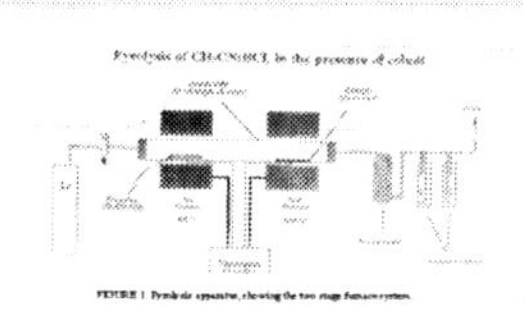

FIGURE 1. Pyrolysis apparatus, showing the two stage furnace system.

RESULTS AND DISCUSSION

TABLA I

SPECTROSCOPY CHARACTERIZATION OF $CH_3CN \cdot BCl_3$

TECHNIQUE	λ AND/OR POSITION	UNITS
IR (KBr)	[illegible]	cm⁻¹
^{1}H-NMR (DMSO-d6)	[illegible]	ppm
^{13}C-NMR (DMSO-d6)	[illegible]	ppm
MS (rel. intensity)	[illegible]	m/z
	[illegible]	m/z
	[illegible]	m/z
	[illegible]	m/z

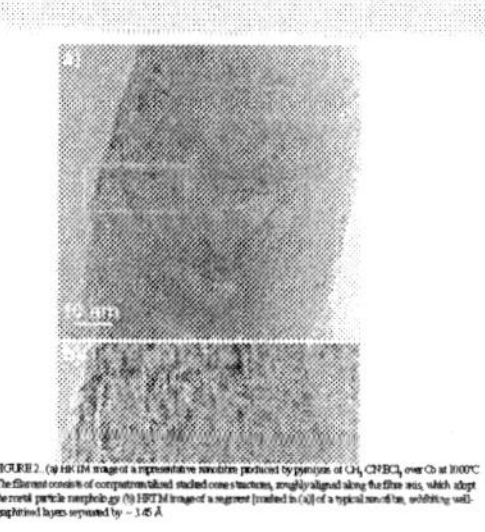

FIGURE 2. (a) HRTEM image of a representative nanofibre produced by pyrolysis of $CH_3CN \cdot BCl_3$ over Co at 1000°C. The filament consists of compartmentalised stacked cone structures, roughly aligned along the fibre axis, which adopt the metal particle morphology. (b) HRTEM image of a segment [marked in (a)] of a typical nanofibre, exhibiting well-graphitised layers separated by ~3.45 Å.

CONCLUSIONS

1 SPECTROSCOPY CHARACTERIZATION OF $CH_3CN \cdot BCl_3$

2 PYROLYTICALLY GROWN BCN NANOFIBRES HAVE BEEN STUDIED AT THE NANOSCALE LEVEL

3 IN SPITE OF THE RELATIVELY LOW TEMPERATURE COMPARED TO OTHER SYNTHETIC ROUTES EMPLOYED PREVIOUSLY

4 THE RESULTING BCN FILAMENTS EXHIBIT INCREASING DIAMETERS AS A FUNCTION OF DISTANCE FROM THE CATALYTIC PARTICLES

FIGURE 3. (a) PTEM image of a representative nanofilament exhibiting two metal particles at the tip. Arrows indicate the location of the high spatial resolution EELS line scans used to obtain chemical concentration profiles of (b)-(d). (b) Concentration profiles of B, C and N across the fibre [see (a)]. (c) Concentration profiles of B, C and N taking grow-th direction of the fibre [see (a)]. (d) Concentration profiles of B, C and O in the catalytic Co particle [see (a)].

REFERENCES

(1) Terrones, M.; Benito, A. M.; Manteca-Diego, C.; Hsu, W. R. O.; Osman, I.; Hare, J. P.; Reid, D. G.; Terrones, H.; Cheetham, A. K.; Prassides, K.; H. W.; Kroto, H. W.; Walton D. R. M.; *Chem. Phys. Lett.* **1996**, 257, 576.

(2) Kohler-Redlich, Ph.; Terrones, M.; Manteca-Diego, C.; Hsu, W. K.; Terrones, H.; Rühle, M.; Kroto, H. W.; Walton, D. R. M.; *Chem. Lett.* **1999**, 310, 459.

(3) Laubengayer, A. W. y Sears, D. S.; *J. A. Chem. Soc.* **1945**, 67, 164.

(4) Nylan, E. y Soloway, A. H.; *J. A. Chem. Soc.* **1959**, 2681.

Are Correct the Planck's Units? Is the Fine-Structure Constant Really Dimensionless?

José De La Luz Montero García*, Jesús Francisco Novoa Blanco[†]

*Institute for Scientific and Technological Information (IDICT); National Capitol, Havana, Cuba.
E-mail: quantumstructure@yahoo.es
[†]The late.

Abstract. The dimensional exhaustive elucidation of matter and the Unified Theory of the Physical and Mathematical Universal Constants allow to understand and to prove the basic mistakes of Planck's Units and dimensionality of the fine-structure constant.

Keywords: Planck's units; fine-structure constant; particle; universal constants; dimension.

Are Correct the Planck's Units?

Our arithmetic-geometric-topological-physical investigation[1] of matter brought as result that the own scale of matter has by base a dimensional octet (PDO) being the primitive dimension characteristic of: A self-assembly longitudinal (three-axial) ($\alpha = 3,3530 \cdot 10^{-34}$ m), time ($\beta = 1,1184 \cdot 10^{-42}$ s), mass ($\theta = 4,5166 \cdot 10^{-7}$ kg), of electric charge ($\Delta = 1,0978 \cdot 10^{-17}$ C), of thermodynamic temperature of energetic hierarchy; ($\gamma = 4,2911 \cdot 10^{31}$ K), of particle ($\phi = 6,8518 \cdot 10^{1}$ p), of grouping ($\Omega = 1,1378 \cdot 10^{-25}$ kmol) and of cycle universal ($\varphi = 2\pi$ rad).

This system precise the Planck's Units and we denominated MKSC.P system, that is: Precised meter-kilogram-second-coulomb. It is proved that: it is possible to obtain analytical-definitionally and to calculate the quantitative value of any and all well-known universal constant (retrodiction) and to know (prediction).

Is the Fine-Structure Constant (α_F) Really Dimensionless?

The dimensional analysis of α_F and the application of the Principle or Theorem of the Unicity of World[1], according to the MKSC.P, are proved has omitted the dimension particle. It is thus, that α_F has strict dimensional character:

$$\left[\alpha_F\right] = \left[2^{-1} \cdot p^{-1}\right] \Rightarrow \alpha_F = (2 \cdot \phi)^{-1} = 2 \cdot (68,5180)\,[p] = 137,0360\,[p].$$

The first and second fundamental particle equivalence $\dfrac{\text{kg} \cdot \text{s}}{\text{m} \cdot \text{C}} \leftrightarrow 2 \cdot p \Rightarrow \dfrac{\theta \cdot \beta}{\alpha \cdot \Delta} = \dfrac{1}{c} \cdot \dfrac{\theta}{\Delta} = 2 \cdot \phi$ and

$p \leftrightarrow m \cdot C \cdot s^{-2} \cdot K^{-1} \cdot 1$. The primitive dimension of particle —in addition to being the basic foundation of the duality wave-corpuscle— conditions the existence of a fundamental global nexus between electromagnetism and gravitation of the Real Material Space (RMS). Thus it

happens: $\dfrac{1}{c} \cdot \left(\dfrac{1}{G \cdot \varepsilon_0}\right)^{1/2} = \left(\dfrac{\mu_0}{G}\right)^{1/2} = 2 \cdot \phi = \dfrac{1}{\alpha_F}$ which has a deep space meaning.

Besides, the hypothetical one and so questioned Dirac's magnetic charge, his famous monopole, whose search until today has been unfruitful, is: $\mu_D = n \cdot (\phi \cdot e) = n \cdot \Delta$; $[e] = \left[C.p^{-1}\right] \Rightarrow \Delta - \phi \cdot e$.

REFERENCES

[1] Novoa, J., Montero, J. (1999). Realize. Quantum Cellular Structural Geometric Theory; Unified Theory of Physical and Mathematical Universal Constants; Program for the Physical-Arithmetic-Geometric-Topological-Dimensional Unification of Matter with its First Fundamental Equations, Laws, Principles and Postulates (Introduction). Havana: Library of Congress (U.S. Copyright Office TXu-911-634). Unpublished.

CP905, *Frontiers of Fundamental Physics (FFP8), Eighth International Symposium*
edited by B. G. Sidharth, A. Alfonso-Faus, and M. J. Fullana
© 2007 American Institute of Physics 978-0-7354-0412-0/07/$23.00

Double Exponential Relativity Theory Coupled Theoretically with Quantum Theory?

José De La Luz Montero García*, Jesús Francisco Novoa Blanco[†]

*Institute for Scientific and Technological Information (IDICT); National Capitol, Havana, Cuba.
E-mail: quantumstructure@yahoo.es
[†] The late.

Abstract. Here the problem of special relativity is analyzed into the context of a new theoretical formulation: the Double Exponential Theory of Special Relativity with respect to which the current Special or Restricted Theory of Relativity (STR) turns to be a particular case only.

Keywords: relativity theory; Lorentz invariant; quantum mechanics; quantum theory; distance.

It is obvious that if we modify Lorentz invariant, everything that follows turns to be modified. Lorentz invariant, is not very clear either because turns to be a market denial of that the variation of all magnitude is, on principle, proportional to itself.

The Double Exponential Invariant of Transformation rectifies this one cardinal mistake of Einstein's own STR[1].

$$E = m \cdot c^2 = m_0 \cdot \left(\exp\left\{ \frac{f_0}{k}\left[e^{k\left(\beta^2 - \beta_0^2\right)} - 1 \right] \right\} - 1 \right) \cdot c^2 \tag{1}$$

This, contrary to Lorentz invariant, allows to understand the principle of the relativity as transition of states and it do not only as process regarding systems of inertial references and the transformations of some in other by virtue of that are equivalent for the description of the processes. Moreover, express that it isn't forbidden to accelerate the system.

The magnitudes f_0 and k, constitute an unexplored field of theoretical (and experimental) research recently revealed by the authors, which opens the doors of relativity to Quantum Mechanics. Where, the attributes of physical bodies don't affect the objective relativism of the dimensions of matter and Real Material Space (RMS).

The value of c comes from the fact of that it is the mathematical ratio between the universal constants of primitive material dimensions of longitude α and time β which are mathematical extreme (related with 4th Thermodynamic Principle).

The value c isn't a limit of "substantial velocities", etc. but a universal invariant of dimensional transformation, that it is linked to the cofactor of dimensional transformation longitudinal-temporal of the matrix de transformation dimensional of world.

Being fundamental variety of RMS the definition of distance or longitudinal interval (ds) between infinitely proximate events:

$$ds = \left[\frac{\alpha^2}{\alpha^7}dx^2 + \frac{\alpha^2}{u^2}dy^2 + \frac{\alpha^2}{\alpha^2}dz^2 + \frac{\alpha^2}{\beta^2}dt^2 + \frac{\alpha^2}{\theta^2}dm^2 + \frac{\alpha^2}{\Delta^2}dQ^2 + \frac{\alpha^2}{\gamma^2}dT^2 + \frac{\alpha^2}{\Omega^2}da^2 + \frac{\alpha^2}{\phi^2}dP^2 + \frac{\alpha^2}{\varphi^2}d\varphi^2 \right] \tag{2}$$

This, of course, questions the legitimacy of assuming the electric charge, for example the one of the electron, as an absolute non-relativistic invariant, an implicit supposition widely used in the diverse formulations of electrodynamics. Not even the Higgs particle is out of this assertion.

REFERENCES

[1] Novoa, J., Montero, J. (2005). Cien Años Después: ¿Quién es realmente c? X Simposio y VIII Congreso de la Sociedad Cubana de Física. Facultad de Física Universidad de la Habana, Mayo 2005.

CP905, *Frontiers of Fundamental Physics (FFP8), Eighth International Symposium*
edited by B. G. Sidharth, A. Alfonso-Faus, and M. J. Fullana
© 2007 American Institute of Physics 978-0-7354-0412-0/07/$23.00

Natural Universal Radiation?

José De La Luz Montero García*, Jesús Francisco Novoa Blanco[†]

*_Institute for Scientific and Technological Information (IDICT); National Capitol, Havana, Cuba._
E-mail: **quantumstructure@yahoo.es**
[†] The late.

Abstract. The Cosmic Microwave Background Radiation is the constant and eternal explosion of quantum cells of the real material space, building up all that exists.

So in accordance with uncertainties considerations it is possible to assert that the Cosmic Microwave Background Radiation is equal numerically the topological constant of real material space.

Keywords: Cosmic Microwave Background Radiation; singularity; Universe; Big Bang.

The Universe is in constant generalized local rotation and, nevertheless it is structured; in effect, as structuring and unstructuring is processes concatenated sequentially and gravitation not is other thing that the mechanical interaction between tie systems space always, Universe pulsating, but locally of way universal, as it does not conceive _Big Bang_ and _Big crunsh_ that sees the singularity where it is the regularity that have not been able to notice[1].

Where everything and any movement or transformation happens through the interaction of the given system to the rest of Material Real Space (RMS) to which belongs and describes. Such interaction always passes through an enormous authentic chain of _Big bangs_ and _Big crunchs_. These last microprocessings are constantly happening but we did not notice then its energetic net result[1] to them almost null and it is characterized by $\Delta G = E_{nerg} - \gamma \cdot \left(\dfrac{E_{nerg}}{\gamma} \right) = 0$.

The transit between cells does not carry net evolution of energy and, therefore, entropy. The colossal basic processes of the space are transits between states of degenerated energy.

The movement, in such sense, turns out to be the transfering with respect to certain system of reference of a disturbance of the own space that is what we denominated object, system, being, body, corpuscle, etc. Such disturbance in its advance, on the other hand front is accompanied of a process of expansion of the cells of the material space with its corresponding transformation towards the body, whereas by the back part the process is the opposed one, the body is compressed towards the space.

This circumstance carries the evident consequence that the space disturbance always undergoes a compaction in the sense of the movement what inevitably it leads to a diminution of it length, an increase of its mass and a slow movements of the time with respect to the given system of reference.

According to considerations of uncertainties, can be affirmed that the Cosmic Microwave Background Radiation is numerically equal to the topological constant of the RMS[1].

The Cosmic Microwave Background Radiation is the Natural Universal Radiations in correspondence with the Principle of the Generalized Dynamic Conception: the matter and any concrete form of its presentation must be understood and be modeled as process and not as act; the state does not interest so much solely nor, as the transition between these.

If at local level always the Natural Universal Radiations are happening rotational processes they are the manifestation of the conservation laws that express the properties of RMS and it has numerical expression in the universal topological constant.

---

[1] Novoa, J., Montero, J. (1999). Realize. Quantum Cellular Structural Geometric Theory; Unified Theory of Physical and Mathematical Universal Constants; Program for the Physical-Arithmetic-Geometric-Topological-Dimensional Unification of Matter with its First Fundamental Equations, Laws, Principles and Postulates. (Introduction), Havana: Library of Congress (U.S. Copyright Office TXu-911-634). Unpublished.

CP905, _Frontiers of Fundamental Physics (FFP8), Eighth International Symposium_
edited by B. G. Sidharth, A. Alfonso-Faus, and M. J. Fullana
© 2007 American Institute of Physics 978-0-7354-0412-0/07/$23.00

Robustness of quantum Grover algorithm against decoherence

Pedro J. Salas and A. Gómez-González

Dpto. Tecnologías Especiales Aplicadas a la Telecomunicación, E.T.S.I. de Telecomunicación, Universidad Politécnica de Madrid, Ciudad Universitaria s/n, 28040 Madrid (Spain)

Keywords: Quantum computation, Grover algorithm.
PACS: 03.67.Lx, 03.67.Pp

Quantum computing has emerged as one of the most challenging fields in physics, both for theoreticians and experimentalists. Two of the most interesting quantum algorithms have been discovered by Shor (1994, to factorize large whole numbers) and Grover (1996). The goal of the Grover quantum algorithm in to find an item in a randomly ordered data base.

Consider an N-item unsorted data base. Any classic strategy searching for an item would require (N-1) steps to find it with certainty (if such an item exists). The reason is that the only way to carry out the search is to analyze each item one by one until the searched-for item is found. Grover algorithm finds the item in $O(\sqrt{N})$ time steps or Grover gate applications, which represents a considerable acceleration in the searching process. The advantage is rooted in applying a cleverly constructed Grover gate to a convenient initial entangled state of n-qubit registers. Bearing in mind that the searched-for item is encoded in the unknown $|k\rangle$ state, the quantum computation runs in order to amplify the a_k coefficient. In the final step, a measure collapse the whole state onto $|k\rangle$ with high probability, solving the problem.

The strength of Grover algorithm is also its weakness: entangled states are very sensitive to decoherence originating from the interaction between the quantum computer and its environment. The objective of this work will be to study the noise effect in the Grover algorithm. To simulate the behaviour of quantum networks we use an independent stochastic error model based on the notion of error locations[1]. At each network location the error is applied at random and independently of other errors in the same or different locations. The evolution error (coming from the free evolution time steps in the quantum network) is introduced by means of the X, Y and Z (Pauli matrices) errors in each qubit and time step with the same error probability $\varepsilon/3$ and no error (I unity operator) appearing with a probability $(1-\varepsilon)$. Noisy one-qubit gates (Hadamard and measurement), have γ error probability at each gate location. Two qubit gates (CNOT) and three qubit gates (Toffoli) have a gate error probability[2] proportional to γ. The results permit us to conclude that the allowed error in the algorithm scales as N-a, with a $\approx$ 1.1. This non-exponential behaviour suggests some degree of robustness, especially when the database is not too large.

[1] E. Knill, R. Laflamme and W. H. Zurek, *Proc. R. Soc. London A* **454**, 365 (1998).
[2] P. J. Salas and A.L. Sanz, *Phys. Rev. A* **66** 022302 (2002) (quant-ph/0207068).

CP905, *Frontiers of Fundamental Physics (FFP8), Eighth International Symposium*
edited by B. G. Sidharth, A. Alfonso-Faus, and M. J. Fullana

PARTICIPANTS

ABEJON ADÁMEZ, MANUEL
ABUBAKAR, BABAGANA
ALFONSO Y GARCIA, JAVIER DE
ALFONSO, MARIA DE LA CONCEPCIÓN
ALFONSO-FAUS, ANTONIO
ALFONSO-GARCIA, MIGUEL
ALMENDRAL, AVELINA
ALONSO, JACINTO JULIO
AMAYA Y GARCÍA DE LA ESCOSURA, JOSÉ MANUEL
ANABALÓN, ANDRÉS
ANTÓN CORRALES, JOSÉ MANUEL
ARRANZ, F.J.
ARTILES FONTANS, MARÍA JESÚS
ARTILES, GLORIA
ASHTEKAR, ABHAY
ASKEROV & FIGAROV, V &FIGAROVA, S
AZCÁRRAGA ARANA, ALVARO
BENITO, JOSÉ MARÍA
BENITO, ROSA MARÍA
BERNABÉU, INÉS
BLANCO, JULIÁN
BOLAÑOS, FRANCISCO
BORONDO, F
BRAVO, CARLOS
BROEKAERT, JAN
BURGUETE AZNAR, VÍCTOR AND WIFE
CABRERA SOLÉ, JOSÉ RICARDO
CAMPOS, M. DE
CARBAJO FUERTES, FERNANDO
CARBONELL, EUDALD
CARBONERO, CRISTINA
CARRASCO GUÍO, PABLO
CASTRO PÉREZ, LUIS
CEBRIÁN, CRISTINA
CHICOT URECH, JOSÉ MANUEL
CHUBYKALO, ANDREW
COLAO, FRANCISCA
DE GENNES, PIERRE GILLES
DE LA PLAZA, SATURNINO
DEGROOTE, EUGENIO
DELGADO BARRIO, GERARDO
DÍAZ-CRESPO FERNÁNDEZ, JUAN RAMÓN
DUQUE, PEDRO
EL NASCHIE, M.S
ELIZALDE, EMILIO
ENDERS, PETER
ESCOBAR ORELLANA, RAFAEL

ESPINOZA, AUGUSTO
ESTÉFAN ARBELÁEZ, MARÍA CECILIA
FERNÁNDEZ DE LA BASTIDA, JOSÉ MANUEL
FERNÁNDEZ PAREDES, MERCEDES
FERNÁNDEZ PUJALTE, FRANCISCO MANUEL
FERREIRO, VICTORIA
FINKELSTEIN, DAVID
FITÉ, ANGEL
FULLANA i ALFONSO, MÀRIUS
FULLANA i ALFONSO, SÒNIA
FULLANA i JORDAN, ARIADNA MARIOLA
GAITE, JOSÉ
GALÁN, PABLO
GALLEGO, MARIA DEL MAR
GARCÍA, MARÍA
GARCÍA, SONIA
GARCÍA DEL RIO, ALVAR
GARCÍA GARCÍA, ELENA
GARCÍA FRÍAS, JAVIER
GARCÍA PALOMO, MIGUEL
GARCÍA RODRÍGUEZ, MARTA
GARCÍA VEGA, SAMUEL
GÓMEZ CASTAÑO, MIGUEL
GONZÁLEZ CERDÁN, PASCUAL
GONZÁLEZ HAYA, ÁNGELA
GONZALO, JULIO A.
GOPEGUI PALACIOS, LUIS
GORADIA, SHANTILAL
GREINER, WALTER J.W
GUIJARRO, JOSÉ FRANCISCO
GULSHANI, PARVIZ
HADLEY, MARK
HARI RAO.
HARTNETT, JOHN
HERNÁNDEZ SÁNCHEZ, SUSANA
HERNANDO PUIEDRA, JOSÉ IGNACIO
HERRERO DEBÓN, ALICIA
HONTANILLA, MARÍA AMPARO
HONTANILLA, MARÍA DE LA CONCEPCIÓN
HONTANILLA, PEDRO
HOOFT, GERARDUS 'T
IBÁÑEZ OLEA, JOSÉ MARÍA
IBÁÑEZ ROMÁN, GONZALO
IZQUIERDO, ANTONIO AND WIFE (MARIA ANTONIA)
JIMÉNEZ ARANA, ÁNGEL
JIMÉNEZ BARRIO, MARÍA
JORDAN i PLÁ, MARÍA TERESA
JORGE HERRERO, PABLO
JOURNEAU, PHILIPPE
JUANAS FERNÁNDEZ, JESÚS

KHADEMI, SIAMAK
KHOLMETSKII, ALEXANDER L.
KLEINERT, H
KOSMIEDER, LOTHAR
KROEGER, H
LAI, C.H
LAÍN HUDITIAN, LILIANA
LANCHAS FUENTES, LAURA.
LARA SÁEZ, ANDRÉS
LIM, S.C.
LINIERS BARREIROS, IGNACIO DE
LLOPIS, RAUSELL, INDALECIO
LÓPEZ CORDOBA, JOSÉ LUIS
LÓPEZ JIMÉNEZ, ÁLVARO
LÓPEZ RUIZ, JOSÉ LUIS
LOSADA, J.C
LUEPKE ESTEFAN, FRANZ.
LUEPKE ESTEFAN, XARLES ERIK
LUQUE GONZÁLEZ, MARÍA LUISA
LLEÓ, ATANASIO
MADORE, JOHN
MANTECA, CONSOLACIÓN
MARTÍNEZ SAITO, MARIO
MARTÍNEZ SAIZ, MARÍA
MASEGOSA FANEGO, ROSA
MATA, MARÍA
MATEO, AMANDA
MIGUEL RUANO, GUILLERMO
MILANS DEL BOSCH, D. GONZALO
MILANS DEL BOSCH., MARÍA
MIZUSHIMA, MASATAKA
MIZUSHIMA, YONEKO
MONCADA, MARTA
MONTENEGRO VILLACIEROS, SANTIAGO
MONTERO, JOSÉ DE LA LUZ
NIETO HERNÁNDEZ, DAVID
NIKULOV, ALEXEY
NOTTALE, LAURENT
OHIRA, TORU
OKON GURVICH, ELÍAS
OLAZABAL DE ESPÍNOLA, AINOA DE
OLIVEIRA, FIRMIN
ORTEGA, JUAN
OSHEROFF. DOUGLAS D.
OTERO DOMÍNGUEZ, MARÍA JESÚS
OUDIH, M. R
OVIEDO SILVEIRA, JAIME.
PALAZZI, PAOLO
PAREDES, JOSÉ ANTONIO
PARRADO, RAFAEL

PASCUAL ALCOVER, JOAN
PEREDA BERLANGA, SERGIO
PÉREZ CASTELL, MANUEL
PÉREZ YUSTE, ANTONIO
PORTAENCASA, RAQUEL
PORTILLO GARCÍA, DAVID
PRADA NOGUEIRA, ISAAC
QUINTANA CABALLERO, IAN
RECIO GARCÍA-CERVIGÓN, GERMÁN
REVUELTA PEÑA, FABIO
REY FERRANDO, HUGO
RODRÍGUEZ PUÑAL, ALEJANDRO
RODRÍGUEZ SOLER, DAVID
ROJAS ESTEFAN, SAMIR R.
RUBBIA, CARLO
RUEDA FUENTES, JOSÉ DAMIÁN
RUIZ DELGADO, MANUEL
SÁEZ MILÁN, DIEGO P.
SALAS PERALTA, PEDRO
SAMPEDRO GÓMEZ, JESÚS MANUEL
SÁNCHEZ DIEZ, FULGENCIO AND WIFE.
SÁNCHEZ RON, JOSÉ MANUEL
SAN ROMÁN ARDANZA, BLANCA
SCHNEIDER, THORSTEN
SEPÚLVEDA, MARÍA ÁNGELES
SHARMA, AJAY
SIDHARTH, B.G.
SIERRA, ALFONSO
SMIRNOV-RUEDA, ROMAN
SORIANO MANZANO-SOLÍS, BÁRBARA
SOTO STAUMONT, ALICIA
STEFANOVICH, EUGENE
SUDARSHAN, E.C.G.
TEROL PIQUERAS, JOSÉ ENRIQUE
TRIVIÑO PARADA, JOSÉ
VERA MEGE, RAFAEL
VICENS, EDUARDO
VILACOBA RAMOS, ANDRÉS
VILLANUEVA, SILVIA ESTER
VILLAMARIZ, MARIA ROSARIO
VILAR, JOSÉ ANTONIO
VILLAR VILLAR, PAULA
YAMALEEV, R.
YAÑEZ VICO, CARLOS

Author Index

WE ESPECIALLY WANT TO THANK
OUR SUPPLIERS

LA QUINTA CATERING, S.L.
Chile, 2 - Las Matas
28290 MADRID
mariamilans@cateringvillareal.com
Tel.: +34 91 360 16 95
Fax : +34 91 429 86 70

DIMAR
Chapinería, 6 D
P. Emp.Ventorro del Cano
28925 MADRID
dimar@dimarcongresos.com
Tel.: +34 91 633 52 80
Fax: +34 91 633 98 62

TUNA DE INGENIEROS NAVALES
DE MADRID
Avda. Arco de la Victoria, s/n
28040 MADRID
tunanavales@hotmail.com
Tel.: +34 91 336 72 08

SALA DE ARTE A.IZQUIERDO
Pº Santa María de la Cabeza, 57
28045 MADRID
www.estudiodearte-aizquierdo.com
Tel. & Fax: +34 91 474 97 98

Tuna Ingenieros Navales
Madrid

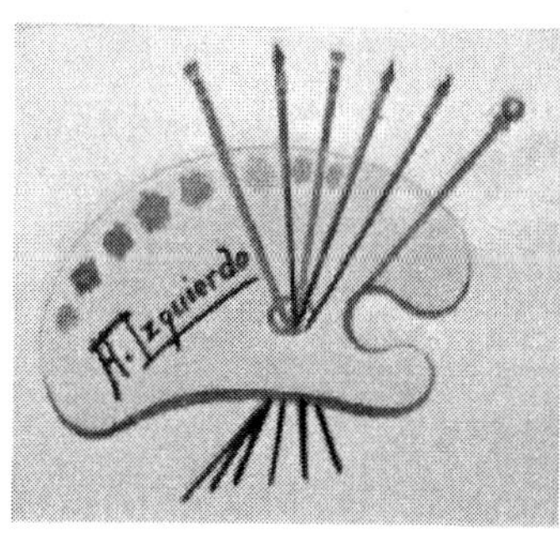

SERGIO PEREDA BERLANGA
C/Real, 16
28450 Collado Mediano
MADRID
spereda@sergiopereda.com
Tel.: +34 636 15 09 54

ESTUDIO DE DANZA MARÍA MATA
C/Núñez de Balboa, 119
28006 MADRID
info@mariamata.com
Tel.: +34 91 564 08 13

Photo S.P.B.

VOCES EN LA RED, S.L.
C/Bravo Murillo, 23   7º C
28015 MADRID
www.vocesenlared.com
Tel. +34 91 291 9996
Fax: +34 902 877 717